INTERNATIONAL MINING FORUM 2011

24-26 NOVEMBER 2011 – BOGDANKA, POLAND

New Techniques and Technologies in Thin Coal Seam Exploitation

International Mining Forum 2011

Edited by

Artur Dyczko
Polish Academy of Sciences, Mineral and Energy Economy Research Institute, Cracow, Poland

Jerzy Kicki
AGH University of Science and Technology, Cracow, Poland
Polish Academy of Sciences, Mineral and Energy Economy Research Institute, Cracow, Poland

Michał Myszkowski
Caterpillar Global Mining, Luenen, Germany

Zbigniew Stopa
Lubelski Węgiel "Bogdanka" S.A., Poland

Andrzej Tor
Jastrzębska Spółka Węglowa S.A., Poland

CRC Press
Taylor & Francis Group
Boca Raton London New York Leiden

CRC Press is an imprint of the
Taylor & Francis Group, an **informa** business
A BALKEMA BOOK

CRC Press/Balkema is an imprint of the Taylor & Francis Group, an informa business

Typeset by Jacek Jarosz, Marek Indyk
Printed and bound in Poland

Published by: CRC Press/Balkema
P.O. Box 447, 2300 AK Leiden, The Netherlands
e-mail: Pub.NL@taylorandfrancis.com
www.crcpress.com – www.taylorandfrancis.co.uk – www.balkema.nl

ISBN10: 0-415-68938-4
ISBN13: 978-0-415-68938-0

International Mining Forum 2011, Dyczko et al. (eds) © 2011 Taylor & Francis Group, London, UK. ISBN 978-0-415-68938-0

Table of Contents

CONSTRUCTION AND MAINTENANCE OF OPENING-OUT HEADINGS FOR PLOW LONGWALLS

Preface

The objective of the book is to demonstrate the up-to-date Polish and international experience in the area of thin coal seam mining. The implementation of the plow technology in Poland is done by young engineers for whom the automated extraction system presents itself as a great challenge pinpointing the direction of future developments – a production mining face manned by a skeleton crew, fully utilizing the potential offered by the information technology. They create new ways and means of running the production processes, which are successfully used together with the new generation plows. Successive generations of young mining engineers entering their careers are equipped with a new type of ability and, most of all, awareness which is characterized by a new understanding of reality, in which "real" is what is on the screen. They are the guarantee that this direction of research and the drive towards minimizing human involvement in the under-ground production chores will receive increasing focus and will remain the area of development. The experiences to-date of three mines (KWK Zofiówka, KWK Jas-Mos, LW Bogdanka S.A.) show that there is a real potential for the Polish companies to join in the research aimed at developing new plow installations. This publication is to sum-up the actions-to date taken both in Poland and all over the world in terms of implementing the plow technology in mining.

Artur Dyczko, Jerzy Kicki, Michał Myszkowski, Zbigniew Stopa, Andrzej Tor (eds)

November 2011

Organization

Scientific Editorial Committee:

Jerzy KICKI
(Mineral and Energy Economy Research Institute of Polish Academy of Sciences)
Zbigniew STOPA
(Deputy President of the Management Board on Production of LW Bogdanka S.A.)
Andrzej TOR
(Deputy President of the Management Board on Production of JSW S.A.)
Michał MYSZKOWSKI
(Caterpillar Global Mining)

Scientific Committee:

Volodymyr Bondarenko, professor (National Mining University of Dnipropetrovsk, Ukraine)
Piotr Czaja, professor (AGH University of Science and Technology, Poland)
Jozef Dubiński, professor (Central Mining Institute, Poland)
Roman Dyczkowsky (National Mining University of Dnipropetrovsk, Ukraine)
Jaroslav Dvoracek, professor (Technical University VSB, Czech Republic)
Pawel Krzystolik, professor ("Barbara" Experimental Mine, Poland)
Garry Litwinsky, professor (Donbass State Technical University of Alchevsk, Ukraine)
Jacek Paraszczak, professor (Laval University, Quebec City, Canada)
Uli Paschedag (Caterpillar)
Anton Sroka, professor (Technical University Bergakademie Freiberg, FR Germany)
Mladen Stjepanowic, professor (University of Belgrade, Republic of Serbia)
Antoni Tajduś, professor (AGH University of cience and Technology, Poland)

Organizing Committee:

Artur Dyczko (Chairman)
Katarzyna Witek (Vice-Chairman)
Jacek Jarosz (Vice-Chairman)
Małgorzata Boksa (Secretary General)
Eugeniusz J. Sobczyk
Piotr Saługa
Michał Kopacz
Dominik Galica
Szymon Sypniowski
Rafał Polak
Michał Kopacz

International Mining Forum 2011, Dyczko et al. (eds) © 2011 Taylor & Francis Group, London, UK. ISBN 978-0-415-68938-0

PLOW TECHNOLOGY AND ITS PERSPECTIVES

International Mining Forum 2011, Dyczko et al. (eds) © 2011 Taylor & Francis Group, London, UK. ISBN 978-0-415-68938-0

Plow Technology – History and Today's State-of-the-Art

U. Paschedag
Caterpillar Global Mining, Germany

ABSTRACT: Plowing is a long-known coal mining method invented in the early 1940s. Many different plow models were designed and trialed underground until in the mid-1990s only two models remained: the Reisshakenhobel® or base plate plow and the Gleithobel® or gliding plow. Horsepower and plow speed have increased steadily over time; however it wasn't until 1989 when the biggest weakness of plow systems was overcome: automated plowing, using electro-hydraulic controls, with defined cutting depths finally allowed plowing to become the most productive mining method for seams below 1.8 m seam thickness. This technology step like the most inventions in plowing came from Caterpillars predecessor Westfalia Lünen. Another mile-stone was yet another invention by Westfalia as the world's first and only really mine-worthy and rugged control system PM 4 was introduced in 1993. Today, three standard plow systems are offered by Caterpillar together with tailor-made plow shields and the currently most advanced electronic shield control system, the PMC®-R. Installed cutting power of up to 2 x 800 kW, coupled with a blazing plow speed of up to 3.6 m/s and world-leading automation capabilities leads to an integrated longwall mining system for medium and thin seams which produces more clean coal at lower cost than any shearer can

KEYWORDS: Plow longwall history, automated plowing with defined cutting depths, installed horsepower, plow speed

1 PLOWING HISTORY

The longwall mining method to extract coal from underground was established in the early 20th century. True mechanization of the coal mining process was only established in the early 1940s; however the first patents for the „slicing cutting" of coal were granted as early as 1912. The following shows the list of the patents granted and some examples of these inventions.

- 1912 – Wissmann (Germany)
- 1917 – Hirst (UK)
- 1922 – Diancourt (Netherlands)
- 1927 – Ducoer (France)
- 1929 – Morgen (USA)
- 1930 – Wilhelm Löbbe receives the patent for the coal plow
- 1935 – Hammel (Germany)
- 1941 – Ibbenbüren mine builds the first coal plow (Inventor: Konrad Grebe)

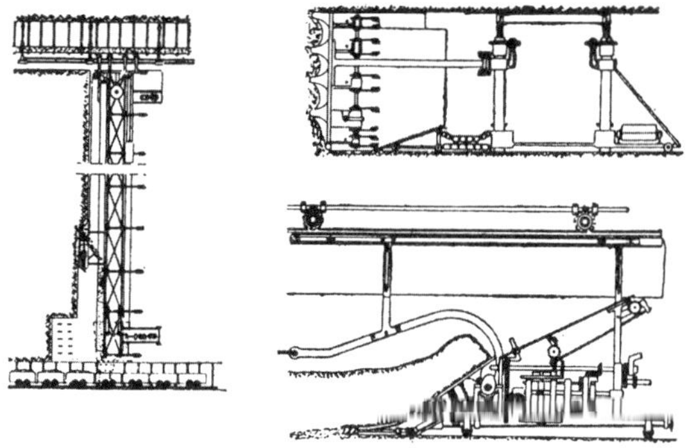

Figure 1. Invention of Wissmann (1912)

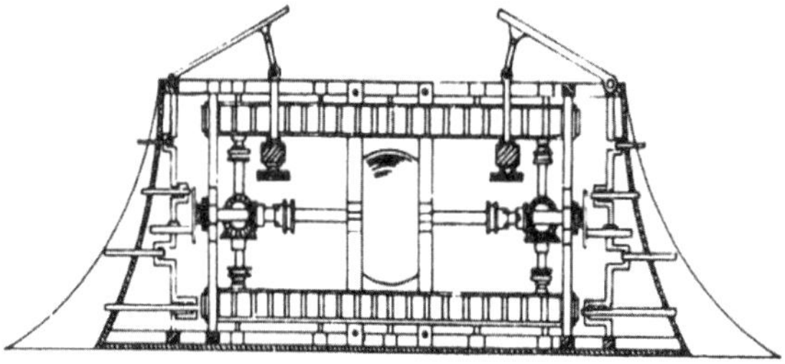

Figure 2. Invention of Diancourt (1922)

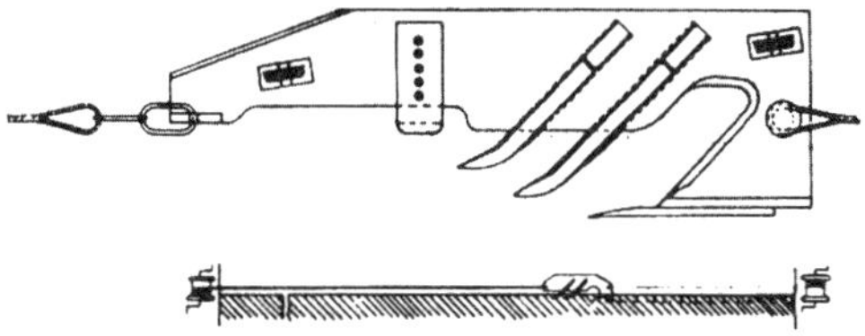

Figure 3. Invention of Ducoer (1927)

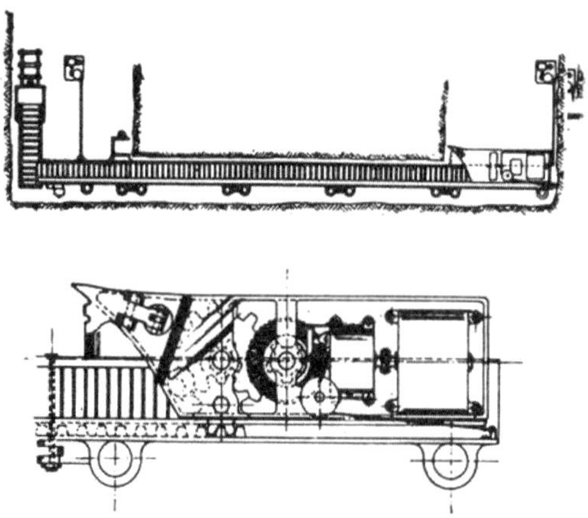

Figure 4. Invention of Morgan (1929)

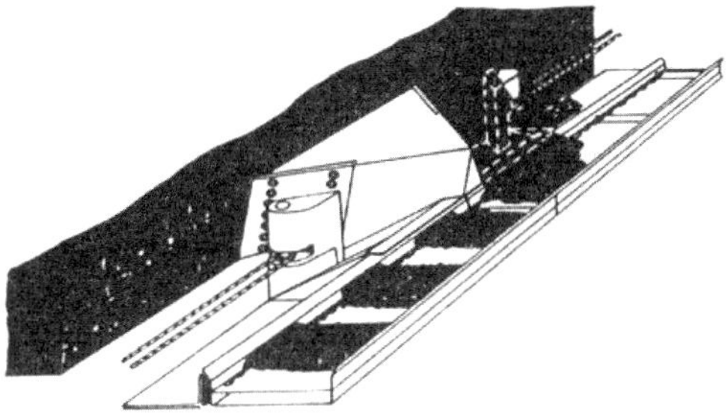

Figure 5. Invention of Grebe (1941)

Konrad Grebe was the inventor of the first built plow, installed at Germany's Ibbenbüren Mine in 1941. The plow cutting principle was already established, the plow body was pulled by a rope back then. The installed power was 2 x 40 kW; the cutting speed was 0.3 m/s.

Figure 6. Konrad Grebe (middle), the inventor of the first plow used at Ibbenbüren Mine in 1941

Figure 7. First Ibbenbüren plow in 1941

Figure 8. First Ibbenbüren plow in 1941

The plow was invented and refined by Wilhelm Löbbe, chief engineer at Westfalia Luenen (a predecessor company of Bucyrus Europe), back in 1937 in an attempt to modernize and mechanize underground coal mining. Löbbe also invented the PANZER-Förderer® or armored face conveyor, which was successfully installed underground in 1942 in Poland for the first time. Löbbe improved plow performance by reducing the cutting depth and increasing the cutting speed. The 'fast plow' was first installed in 1949 at Friedrich-Heinrich mine in Germany. With an installed power of only 2 x 40 kW, it achieved a daily output of over 1,000 tonnes for the first time in 1950.

Figure 9. Wilhelm Loebbe (1890–1950) was chief engineer at Westfalia Lünen, predecessor of DBT, Bucyrus and now Caterpillar

In fact this first really successful plow was so popular that the German government decided to publish a special stamp for it.

Figure 11. Löbbe plow stamp (translation: "The Loebbe plow - helper of the miner. The miner – helper of the human race.")

Figure 10. Loebbe Plow System 1950

From this point on, plow development picked up speed. In 1956, Westfalia Lünen and later the Beien Company built the first plow systems with a gob-side plow guide (prototype of the 'Reisshakenhobel®' or base plate plow). In 1960, Westfalia Lünen installed a plow without base plate but for the first time guided on gliding ramps at the NV Nederlandse Staatsmijnen in Heerlen, Netherlands (prototype of the Gleithobel® plow). 1963 saw the first underground installation of the first real Gleithobel plow, type I/26 at the Prosper III/IV, Graf Bismark und Westerholt mines in Germany. Between 1960 and 2003 there were 14 further plow models just for the Gleithobel alone.

The following plow models were pulled by a rope:

- Ibbenbüren mine: Einheitshobel
- Demag Company: Stufenhobel
- Brüninghaus Company: Schnellhobel
- Westfalia Lünen: Schnellhauer „Hannibal-Hauer".

After 1947, all plows were pulled by chains. The following shows a list of all these plows:

- Westfalia Lünen: Löbbe-Hobel, Reisshakenhobel®, Gleithobel®, Gleitschwerthobel® *(furthermore Anbauhobel, Prismenhobel, Tandemhobel, Vibrationshobel, Drehhobel, Pottshobel, Pulthobel, Tischhobel, Hydrohobel, Steilhobel)*
- Brand: Kettenstufenhobel
- Eickhoff: Universalhobel

- Beien: Megahobel, Rampenhobel, Schlaghobel, Raupenhobel, Gelenkhobel
- Klöckner Becorit: Universalhobel, Schlagkopfhobel
- GHH: Hydrohobel Jet-Miner
- Halbach&Braun: Kompakthobel, Kompaktrollenhobel, Schwertrollenhobel
- Heintzmann: CLM.

The following pictures show some of these plows [1].

Figure 12. Westfalia Anbauhobel (left) and Eickhoff Universalhobel (right)

Figure 13. Klöckner-Becorit plows (left) and Heintzmann CLM plow system (right)

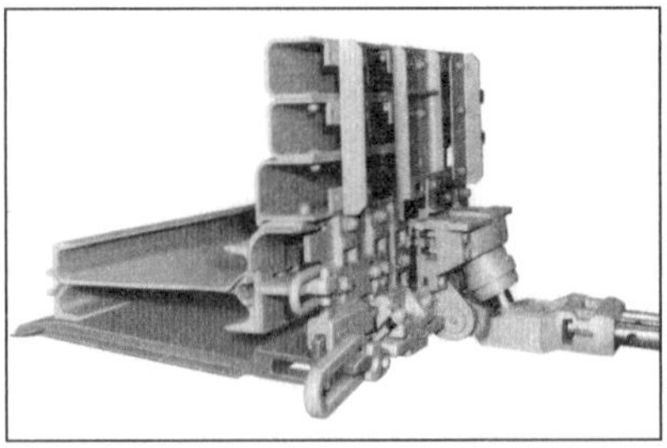

Figure 14. Westfalia Gleitschwerthobel®, combining the Gleithobel® and the Reisshakenhobel® (base plate Plow) principles

Besides these German companies also some others from Japan, Russia, Czech Republic and Poland developed plows, always based upon one or more of the above mentioned models. Until 1956, Westfalia Lünen alone delivered 248 Löbbe plow systems. As per an investigation in 1965, the German coal industry already used 531 plow systems from various suppliers; the vast majority of which came from Westfalia Lünen. Ever since the introduction of the Löbbe plow, Caterpillar and its predecessors Bucyrus, DBT, Westfalia Becorit and Westfalia Lünen has installed well more than 2000 plow system in many countries around the world. They were and are used for the following minerals: hard cola, lignite coal, phosphate, bauxite.

In addition to the conventional cutting type plows, there were also several attempts to activate plows through an activated cutting process. The following shows two of these models.

Figure 15. Activated plow from the Beien company (activation of bits through chain detour through the plow body)

Figure 16. Activated plow from the GHH company (high-pressure water jets pre-cut the coal behind the mechanical bits)

The activated plows were investigated and partly developed into trial machines because the older plow systems had low horsepower and no automation, so the cutting of hard coal or the cutting through rock was basically impossible. However all of the trials that were performed did not result in a machine that provided satisfactory production figure, furthermore such machines showed much less availability and maintenance was complicated. As incremental plowing with defined cutting depths was invented by Westfalia Lünen in 1989, coupled with more and more horsepower available, these developments were finally all abandoned in the early 1990s.

By the mid-1980s installed power had drastically increased to 2 x 250 kW. To match this power the plow chain size increased also to 34 mm diameter which allowed plow speeds of 1.8 m/s to be achieved. By the 1990s, installed cutting power had increased again to first 2 x 315 kW using a plow chain diameter of 38 mm, later in that decade 2 x 400 kW with the maximum plow speed finally reaching 3 m/s.

Figure 16. Schlagkopfhobel of Bergbau-Forschung GmbH (activated through hydraulic pick hammers on both sides of the machine which was pulled by a plow chain)

Working principle and early achievement of cutting depth

The working principle of plows has not really changed since its invention: the plow guide is attached to the face conveyor; the plow body is pulled (first by ropes, later solely by an endless chain) in both directions from the face end drives. The conveyor drives are also located at the face ends. The plow body "slices" off the coal from the seam and loads it into the face conveyor through its shape.

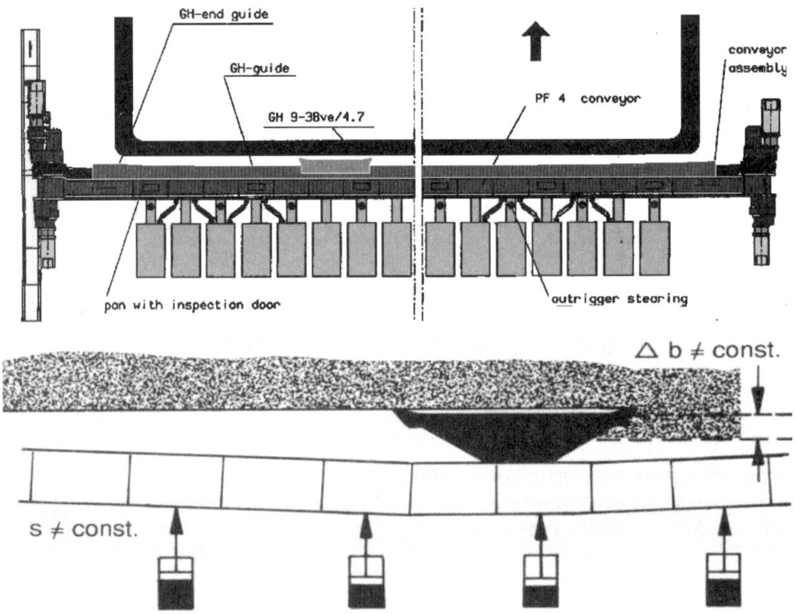

Figure 18. Plow system setup and constant low-pressure push principle

After the plow has passed by the face conveyor and the attached plow guides are pushed forward, which happened until the late 1980s / early mid-1990s by constant low pressure push from stand-alone push beams and later from roof supports of different kinds, from 1970 on shield supports. This was leading to inconstant cutting depths and an undulated face which had to be corrected very frequently. Also production was limited as the average cutting depth was lower than the one intended to be cut per plow pass.

Further to the face end drives, plow systems are characterized for quite some time by a horizon control system called "outrigger steering" device, allowing the operator to adjust the plow cutting to be straight forward, uphill or downhill by changing the angle of the face conveyor pan. Furthermore, also for quite some time so-called "block anchorage" system are being used for two purposes: they keep the AFC from creeping and at the same time the keep enough mechanical play between the individual line pans to allow them to be pushed over easily.

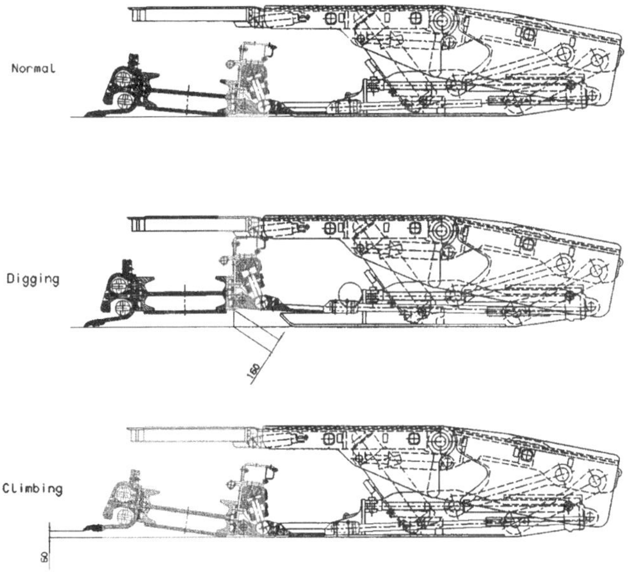

Figure 19. Outrigger steering plow horizon control system

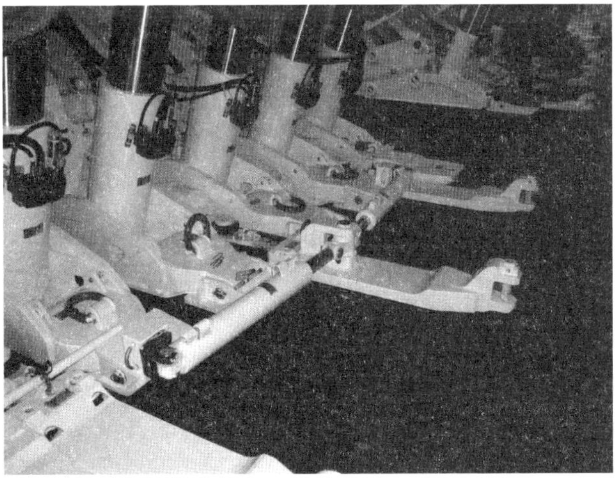

Figure 20. Modern Caterpillar block anchorage system

Remaining plow systems after 1995

Many of the previously mentioned plow models disappeared over the years due to their disadvantages, technical inefficiencies, as the companies designing and manufacturing them went into bankruptcy or merged with other companies. The last of these mergers was the one between Westfalia Becorit, Hemscheidt and Halbach&Braun in December of 1994, forming DBT Deutsche Bergbau-Technik under the ownership of RAG AG. Based upon an in-depth look at the advantages and disadvantages of the existing four plow models, it was decided to continue only using Westfalia's Reisshakenhobel® or base plate plow system S4N and Westfalia's Gleithobel® plow system 9-38 ve[(2)].

Figure 21. Standard plow systems at DBT Deutsche Bergbau-Technik in 1995: Reisshakenhobel® or base plate plow S 4N (left) and Gleithobel® 9-38 ve 5.7 (right)

Automated plowing with defined cutting depths

The breakthrough invention (outside of ever more horsepower) was the invention of automated plowing with defined cutting depths. The use of electro-magnetic valves and reed rod stroke measurement sensors inside the ram cylinders coupled with electronic control units in every shield finally allowed to overcome the biggest weakness of previous plow systems. Now, the AFC with the attached plow guide could be pushed over with high pressure after the plow body had passed by a predefined stroke equivalent to the next cutting depth.

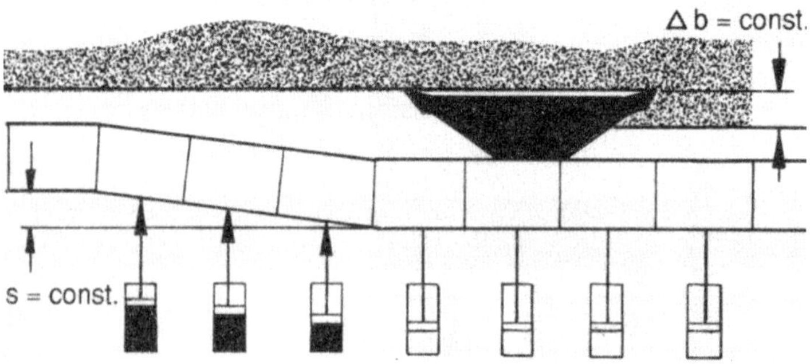

Figure 22. Automated plowing with defined cutting depth and high-pressure push

This method was successfully applied for the very first time in the USA, when Westfalia Lünen delivered two automated plow systems to U. S. Steel's Pinnacle mine in West Virginia, USA. While the automation algorithms installed by Westfalia into their third-party sourced electronics worked very successfully, the reliability of these electronics was very poor. It took until

1996, when the Pinnacle installed the company's self-developed PM 4 electronics to finally make the last step towards reliable automation. This, the world's first automated plow system, was upgraded with regards mainly to installed horsepower (at the end 2 x 400 kW) several times until in 2010 the then new owner Cleveland Cliffs ordered a complete new automated plow system from then Bucyrus.

Today, the main requirements for any modern high-performance plowing are remote control and a precisely adjustable cutting depth which produces a straight face line. These are easily achieved by incremental plowing, a feature unique to Caterpillar plows. Caterpillar incremental plowing, which is precisely controlled electronically, ensures that the plow cuts to an exactly defined depth, regardless of coal hardness, seam structure and presence of rock intrusions. This results in balanced loads on conveyors and steady advance through the seam. The AFC/plow system and shield supports can be positioned exactly where they are needed (snakes, faults etc.). The system is self-correcting for over- or underpush, maintaining a straight face regardless of conditions. In fact, as a result of automation, Caterpillar plow systems now cut as straight a face as shearers do. Today, the electrohydraulic system of choice is Caterpillar's PMC®-R system, which was developed as the successor of the long-time pacesetter in the industry, the PM 4, in 2004.

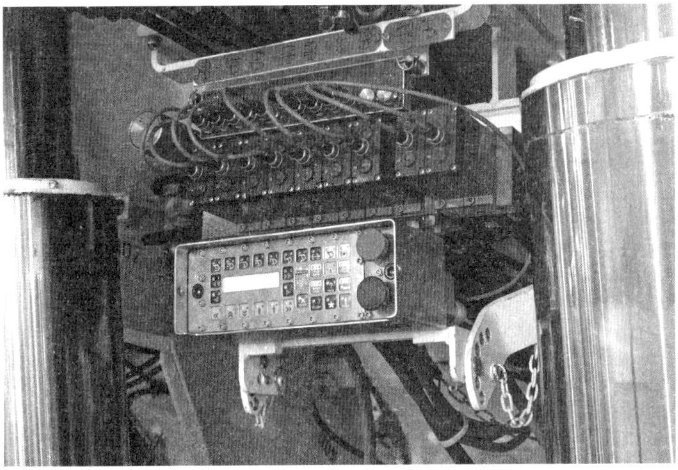

Figure23. Caterpillar's PMC®-R electrohydraulic shield control system

Figure 24. Underground longwall control cabin with face visualization

Caterpillar's predecessor DBT further developed these plow systems in various steps since 2000, adjusting to the demand of increased horsepower for cutting and increased productivity. Caterpillar now offers two types of plows with three different models: The Reisshakenhobel® or base plate plow RHH 800 and the Gleithobel® (GH or gliding plow) with two different models; the GHH 800 and the top-of-the-line GH 1600.

At the entry level, the RHH 800 base plate plow cuts seams 0.8–1.6 m in soft to hard coal has installed power of 2 x 400 kW, a maximum plow speed of 2.5 m/s and a maximum cutting depth in softer coal of 150 mm with a face length of up to 300 m. A 38 mm or 42 mm chain may be used.

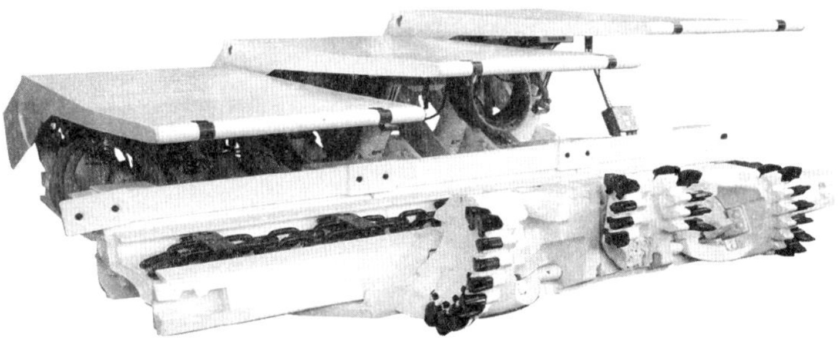

Figure 25. Caterpillars RHH 800 base plate plow system

The Gliding Plow GH 800 is designed for slightly thicker seams of 0.9–2.1 m. With the same installed power as the RHH 800, it achieves a higher maximum plow speed of 3.0 m/s and a greater cutting depth of 180 mm. The GH 800 also cuts soft to hard coal with a face length of up to 400 m. This system uses a plow chain with a diameter of 38 mm. An updated version of this successful model is underway.

Figure 26. Caterpillar's GH 800 plow system underground

Top of the range is the Gliding Plow GH 1600. Designed for mining medium to extremely hard coal in seams of 1.1–2.3 m, the GH 1600 has a maximum installed power of 2 x 800 kW, allowing it to achieve a maximum plow speed of 3.6 m/s with a maximum cutting depth in softer coal of 250 mm with a face length of up to 400 m, limited only by the maximum installable

AFC power. This plows system uses a plow chain with a diameter of 42 mm. In Germany, this next-generation plow system first installed in 2003 achieved a productivity of more than twice the amount of coal ever mined from a 400 m face on a 1.5 m seam of extremely hard coal with up to 3,500 t/h.

Figure 27. Caterpillar's GH 1600 plow system for maximum productivity

The height of all Caterpillar plow bodies (turret head) is easily and infinitely adjustable with a worm gear within a certain range. The plow body height can also be adjusted in larger increments by simply installing or removing auxiliary bit blocks. This allows for quick an easy adjustment to varying seam thicknesses, minimizing both the cutting of adjacent rock and the associated preparation costs.

Overload protection for prevents damage resulting from shock loads to drive components, minimizing chain failures, extending component life, and allowing for a quick restart. Caterpillar's planetary UEL overload protection system contains an integrated multi-disc clutch set to slip at well below the breaking force of the chain. If the chain stalls, the system decreases gearbox torque and shuts down the plow drives before a dangerous peak chain force is reached. This is achieved through yet another electrohydraulic control system. Finally, modern Caterpillar plow shields made out of high-strength steel for long life are available to make up for a truly integrated plow system.

2 REFERENCE PROJECTS

2.1 Pinnacle Mine

In 1989, US Steel's Pinnacle mine became the world's first automated longwall face that eliminated the need for an operator at the face. After conducting an extensive evaluation into the most efficient mining method for low seam coal, Cliffs Natural Resources Inc. opted for an automated Caterpillar plow longwall system, concluding that it was the safest, most cost-effective and most environmentally friendly way to extract coal from the low seam (Pocahontas No. 3) at the Pinnacle mine.

The 2010 installation has 2 x 600 kW plow drives, large P-45 UEL, a 42 x 137 mm plow chain, a plow body rated at 1.6 MW, the Caterpillar 2003 heavy-duty bit system, heavy-duty plow guides, robotically welded plow guides, PMC®-D plow control system, more plow steering cylinders. The AFC installed power is increased to 2 x 600 kW, with a higher AFC speed of 127 m/min. For Cliffs, the unveiling marks the start of a new era in low-seam productivity and safety. For Caterpillar, it marks the kick-off for a nationwide promotion of this highly advanced and efficient technology which is ready to assume a very important role in North American mining going forward. With its new mining system, Cleveland Cliffs will attempt to break the current world record for low seam longwall mining, set at the same Pinnacle mine in the late 1990ies with nearly 23,000 clean metric tons of coal in one day.

Figure 28. Surface compatibility test for the Cleveland Cliffs plow system in 2010 in the USA

2.2 Bogdanka

In March 2010, Polish mine operator Lubelski Węgiel Bogdanka S.A. commissioned the automated Caterpillar GH 1600 plow designed for high productivity in very hard coal and the system of choice for seams up to 2.3 m at its Bogdanka mine. The system features 2 x 210/630 kW (two-speed motors) for the plow system and 2 x 800 kW for the AFC system.

Within months of delivery, the new longwall had far exceeded the production target of 10,000 tonnes per day specified in the invitation to tender – and even set a new European record for plow faces as well as a new world production record for plow longwalls in mines using arches to support their entries. On August 25, 2010 the Caterpillar longwall achieved a daily output record of 16,894 metric tonnes. During only 10.5 hours of operation, the plow made over 300 passes along the 250 meter panel, traveling a total distance of 76 kilometers, an average of 120 m/min. The plow face advanced a total of 21.6 m with a cutting height of only 1.5 m – an achievement impossible for any shearer in such seam heights. This is the result of huge advances in plow automation and technology especially over the past 10 years.

The Caterpillar plow longwall installed in Bogdanka is working in automatic mode, allowing a significant reduction of staff required at the face during operation. Incremental plowing is used, cutting to an exactly defined depth, regardless of coal hardness, seam structure and faults. Horizon control allows exact control of the plow angle, ensuring that the plow does not dig or climb unless required by the seam geology.

Figure 29. Bogdanka plow GH 1600 underground

Polish mining industry had abandoned plow technology in the mid-1990s because of its low productivity caused by low horsepower and the lack of automation, but the success of Caterpillar' latest plow systems is causing a renaissance. JWR's Zofiowka mine was the first to buy a Caterpillar plow system (it now has two sets in operation); Bogdanka has already purchased another plow longwall system for delivery in 2012 for its new Stefanów panel. And at Bogdanka alone, Caterpillar plow technology will increase recoverable reserves by some 100 million tonnes.

3 SUMMARY

Automated plowing is the most productive longwall mining method for seams below appr. 1.8 m cutting height. High installed cutting power and plow speed allow a maximum capacity of 3,500 t/hour. Caterpillar plow systems allow clean extraction of low and medium coal seams with minimum mining of adjacent rock that reduces costs associated with preparation and refuse disposal. Incremental plowing with defined cutting depths and full automation without any operator being required at the face, combined with cutting power of up to 2 x 800 kW, ensures high productivity even when extracting hard coal. With the ability to operate in seams as low as 0.8 m plows can therefore mine seams previously thought to be inaccessible. And from this minimum seam height up to 1.8 m, the Caterpillar plow system easily outperforms any shearer.

Safety is enhanced by lower methane concentrations (due both to the larger size of the cut product and greater distribution across the face created by the higher plow transit speed) and the fact that – in contrast to the shearer – the plow can operate completely unattended.

And finally, Caterpillar plows are engineered and manufactured by the world leader in plow technology – with continuing research and development of all components to ensure maximum productivity, service life and return on investment.

REFERENCES

Archives of Hans Dreher.

Heinz-Werner Voß/Manfred Bittner: "Status of longwall mining of medium and thin seams in Germany". Publication in Glückauf Mining Reporter 1-2003; pages 20-27.

International Mining Forum 2011, Dyczko et al. (eds) © 2011 Taylor & Francis Group, London, UK. ISBN 978-0-415-68938-0

Plough Technique at LW Bogdanka S.A. – Present State and Prospects of Development

Z. Stopa
Lubelski Węgiel Bogdanka S.A., Poland

ABSTRACT: The article discusses the experiences which Bogdanka mine has gathered to date in the field of implementation of the plough technique. It presents technical specifications of the first mechanical plough system purchased by Lubelski Węgiel Bogdanka S.A. as well as the location, the geological and mining conditions, and the full production output reached by the research wall 1/VI in deposit 385/2 which was equipped with that system in 2010. The article discusses the second wall 7/VII/385, equipped with that system, launched at the end of September/beginning of October 2011 in Stefanów field. It presents changes which were implemented in that system and in the technology of mining the second wall as well as the plans of the mine to purchase a second system, the delivery of which is forecast to take place in 2012.

KEYWORDS: underground mining, hard coal, mining systems, plough mining technique, Lubelskie Zagłębie Węglowe

1 INTRODUCTION

The underground mining operations conducted for almost 30 years in the Lublin Coal Basin (Lubelskie Zagłębie Węglowe) by the mine of Lubelski Węgiel Bogdanka S.A. has been based on the shearer technique. That technique requires shearer operators to constantly attend the longwall area and move to follow the operated machine. Such constant presence of operators in the wall as well as the size of the shearer and the armored face conveyor, which constitutes the shearer travel track, results in the shearer technique, applied in such a version, being capable of being used in walls that are at least 1.6 m high. During the mining operations conducted to date, Bogdanka mine has fine-tuned the shearer-based technology of wall mining to a satisfactory level. This allowed the mine to reach high concentration of output, high efficiency and favourable economic results when mining fields that are more than 1.6 m thick. In the case of Lubelski Węgiel Bogdanka S.A., in seams that are 2.0-2.5 m thick, this technique allows to reach the level of daily output from one wall up to 20,000 tonnes of coal yield per day and, in the case of walls which are 1.6-2.0 m thick, up to 15,000 tonnes per day.

Implementing the plough technique in Bogdanka mine is expected to ensure that the following two goals will be achieved simultaneously:

- to enable commercial mining of fields thicker than 1.2 m, which will allow expanding the industrial coal resources by adding those located in fields which are 1.2 to 1.6 m thick,
- to improve the yield quality by limiting the amount of gangue being mined, by pure mining.

As far as resources are considered, it should be emphasised that many millions of tonnes of coal deposited conveniently, not far from the shafts, could not be mined only because the thickness of the deposit located there is slightly lower than that mined with the current shearer technique, which is over 1.6 m thick. This particularly applies to the seam 385/2, mined since 1997, but also to prospective seams 389 and 391. Mastering the ability to conduct commercial mining of the seams as thin as 1.2 m will allow to increase the mineable reserves and prolongs the time of mining of the individual aims.

The second essential goal is the purity of mining. In low mining walls, there is a tendency – natural as well as enforced by geometric dimensions of the wall equipment – to increase the wall height, most often by deposit roof ripping. This leads to deterioration of the yield quality. Therefore, it is the plough mining technique which offers the opportunity of pure mining of low thickness seam sand avoiding mining any stone.

The implementation of the coal mining plough technology has been recognized as an undertaking of strategic importance for the development of Bogdanka mine and has been consistently carried out for several years. The experience which Bogdanka mine has gathered to date regarding the implementation of the plough technique comprises:

1) formulating assumptions, contracting and purchase of the first plough system,
2) field panelling, equipping, and mining the first plough wall 1/VI in deposit 385 in Nadrybie in 2010,
3) field panelling, re-equipping the first plough system for wall 7/VII in deposit 385 in Stefanów, and launching it at the end of September/beginning of October in 2011,
4) formulating requirements, organisation of a tender, and contracting the second plough system for wall 2/VI in deposit 385 in Nadrybie, with accomplishment planned in 2012.

2. PREPARATION OF WALL 1/VI IN SEAM 385/2 AND EQUIPPING IT WITH THE FIRST PLOUGH SYSTEM

2.1. Preparatory works performed to launch the first plough wall

In order to determine the mineability of thin deposits with the plough technique in the conditions of Bogdanka mine, experts of the German DMT performed coal mining tests in deposit 385/2 in 2007. The result of those tests showed that the coal from seam 385/2 was well mineable and most suitable for plough mining.

The following research and development works were performed in order to prepare a decision on the possibility of using the plough technique in the conditions of Bogdanka mine and to develop rational technical solutions:

– "*Analysis of the development of plough technique worldwide and the possibility of its implementation at Lubelski Węgiel Bogdanka S.A.*", August 2008 [3], concerning the examination of the status of the technique worldwide and a technical and commercial analysis of this venture.
– "*Formulating optimum support specifications of the mechanized wall support system for the first plough system for mining thin deposits in the geological and mining conditions of Bogdanka mine*", August 2008 [4], concerning the selection of support specifications of the wall support system.
– "*Design of the gallery support system frame for the purpose of efficient mining by the plough technique in the conditions of Lubelski Węgiel Bogdanka S.A.*", July 2008 [5], concerning the optimisation of solutions employed in the longwall gallery support.

In autumn 2008, a tender procedure was started to purchase a new mechanical plough system with the required output determined at the level of 10,000 tons of coal yield per day. Finally, Lubelski Węgiel Bogdanka S.A. and BUCYRUS DBT Europe GmbH signed a contract on 26 March 2009 to deliver a new longwall plough system, intended to work in the mining plant of LW Bogdanka, with the delivery time of up to 12 months. The actual contract delivery time was faster. As soon as in the second half of November 2009, the compatibility test of the plough system, produced by BUCYRUS for Bogdanka mine, was carried out at the Manufacturer's premises. Following the positive outcome of the test, the deliveries started, being mostly completed by the end of 2009 and finalized at the end of 2010.

2.2. Geological and mining conditions of the first plough wall

The first plough wall in Bogdanka mine was situated in a very well selected location. It was the first wall in field VI in seam 385/2 in Nadrybie (longwall symbol: 1/VI/385), not far from the shaft. Considering the research nature of the work, the wall length was fixed to be 250 m, whereas the panel length, which resulted from geometric dimensions of field VI, amounted to 1.750 m.

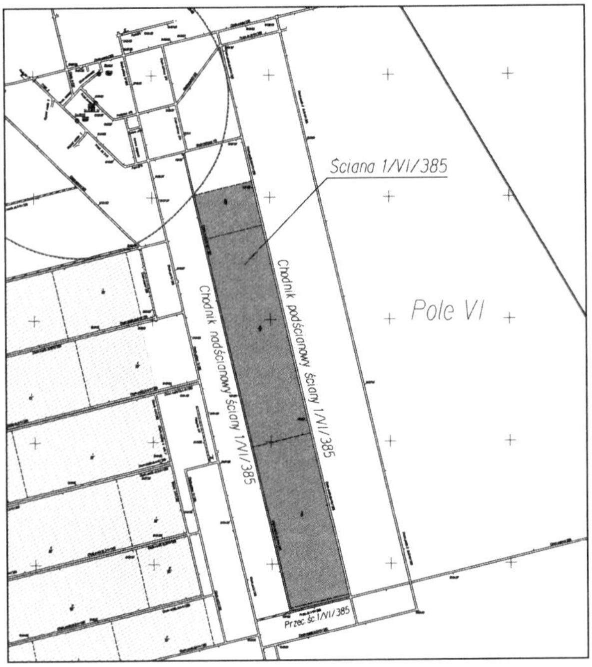

Figure 1. Location of the first plough wall at Lubelski Węgiel Bogdanka S.A.

Notations: Ściana 2/VI/385 – *Wall 2/VI/38;* Pole VI – *Field VI;* Chodnik nadścianowy ściany 1/VI/385 – *Wall 1/VI/385 tailgate;* Ściana 1/VI/385 – *Wall 1/VI/385;* Chodnik podścianowy ściany 1/VI/385 – *Wall 2/VI/385 headgate;* Chodnik podścianowy 2 ściany 2/VI/385 – *2 wall 2/VI/385 headgate*

The preparatory works for the first plough wall in Bogdanka mine comprised:

- tailgate, 1,930 m long,
- headgate, 2,040 m long,
- a 250 m long wall start-up cross-cut, made by reconstructing a part of the eastern gallery 4/385.

Splitting up field VI of deposit 385/2 for the planned plough wall 1/VI/385 was completed in the period between May 2008 and September 2009, by means of a AM-75 roadheader with a VBG-18 drilling and bolting machine attached.

The average monthly progress distances of the headgate heading of wall 1/VI/385 with the ŁPSC 12/S support system with bolting of roof, and tailgate of wall 1/VI/385 with the ŁPSC 11/S support system with roof bolting, were about 350 m. The frame support system used V36 section bar, made of S480W steel, with inter-frame spacing of 0.9 m. Moreover, the rock mass was reinforced during excavation by means of 2.7 m long steel bolts between the support system frames. The support system frames were reinforced by means of bolt horseheads with high bolting for the purpose related to maintaining a propless crossing with the longwall.

The 250 m long start-up cross-cut of wall 1/VI/385 was made with the ŁPrP/K 13 support system within its full height, also with the bolt support system.

The geological and mining conditions within the area of the wall field for the first plough wall were as follows:

- Parameters of seam 385/2 in the mining area of wall 1/VI:
  - seam thickness – 1.4–1.7 m,
  - seam slope – 0–2.5° westwards,
  - mineability index – 0.75–1.2,
  - coal compressive strength – 8–19 MPa,
- Existing natural hazards:
  - methane – 1st category,
  - explosion of coal dust – class B,
  - fire – IV group of auto-ignitability
  - water – 1st degree.

The roof of the seam contained alternating layers of siltstone, mudstone and sandstone. The compressive strength of the 3-meter package of roof layers, determined in longwall galleries, was between 24.4 to 40.7 MPa.

The deposit floor was built of alternating layers of siltstone and mudstone, with some local thin layers of sandstone. The compressive strength of the 3-meter package of roof layers, determined in longwall galleries, was between 12.9 to 31.8 MPa.

There were no disturbances, such as faults or flexures, within the panel of wall 1/VI/385.

2.3. Technical specifications of the first plough system

Detailed technical specifications of the first plough system were as follows:

- mechanized support system:
  a) Outer section 1480/2300-2×3619-1950:
    - range of operation – 950–2000 mm,
      – 1480–2300 mm (with a 270 mm extension),
    - support system width – 1700–2200 mm,

  b) Linear section 950/2000 - 2×3619-1750:
    - range of operation – 950–2000 mm,
    - support system width – 1750 mm,
- GH 1600-1 slide plough:
  - height of mine ability – 980–2010 mm,
  - tower output stroke – 0–300 mm,
  - mining head range – 0–210 mm,
  - chain – 1 x 42×137 mm,
  - speed – 0.98–2.94 m/s,
  - engine– – 2 x 210/630 kW, 3300 V,
- PF 4/1032 wall conveyor with face side discharge:
  - chain speed – 1.52 m/s,
  - max. output – 3000 t/h,
  - chute length – 1750 mm,
  - engines – 2 x 800 kW, 3300 V,
- PF 4/1132 wall bottom conveyor with SK 1111 crusher:
  - chain – 2 x 34 x 126 mm,
  - chain speed – 2.03 m/s,
  - max. output – 3000 t/h,
  - engine – 1 x 400 kW, 3300V,
- transmission systems of:

- the main wall conveyor drive with plough drive produced by BUCYRUS,
- the wall bottom scraper conveyor produced by BUCYRUS,
- the auxiliary drive of the UPN plough system produced by Sigma S.A. for the wall conveyor auxiliary drive transmission along with the plough drive.

2.4. Equipment of plough longwall 1/VI/385

Building up the section in plough wall 1/VI/385 started in the middle of January 2010. In order to set up sections of the mechanised support system in the coal deposit while installing equipment, the start-up cross-cut height was reduced from 4.4 m to 2.5 m by providing a stone bed up to the deposit floor level.

The mechanised support system for the start-up cross-cut was transported by means of LZH 60.D.0 or Bewex 70 overhead gasoline train and a 240 kN modular carrier set which moved on a rail track made of a 155 double-tee section bar.

The task of the brigades which built up sections in the start-up cross-cut was to: move sections and set them up properly, connect the mechanised support system section to the conveyor, build up elements of the artificial roof, disassemble elements of arch support system of the cross-cut, set cribs on straight pieces of the artificial roof, and expand the mechanised support system section. The first to be assembled was the route of the PF-4/1032 wall conveyor and then came sections of the mechanised support system. The first section was put into the start-up cross-cut on 19 January 2010; the last 141 section was set up on 13 February 2010.

3 PRODUCTION OUTPUT OF THE LONGWALL 1/VI IN SEAM 385/2

The plough wall 1/VI385 in seam 385/2, first in the history of Lubelski Węgiel Bogdanka S.A, was launched on 23 March 2010.

It took two days to transfer the support system sections from the start-up cross-cut to the deposit, i.e. to reach full supportability of the longwall support system.

The stage of the preliminary start-up of wall 1/VI/385 was conducted in full cooperation with the representatives of BUCYRUS; this period of time was spent on adjusting the parameters of the PMC-R and PMC-D system to the local operational conditions, i.e. the plough operation automation parameters (check points required to position it, slow down and make U-turns) and the section operation parameters (operational pressures, lead times of present automatic functions, setting up the "elephant pace", sequential sprinkling, cooperation of the PMC-R system with PMC-D, etc.) as well as setting up the parameters of the Integrated Control System of the Mining System in cooperation with the representatives of Elgór+Hansen Sp. z o.o.

The first months of mining wall 1/VI/385 allowed the employees working on it to learn in practice about the new technical solutions used in the automated plough wall. The operators of the continuous mining system must be experienced in order to work with the visualisation software, such as VShield and VPlough, or the Integrated Control System of the Mining System, and work during wall mining is the only way to get proper qualifications. Control and automation are one of the most important factors of the high-efficiency plough technique. They are based on electro-hydraulic control of the wall support system. Therefore, detailed measurement of the conveyor transmission distance makes it possible to accurately determine the depth of cutting. Furthermore, control and automation enable moving the mechanised support system sections without the involvement of operators.

The main station which controlled the entire plough system was situated in the tailgate of longwall 1/VI/385. The operator was continuously receiving information about all important data related to the system's operation, such as the position of the plough head and the direction of its movement, the position of the conveyor and the target route of the longwall face line, the output of each shifter for a preset position of the support system section against the conveyor as well as the current pressure in the support system hydraulic props.

Fine tuning of the plough system's control and automation software as well as adjusting the optimal parameters of the unit to the geological and mining conditions which prevailed in the wall allowed, after more than one month after launching the system, to reach the output of 6.000 tonnes per day. The production output results (wall mining progress, daily output) achieved by the plough system and the efficient operation time were increasingly better during the subsequent month of operation, reaching the planned output at the expected level of 10,000 tonnes of yield per day within three months after start-up.

On 15-17 June 2010, a test was performed to check the correctness of equipment operation (the so-called efficiency test), which confirmed that the minimum daily output, which had been required and declared by BUCYRUS to be 10,000 tonnes of yield, assuming the wall operation time of 20 hours per day and the efficient operation time of the plough of 10 hours per day, was reached. The detailed production results for the efficiency test duration period are presented in Chart 1.

Chart 1. Production results for the plough system efficiency test duration period

Date	Wall mining progress	Seam height	Yield volume	Yield mass	Plough operation time
	m	m	m^3	t	h ; min
15/06/2010	14.40	1.57	6,480	10,692	10 ; 00
16/06/2010	14.95	1.55	6,540	10,792	10 ; 00
17/06/2010	15.80	1.55	6,992	11,546	10 ; 00
25/08/2010	21.80	1.66	10,239	16,894	11 ; 22
26/08/2010	20.10	1.68	9,493	15,664	9 ; 58

After two months of further operation of the system, when a significant part of the wall panel had already been mined, another wall efficiency test was performed. On 25 August 2010, the record daily output of 16,894 tonnes was reached.

On 25 and 26 August, plough wall 1/VI/385 recorded the maximum results (chart 1):
August 25: – wall mining progress – 21.8 m, mining output – 16,894 t,
August 26: – wall mining progress – 20.1 m, mining output – 15,664 t.

During the later period, longwall 1/VI/385 was mined according to the schedule, according to which wall would run to its end by the end of October 2010.

Chart 2. Complete production results achieved by wall 1/VI in deposit 385/2

Parameter		2010							
		III	IV	V	VI	VII	VIII	IX	X
Wall mining progress [m]	monthly	12.0	187.2	238.3	238.1	313.3	303.9	252.3	174.3
	monthly cumulative	12.0	199.2	437.5	675.6	988.9	1,292.8	1,545.1	1,719.4
	daily average	4.2	8.1	10.8	10.8	11.6	11.7	10.1	8.3
	daily maximum	5.5	13.5	15.4	15.8	17.0	21.8	18.3	16.6
Gross output [thousand t]	monthly	9,063	135.7	175.5	177.5	214.1	217.6	164.1	111.0
	monthly cumulative	9,063	144.8	320.3	497.8	711.8	929.5	1,093.6	1,204.6
	daily average	3,021	5,901	7,978	8,067	7,928	8,371	6,565	5,287
	daily maximum	3,988	8,960	12,905	11,546	11,854	16,894	11,729	9,549
Plough operation time [h : min]	monthly	8 : 45	112 : 34	123 : 47	136 : 23	154 : 49	145 : 16	127 : 27	100 : 40
	monthly cumulative	8 : 45	121 : 19	245 : 06	381 : 29	536 : 18	681 : 34	809 : 01	809 : 01
	daily average	2 : 55	4 : 54	5 : 38	6 : 12	5 : 44	5 : 35	5 : 06	4 : 48
	daily maximum	3 : 51	8 : 04	8 : 27	9 : 32	7 : 35	11 : 22	8 : 48	7 : 09

The data has been sourced from controller reports.

The production results of the first plough wall in Bogdanka mine achieved over the entire panel length (excluding March and November, that is the start-up month and the closure month) are presented in Chart 2 and graphically on Figures 2, 3 and 4. Figure 2 presents the daily mining progress of the wall, Figure 3 – the daily gross output, and Figure 4 – the daily plough operation time.

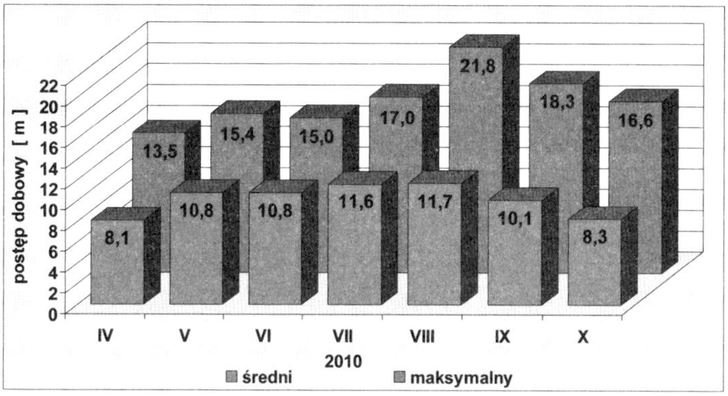

Fig. 2. Daily mining progress of plough wall 1/VI/385

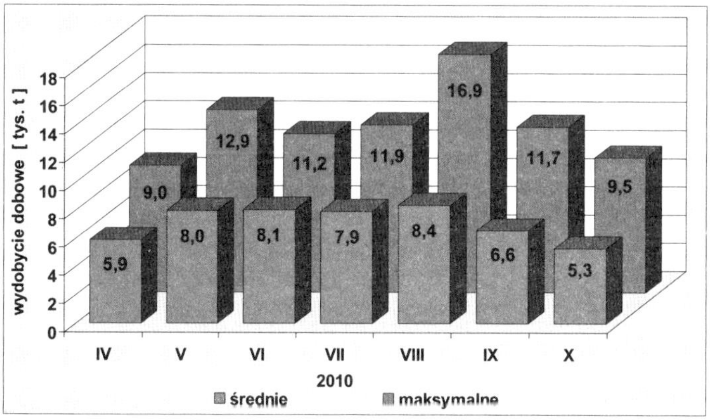

Fig. 3. Daily mining output of plough wall 1/VI/385

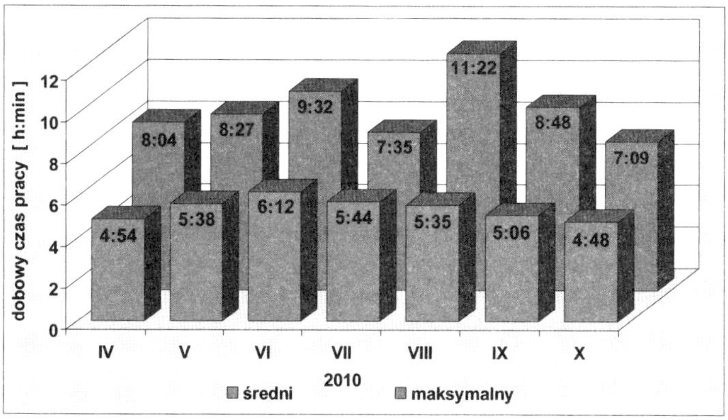

Fig. 4. Daily time of plough operation in wall 1/VI/385

Average production results generated by plough wall 1/VI/385 from all over its panel length were as follows:

- average daily progress – 10.4 m,
- average daily gross output – 8.2 thousand t,
- average monthly progress – 247.4 m,
- average monthly output – 195.4 thousand t,
- average daily time of the plough operation – 5 h 23 min,
- total wall panel length – 1,750 m,
- total wall output – 1,382 thousand t.

The calculations of average values did not take into account the wall start-up month – March (progress: 12 m), and the wall closure month – November (progress: 6.5 m).

4 PREPARATION, EQUIPMENT AND LAUNCHING OF THE SECOND WALL EQUIPPED WITH THE FIRST PLOUGH SYSTEM

Longwall 1/VI/385 was an exploratory wall. Situated not far from the shaft, it offered favourable geo-mechanical conditions, short length and small panel length as compared to the conditions prevailing in Bogdanka mine. The second site where the first plough system is used is wall 7/VII in same deposit 385/2, in Stefanów field. This wall already has a full length of 300 m and a very large panel length of over 5 km. The seam is 1.2-1.5 m thick. Longwall 7/VII/385 was launched at the end of September/beginning of October 2011, after skip shaft 2.1 in Stefanów had been opened for mining operation on 30 September.

4.1 Working for wall 7/VII/385

Considering the very large panel length, more intense convergence of longwall galleries and lower thickness of the deposit against the expectations of higher output volume, challenges for the second longwall are not lesser than they were in the case of the exploratory wall.

Working for longwall 7/VII/385, which comprised driving about 12 km of galleries, was completed in November 2010. The location of wall 7/VII/385 is presented in Figure 5.

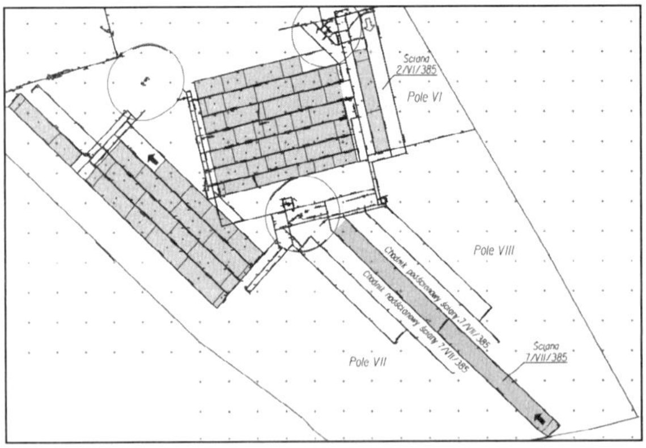

Fig. 5. The location of wall 7/VII/385 in Stefanów field, a second wall equipped with the first plough system

Notations: Ściana 2/VI/385 – *Wall 2/VI/38;* Pole VI – *Field VI;* Chodnik nadścianowy ściany 1/VI/385 – *Wall 1/VI/385 tailgate;* Ściana 1/VI/385 – *Wall 1/VI/385;* Chodnik podścianowy ściany 1/VI/385 – *Wall 2/VI/385 headgate;* Chodnik podścianowy 2 ściany 2/VI/385 – *2 wall 2/VI/385 headgate;* Pole VII – *Field VII;* Chodnik nadścianowy ściany 7/VII/385 – *Wall 7/VII/385 tailgate;* Ściana 7/VII/385 – *Wall 7/VII/385;* Chodnik podścianowy ściany 7/VII/385 – *Wall 7/VII/385 headgate;* Pole VIII – *Field VIII*

4.2 Re-equipping the system and launching longwall 7/VII/385

The period of preparing longwall 1/VI/385 for closure ended in the first decade of November 2010. First sections of the longwall support system were transported out of the longwall at the end of November. Since the date of launching longwall 7/VII/385 depended on finishing of the construction of shaft 2.1 in Stefanów and the date of obtaining the permit to operate the skip shaft, no decision was made to transport the sections to the new longwall and build them in. The sections transported to Stefanów were stored in the tailgate of longwall 7/VII/385, where they underwent the required inspection. A wall conveyor was built up in the wall start-up cross-cut. Sections were built up in the cross-cut of wall 7/VII/385 in June and completed in July. The wall had practically been ready to be launched since August. The decision to launch the wall was finally made at the end of September and resulted from the date when the Mining Authority for Audit Inspections of Electromechanical Devices in Katowice gave its approval for the operation of skip shaft 2.1 in Stefanów.

4.3 Important design modifications of the system, and the technical and organisational changes introduced in longwall 7/VII/385

Longwall 7/VII/385, a second longwall equipped with the first plough system, is essentially different from the exploratory longwall. The length was increased from 250 m to 300 m (having obtained a deviation permit from the President of WUG (Polish State Mining Authority)) and the panel was increased by about 3 times, i.e. to 5.070 m. Several significant design modifications were also introduced in the system. The most important of them include:

1. Change of the expansion and shift station in the bottom wall gallery. The existing station has been replaced with the solution provided by SIGMA, used in the top wall gallery and used to date in shearer walls, which is more suitable for the specific conditions prevailing in Bogdanka mine.
2. Employment of a prototype machine for self-loading of the yield at the longwall and tailgate crossing. This machine does the work of floor loaders, which have to date been loading the yield ejected out by the plough to longwall galleries onto means of haulage.
3. Employment of a prototype self-propelled support system, designed for mechanising wall cavities by the both longwall galleries.

The idea of the yield self-loading machine was developed in cooperation between the engineering staff of the mine and SIGMA, which produced the design and the prototype of the machine. The first machine of this kind, built up already during the stage of installing the wall equipment, caused many controversies. However, the first meters of longwall mining progress proved its high functionality, which manifests itself not only by good and efficient loading of yield within the pouring area, but also the significant improvement of labour safety in the longwall crossing area. As a matter of fact, this machine tightly fences off the longwall inlet, which makes it possible to work safely in the gallery ahead of the wall face (e.g. while disassembling the sidewall elements of frames) when the plough is in operation. As a result, a decision has been made to contract a similar machine for the crossing of the longwall and thetailgate.

Works to develop the self-propelled support system for wall cavities in cooperation between the engineering staff of the mine and of the company SIGMA were conducted simultaneously. That support system is expected to replace the individually made support systems, which should reduce the labour consumption at work, extend the efficient time of the plough operation, and improve the stability of the roof in the cavities. The first such machine was built up in the bottom cavity at the crossing with the headgate on 23 October of the current year, i.e. when the wall obtained about 100 m panel length.

Other modifications in longwall 7/VII/385, as compared to the exploratory longwall 1/VI/385, performed during the initial period of its run and are worth noting, include:

a) situating the plough control station on the ground surface,

b) change of the geometrical shape of the wall front from curvilinear to rectilinear,
c) changes in the area of controlling the supportability of sections while they are in process of being moved,
d) starting works to implement drive bolting stations in longwall galleries which take over transverse loads.

5. SECOND PLOUGH SYSTEM FOR LUBELSKI WĘGIEL BOGDANKA AND SELECTION OF THE FIRST WALL FOR THAT SYSTEM

According to the coal mining schedule for Lubelski Węgiel Bogdanka, a second longwall is planned to be launched in mid 2012, from which coal will be mined through the new skip shaft in Stefanów. Considering the thickness of the deposit and the positive results achieved with the first system, it was decided that this will also be a plough system. Therefore, activities were initiated in 2011 to contract a second plough system for the needs of Bogdanka mine. The system will be used for longwall 2/VI in seam 385/2, situated next to exploratory wall 1/VI/385, mined in 2010. The geological and mining conditions of that longwall are therefore similar, whereas it will have an expanded panel length of about 2,200 m and will be mined in the opposite direction with haulage in shaft 2.1 in Stefanów.

As a result of the tender procedure started at the beginning of 2011, a contract was signed on 31 May 2011 with Bucyrus Europe GmbH with its registered office in Lünen (Germany) (presently: Caterpillar Global Mining CAT) to deliver the second longwall plough system for the amount of approximately PLN 163 million. Delivery time – up to 14 months since the contract signature date.

Certain modifications will be implemented in the second plough system as compared to the first system dated 2010, as a result of the acquired experiences or the technical progress (also those which are already being introduced in the first system for the second longwall), for example:
- expansion and shift station in the headgate has been replaced with the solution provided by SIGMA,
- employment of frequency converters for the 3.3 kV voltage to power the plough and the longwall armored face conveyor, replacing the multi-breaker switches,
- employment of the coal self-loading solutions for the longwall inlet in case of getting a positive evaluation of the prototypes tested in longwall 7/VII/385.

The target model of mining operations in Bogdanka mine assumes that full mining output will be conducted, coming from four walls which are simultaneously active, including two plough longwalls and two shearer longwalls.

6 SUMMARY

1. Implementing the plough technique in Bogdanka mine is expected to ensure that two goals will be achieved simultaneously:
   - to enable commercial mining of seams thicker than 1.2 m, which will allow expanding the mineable reserves by adding these located in seams which are 1.2 to 1.6 m thick,
   - to improve the yield quality by limiting the amount of gangue being mined during the process.
2. The experiences which Bogdanka mine has gathered to date regarding the implementation of the plough technique comprise:
   - formulating assumptions and purchasing the first plough system,
   - field panelling, equipping and mining the first plough longwall 1/VI in seam 385 in Nadrybie in 2010,
   - splitting up the field and re-equipping the first plough system for longwall 7/VII in deposit 385 in Stefanów and launching it at the end of September/beginning of October 2011,

- formulating requirements and organising a tender, finalised with contracting the second plough system for Lubelski Węgiel Bogdanka, with the delivery date in 2012.

3. The first plough system, which was purchased in 2009, enabled in 2010 to launch longwall 1/VI in seam 385/2, which reached the target output of ten thousand tonnes/day almost without a problem, with the seam thickness of about 1.6 m. The record daily output was almost seventeen thousand tonnes, with the longwall progress of 21 m.
4. The full panel length for longwall 1/VI/385, amounting to 1.750 m, was mined in the period of March to November 2010. The average parameters achieved in course of the wall mining operations in the months of April to October 2010 were as follows:
   - average daily progress – 10.4 m,
   - average daily gross output – 8.2 thousand tonnes,
   - average monthly progress – 247.4 m,
   - average daily time of the plough operation – 5 h 23 min,
   - total wall output – 1,382 thousand tonnes.
5. Particular emphasis should be put on the high potential in terms of efficiency and high efficiency reserves which the plough system offers and which can be used in subsequent walls. The average daily time of the plough operation is only about 5.5 h/day, while the maximum time is 11 h/day.
6. Another serious challenge with regard to the implementation of the plough technique in Lubelski Węgiel Bogdanka is the second plough longwall 7/VII/385, 300 m long, with the panel length of 5 km and seam thickness of 1.2–1.5 m, which was equipped with that system. That wall was launched at the end of September/beginning of October 2011 in Stefanów field, and required that the new skip shaft 2.1 be opened for mining operations prior to that.
7. In longwall 7/VII/385, new solutions were introduced in the area of self-loading of coal, ejected by the plough into longwall galleries, and in the way of building up and running gallery cavities. Changes in the technology of mining the wall have also been tested. Those modifications are subject to a detailed evaluation and will successively implemented.
8. Bogdanka mine contracted the second plough system in mid-2011, with the delivery date in 2012, when the second longwall is planned to be launched along with the output transportation through shaft 2.1 in Stefanów. This will be longwall 2/VI/385, situated in Nadrybie, nearby exploratory wall 1/VI/385, but mined in the opposite direction.
9. The target model of mining operations in Bogdanka mine assumes that full mining output will be conducted, coming from four longwalls which are simultaneously active, including two plough longwalls and two shearer longwalls.

REFERENCES

Kicki J. et al. 2008. Analiza stanu techniki strugowej na świecie i możliwości jej wdrożenia w Lubelskim Węglu Bogdanka S.A. [*Analysis of the development of plough technique worldwide and the possibility of its implementation at Lubelski Węgiel Bogdanka S.A.*] Work prepared by the Mineral and Energy Economy Research Institute of PAN in Krakow. Unpublished. (In Polish)

Kwiatkowski K., Leś A. 2010. Wdrożenie pierwszej zautomatyzowanej ściany strugowej w Lubelskim Węglu Bogdanka S.A. [*Launching the first automated plough wall in Lubelski Węgiel Bogdanka S.A.*] XII Conference of the Management Board of SITG – Mining Yesterday and Today. Ustroń. (In Polish)

Kwiatkowski K. 2011. Wykonywanie i utrzymywanie chodników przyścianowych dla ścian strugowych w kopalni Lubelski Węgiel Bogdanka S.A. [*Making and maintenance of longwall galleries for plough walls in the mine of Lubelski Węgiel Bogdanka S.A.*] IMF 2011 conference materials in Bogdanka. (In Polish)

Prusek S. et al. 2006.Projekt obudowy wyrobisk chodnikowych dla możliwości efektywnego zastosowania techniki strugowej w KWK Zofiówka. [*Design of a support system for gallery workings with regard to the possibility of using the plough technique in KWK Zofiówka*] GIG work documentation no. 411 56205-150. Unpublished. (In Polish)

Rajwa S. et al.. 2008. Opracowanie optymalnych parametrów odpornościowych zmechanizowanej obudowy ścianowej dla pierwszego kompleksu strugowego do wybierania cienkich pokładów w warunkach geologiczno-górniczych kopalni Bogdanka. [*Formulating optimum resistance specifications of the mechanized wall support system for the first plough system for mining thin deposits in the geological and mining conditions of Bogdanka mine*] Work prepared by the central Mining Institute in Katowice. Unpublished. (In Polish)

Rotkegel M. et al. 2008. Projekt odrzwi obudowy chodnikowej dla efektywnego prowadzenia eksploatacji techniką strugową w warunkach LW Bogdanka. [*Design of the support system frames for the purpose of efficient mining operation with the use of the plough technique in the conditions of LW Bogdanka*] GIG work documentation no. 581 27718-151. Unpublished. (In Polish)

Stopa Z. 2011. Doświadczenia Lubelskiego Węgla Bogdanka S.A. w zakresie wdrażania technologii strugowej. [*Experiences of Lubelski Węgiel Bogdanka S.A. in implementing the plough technology*] Materials of the Conference: School of Underground Mining. (In Polish)

International Mining Forum 2011, Dyczko et al. (eds) © 2011 Taylor & Francis Group, London, UK. ISBN 978-0-415-68938-0

Operational Experiences of Automated Plow Systems in Tiefa, China

D. X. Tang
Caterpillar Global Mining Greater China & Korea
Caterpillar (China) Investment Co., Ltd.

ABSTRACT: There are eleven Cat (former Bucyrus/DBT) automated plow systems in China. Tiefa Coal Mining (Group) Co., Ltd. is not only the first end user but also has the largest fleet of the Cat automated plow system. Over the past decade it employed this fleet to effectively recover its valuable low seam reserves, which extended the service life of its old mines that otherwise could have been closed and resulted in the unemployment of thousands of miners. Meanwhile Tiefa gained a lot of successful experiences in using the automated plow systems and shared them with other Chinese coal producers. This paper will describe how the problems were approached and successfully resolved by the Tiefa and will reflect China's approach to effective resolution of thin seam coal extraction.

KEYWORDS: automated plow, low seam, operational experiences

1 INTRODUCTION

Tiefa Coal Mining (Group) Co., Ltd. is a state owned coal producing company located at Diaobingshan City, Liaoning Province, in the northeast of P. R. China. It has eight mines operating in its three coal fields with a total field area of 618.43 KM^2 and total proven coal reserves of 2.3 billion tons, 26% of which are low seams. Meanwhile it is expanding its coal production by purchasing coal reserves in Inner Mongolia and establishing a joint venture mine in Shanxi province, which has high quality coking coal low seam reserves. Tiefa has a number of concerns regarding how to handle these low seam reserves. First of all, using traditional mining method, such as drilling and blasting, is low productivity and unsafe. Secondly, to stop mining these low seam reserves will result in the waste of these reserves because it will be very difficult to recover them in the future if thick seams underneath them are mined first using the current longwall method, which is not allowed by Chinese central government. Thirdly, the service life of some of its coal mines will be greatly shorten because their thick seam reserves have been mined out. Take its Xioaqing Mine for example, it would be closed very soon if low seam reserves are not mined and thus would make its three thousands employees lose their jobs.

The automated plow system provided Tiefa with a solution for mining its low seam reserves. In 2000 the first automated plow system was imported from former DBT and was successfully used at its Xiaoqing Mine. Since then Tiefa has invested heavily to import four automated plow systems. The contract for importing its fifth automated plow system has been signed, which will be delivered in early 2012 and will be employed at its joint venture mine at Shanxi province.

2 TIEFA APPROACH TO AUTOMATED PLOW SYSTEM

The history of using the plow system in low seams of China is not short. As early as 1964 some Chinese mining equipment manufacturers started developing the plow equipment together with coal mines and research institutes and were able to supply a couple of types of plow equipment in the late 1990's. In 1980 two sets of Westfalia Becorit 8/30 plow were imported and employed respectively at Qishan Mine and Jiahe Mine of Xuzhou Coal Mining Company, Jianshu Province. In 1993 one set of Halbach & Braun KHS-2 compact plow was put into operation at Datong One Mine of Songzao Coal Mining Company, Sichuan Province. However they equipped with low installed horsepower motors and were not automated and therefore could not achieve a high productivity.

In 1997 Tiefa began consideration of applying an automated plow system after learning of successful operation experiences in the USA and Germany. A feasibility study group comprised of Tiefa, a research institute, and a local shield supplier conducted intensive investigations to evaluate the feasibility of importation and applying the most technological advanced system available in the Tiefa mines. The group concluded that the DBT automated plow system would be an ideal candidate for Tiefa's low seams in terms of technology and quality. After further evaluations and discussions, it was agreed to use the roof supports including face end roof supports supplied by a local manufacture equipped with DBT PM4 electro-hydraulic control system, advancing rams and hydraulic valves. In addition, it was decided that one PM4 E/H control unit should control three roof supports instead of only one. Later it was called "Tiefa Model" and was adopted in other DBT automated plow systems imported to China.

The contract for importing the first automated plow system was signed in December of 1999. Thereafter three more systems were purchased (Table 1). Another system was ordered recently for one joint venture mine of Tiefa at Shanxi Province and will be delivered in early 2013. All these five systems took so called "Tiefa Model". Nevertheless, this does not mean that each system is a simple repeat of the previous systems. Based on the experiences gained in utilizing previous systems and the available cutting-edge technology, the improvements were made for each new system, mainly in the following aspects:

a) Local shields

Both face end shields and face shield were supplied by the local suppliers in order to save the equipment investment. The interfaces between the local shields and the imported ram cylinders with reed rod, outrigger steering cylinders, E/H control units, valve blocks, and pressure sensors, are always the key topics during the design liaison meetings and the prototype tests conducted at the workshops of local shield suppliers to ensure a better performance of each system.

- Face end shields

  The face end shields are employed at both main gate end and tail gate end to protect the miners working at two ends and to push the plow and AFC drives and BSL. All face end shield have four legs. The width of the Tiefa 1 face end shields is 1880 mm with the support resistance being 9900 kN, which were proven later to be too heavy and large. Starting with Tiefa 2 system, 6200 kN became a standard support resistance of the face end supports. The support density can reach above 0.64 MPa while the contact pressure at base is 1.8 MPa. The hydraulic stroke is 600 mm. The face end shields of Tiefa 1 are manually controlled while those of Tiefa 2 and 3 are upgraded to PM4 E/H controlled. PMC-R instead of PM4 was ordered for Tiefa 4 with 12 functions being available.

- Face shield

  All face shields of the four systems have 2 double telescopic legs. The contact pressure at base was reduced from 2.5 MPa of the Tiefa 1 to 1.77 MPa of the Tiefa 2 and 3 and furthermore to 1.33 of the Tiefa 4 to suit soft floor. The auto-controlled water spray was added on the AFC spill plates of the Tiefa 3. In the Tiefa 4 the water spray nozzles were moved to the canopies of the face shield and controlled by PMC-R. The functions of the E/H control were increased from 4 in the Tiefa 1 to 6 in the Tiefa 2 and 3 and further to 12 in the Tiefa 4. The shield closed height is always the most difficult in the design and manufacture of low seam shields. The closed height of the Tiefa 1 shields is 900 mm. It was lowered to 800 mm for the Tiefa 2 and 3 and further to 600 mm for the Tiefa 4. This height is expected to be reduced to 550 mm for the upcoming Tiefa 5 for the joint venture mine in Shanxi with the support resistance being 4400 kN.

b) AFC
The installed power was increased from 2x315 kW to 2x400. Accordingly the face length was increased from 200 m to 245 m. The cross-frame drive frames were adopted in the Tiefa 2 and 3 for the better performance in flat seams.

c) Plow
The installed power was increased from 2x315 kW to 2x400 kW. Meanwhile the plow body travelled faster, from 0.88/1.76 m/s to 0.96/1.92 m/s. The GH9-38Ve/5.7N plow body with a minimum height lower than that of GH9-38Ve/5.7 was ordered for the Tiefa 4 to prepare for the lower seams.

d) Plow chain protection
S-15-LK plow gearboxes were installed in the Tiefa 1 and 2. Their mechanical chain protection was difficult to calibrate and did not perform well. As a result, the plow chain breakages occurred from time to time. The P-30 UEL gearboxes with IKM were used as a substitute in the Tiefa 3 and 4 and achieved a better result. The Tiefa 2 was also upgraded to IKM. The P-30 UEL gearboxes with PMC-D/V will be included in the Tiefa 5.

Table 1. Configurations of automated plow systems of Tiefa

System		Tiefa 1	Tiefa 2	Tiefa 3	Tiefa 4
Face Length (m)		207	215	230	245
Face end shield	Model	ZZ9900/17/30	ZT6200/18/32D	ZT6200/18/32D	ZT6200/18/32D
	Legs	4	4	4	4
	Height (mm)	1700~3000	1800~3200	1820~3200	1800~3200
	Width (mm)	1880~2130	1420~1590	1440~1610	1420~1590
	Shield center (mm)	2000	1500	1500	1500
	Pump pressure (Mpa)	31.5	31.5	31.5	31.5
	Support resistance (kN)	9900	6200	6200	6200
Face shield	Model	ZY6400/09/20D	ZY5200/08/18D	ZY5200/08/18D	ZY4800/06/16.5D
	Legs	2	2	2	2
	Height (mm)	900~1800	800~1800	800~1800	600~1525
	Width (mm)	1440~1600	1440~1600	1440~1600	1440~1600
	Shield center (mm)	1500	1500	1500	1500
	Pump pressure (Mpa)	31.5	31.5	31.5	31.5
	Support resistance (kN)	6400	5200	5200	4800
AFC	Model	PF2.30/732	PF3/822	PF3/822	PF3/822
	Installed power (kW)	2x315	2x400	2x400	2x400
	Discharge	Overhead	Cross-frame	Cross-frame	Overhead
Plow body	Model	GH9-34Ve/4.7	GH9-38Ve/5.7	GH9-38Ve/5.7	GH9-38Ve/5.7N
	Height (mm)	800~1675	880~1645	880~1645	800~1585
	Installed power (kW)	2x160/315	2x200/400	2x200/400	2x200/400
	Speed (m/s)	0.88/1.76	0.88/1.76	0.96/1.92	0.96/1.92
Electrical	Control unit	central control	central control	central control	central control
	Plow chain protection	Mechanical	Mechanical	IKM	IKM
	Shield control	PM4	PM4	PM4	PMC-R

3 PERFORMANCE OF AUTOMATED PLOW SYSTEMS

The first automated plow systems was put in operation at Xiaoqing Mine in early 2001. Until 2009 nineteen panels have been finished using the four plow systems at the three mines of Tiefa (Table 2).

Table 2. Faces mined by the automated plow systems

	Plow system	Mine Name	Face Name	Face length (m)	Mining height (m)
1	Tiefa 1	Xiaoqing	W1E-703	200/150	1.34
2	Tiefa 1	Xiaoqing	W1W-712	200	1.46
3	Tiefa 1	Xiaoqing	W1E-702	200	1.34
4	Tiefa 1	Xiaonan	W3-409	165	1.7
5	Tiefa 1	Xiaoqing	W1E-701	200/150	1.35
6	Tiefa 1	Daming	E1S-704	196	1.5
7	Tiefa 1	Daming	E1S-703	200	1.7
8	Tiefa 1	Daming	E1S-705	200	1.6
9	Tiefa 1	Daming	E1S-702	200	1.6
10	Tiefa 1	Daming	E1S-702	155	1.7
11	Tiefa 2	Xiaonan	W3-410	215	1.8
12	Tiefa 2	Xiaoqing	W2-704	215/170	1.67
13	Tiefa 2	Xiaoqing	W2-705	225	1.5
14	Tiefa 2	Xiaoqing	W2-708	215	1.5
15	Tiefa 2	Xiaoqing	S1W-701	215	1.7
16	Tiefa 2	Xiaoqing	W1E-705	220	1.2
17	Tiefa 3	Xiaoqing	W2-719	230	1.8
18	Tiefa 3	Xiaoqing	W1E-704	230	1.3
19	Tiefa 4	Xiaoqing	W2-713	207	1.4

As a typical automated plow application in Tiefa described below, Panel W3-409 of Xiaonan Mine was successfully mined using Tiefa 1 plow system October of 2002 through February of 2003.

- *Mining conditions:*
  Panel width: 165 m
  Panel length: 924 m
  Ave. seam thickness: 1.7 m
  Coal hardness: f=2~3
  Face inclination: 3 ~10 degree
  Friable roof: mudstone, ave. 0.4 m
  Immediate roof: fine sandstone, sandstone, ave. 8.4 m
  Friable floor: mudstone, ave. 0.1 m
  Immediate floor: sandstone, ave. 2.9 m
  Overburden depth: 407~507 m
  Total panel reserves: 580,000 tons
- *Longwall equipment:*
  Plow system: Tiefa 1(see Table 1)
  BSL: SZZ764/200
  Crusher: PEM650x1000
- *Selected cutting depth:*
  Head to tail: 30 mm
  Tail to head: 50 mm
- *Outputs:*
  Ave. daily output: 4852 tons

Best daily output: 6658 tons
Ave. monthly output: 142,200 tons
Best monthly output: 145,600 tons

Xiaonan's experiences show that under certain conditions a shearer longwall can also come relatively close to that performance. However, the automated plow longwall enjoyed obvious economic benefits besides the improvement in safety and working environment. First of all it used less manpower (Table 3 and 4) due to its automation and thus its manpower efficiency was improved from 90.5 tons/man-shift to 140.1 tons/man-shift. Secondly, the total direct operation costs of the panel (consumable & wear parts, labour, materials, grease, and equipment rental) were reduced by 2.2 million Chinese Yuan. Thirdly, if a shearer was used for this face, 600 mm roof /floor rock would have been cut in order to get enough space for the operator's travelling along the face, which would have increased rock content in coal by 15%. The better quality coal produced by the plow increased the profit by 12 million Chinese Yuan.

Table 3. Manpower Organization Shearer Longwall

Job title	Shift				
	Maintenance	Production 1	Production 2	Production 2	Subtotal
Forman	1	1	1	1	4
Shift head	1	1	1	1	4
Shearer operator		2	2	2	6
Shield operator		3	3	3	9
Face end worker		4	4	4	12
Electrician	5	1	1	1	8
Mechanic	10				10
Backup		3	3	3	9
Total	17	15	15	15	62

Table 4. Manpower Organization Plow Longwall

Job title	Shift				
	Maintenance	Production 1	Production 2	Production 3	Subtotal
Forman	1	1	1	1	4
Shift head	1	1	1	1	4
MCU operator		2	2	2	6
Face end worker		4	4	4	12
Electrician	4	1	1	1	7
Mechanic	7				7
Backup		2	2	2	6
Total	13	11	11	11	46

4 GATE ENTRIES, SETUP ROOM AND RECOVERY ROOM

Gate entries

Compared with a shearer longwall, a plow longwall requires wider head & tail gate entries because its all drives are in the gate entries. However the wider entries are more difficult to support and maintain. To solve this problem Tiefa has tried every means to prepare the entries wide enough to meet the requirements of the plow systems. Take Panel W3-409 of Xiaonan for example again, the dimensions of its rectangular headgate entry were 5.6 m x 2.5 m while the dimensions of its rectangular tail entry were 5.0 m x 2.5 m. Seven Φ22 mm x 2200 mm fully grouted resin bolts per row, 800 mm apart and 800 mm between rows, plus two Φ15.5 mm x 5300 mm (pattern 2000 mm x 2400 mm) cable bolts were used to support the roof of the head gate entry. In the tail gate entry, five Φ22 mm x 2200 mm fully grouted resin bolts per row, 900 mm apart and 900 mm between rows, plus two Φ15.5 mm x 5300 mm (pattern 2000 mm x 2400 mm) cable bolts were used for the roof. In both the head and tail gate entries, Φ22 mm x 2000 mm resin bolts (pattern 800 mm x 800 mm) plus 80 mm x 120 mm mesh were installed on each rib side (Figure 1 and 2). The capacities of the roof bolt, cable bolt, and rib bolt were required larger than 120 kN, 180 kN, and 50 kN respectively.

The above design effectively controlled the entry deformation and thus ensured the success of the plow longwall.

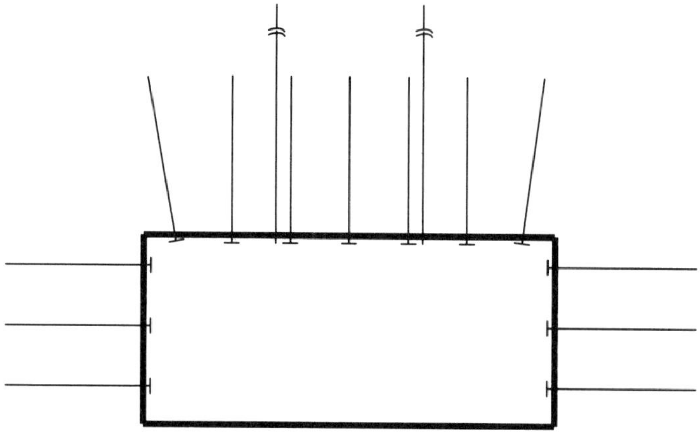

Figure 1. Head gate entry of plow panel

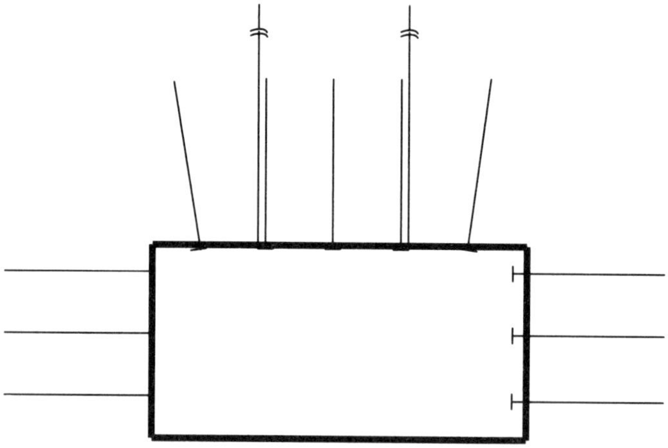

Figure 2. Tail gate entry of plow panel

Setup room

The dimensions of a setup room shall meet the requirement of the transportation and lifting of the face shields during the face installation. In Tiefa's practice, a plow setup room normally was 7.0 m wide and 2.3~2.4 m high and developed along the floor. Roof rock needed to be cut in order to reach the required height. To keep the shields contact the roof well and enter the seam smoothly 1.9 m wide roof on face side was developed inclined towards the face. Nine Φ22 mm x 2200 mm fully grouted resin bolts per row, 800 mm apart and 800 mm between rows, plus two Φ15.5 mm x 5300 mm (pattern 2000 mm x 2400 mm) cable bolts were used for roof support. Two Φ22 mm x 2200 mm bolt were installed on the face side rib and the gob side rib (Figure 3).

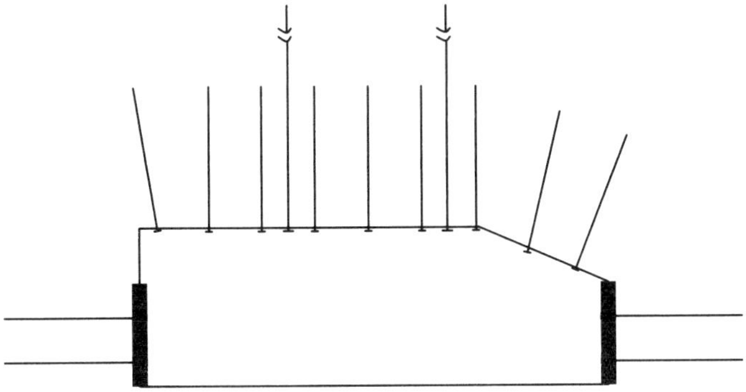

Fig 3. Setup room for plow longwall

5 CONCLUSIONS

The Cat automated plow systems helped Tiefa to achieve significant benefits. First of all, numerous million low seam reserves have been recovered. Secondly, the service life of the Tiefa coal mines is importantly extended. Assuming the total production capacity of Tiefa remains 20 million ton per year in the future, 620 million tons of low seam reserves will make increase the service life of Tiefa coal field by 31 years. Thirdly, the working environment and safety of Tiefa miners have been improved considerably due to an application of automated plow systems from Caterpillar.

Chinese experiences show that although under certain conditions a shearer longwall can come close to the performance of a plow; the automated plow longwall enjoys obvious economic benefits besides the improvement in safety and working environment. It uses less manpower due to its automation, delivering higher efficiency and has significantly lower operating costs. Furthermore, through the application of plows much cleaner output can be reached, as an operation of shearers requires an additional roof/floor cut of several hundred millimetres.

Low seam reserves account for 20% of the total coal reserves in China. But the coal produced from the low seams contributes only 10% of the total production of China because of the low efficiency of the low seam mining without suitable equipment. Encouraged by Tiefa's successful experiences of utilizing Cat automated plow systems, 6 Chinese coal companies have imported Cat automated plow systems, making the total number of Cat automated plow systems reach 11. More potential end users are discussing with Caterpillar in order to apply Cat automated plow systems including more powerful GH1600 to their low seam mining.

International Mining Forum 2011, Dyczko et al. (eds) © 2011 Taylor & Francis Group, London, UK. ISBN 978-0-415-68938-0

Past Experiences and Further Expectations with Regard to Application of Plow Technology at Jastrzębska Spółka Węglowa S.A.

A. Tor
Jastrzębska Spółka Węglowa S.A., Jastrzębie-Zdrój, Poland

ABSTRACT: The paper presents benefits that result from application of plow technology to extraction of thin seams as compared to conventional extraction with use of longwall shearers. The up-to-date experience and scope of further plans associated with coal extraction with use of plow systems at JSW S.A. are outlined with practical outcome acquired during past operations. Application of the plow technology with continuous improvements against the background of gained expertise and consequential innovative solutions are crucial for JSW S.A. due to substantial portion of thin seams in overall resources of comprised coal mines.

KEYWORDS: Hard coal mining, thin seam, plow technology

1 INTRODUCTION

Shearer-based longwall systems still remain the most common coal extraction method used in Polish coal mining industry. The policy aimed at achievement of profitability and implemented at Polish coal mines prefers extraction of thick and medium coal seams (with the height above 1.5 m). It leads to unidirectional technical development of mining industry where new designs are limited to machinery and equipment inherently dedicated to the needs of shearer-extracted longwalls. The problem how to mine coal from thin beds has been neglected and considered as a secondary issue, both due to limited availability of modern machinery and equipment capable of operating under such mining conditions and nearly no demand for such facilities from coal mines.

The long-term experience provides the proof that the shearer system enables efficient extraction form coal beds if their thickness is more than 1.5 m. Mining from seams with the lower thickness entails a number of technical problems that significantly affect productivity of shearers, economic effects and occupational safety of personnel. These hindrances include:

- dramatically confined space within the longwall areas that makes the work extremely burdensome with very little comfort and involves a number of safety hazards to the numerous staff,
- the need to adjust cutting speed of a shearer to physical abilities of its operators to follow its movements within very confined and low excavations, which leads to drop of the shearer productivity,
- no possibility to install loaders on cutting drums of a shearer, which makes it more difficult to load coal onto the AFC,

– prolonged downtimes in case of possible failures, both of the shearer and the powered roof support units or the AFC /BSL system,
– extraction of coal-adjacent waste rocks by cutting gangue from the roof or the floor, loses due to haulage of contaminated coal to the surface, coal separation and gangue storage.

Coal seams extracted by coal mines owned by JSW S.A. comprise a significant portion of coal that is deposited in beds with height below 1.5 m. Since demand for coking coal offered by JSW S.A. remains very high and structure of available resources shows exhaustion of thick seams, the possibility to efficiently extract coal from coal beds with thickness less than 1.5 m has become one of the urgent priorities for further operation and growth of the company. Such seams present a significant portion of resources available to such coal mines as "Pniówek" as well as "Zofiówka" colliery within the "Borynia-Zofiówka" coal mine. In addition, efficient use of coal resources deposited in thin seams shall be beneficial to substantial prolongation of lifetime expected for coal mines within JSW S.A.

2 PLOW TECHNOLOGY

Prior to making decision about implementation of the plow technology at Jastrzębska Spółka Węglowa S.A. the Company's management and engineering staff held a series of meetings with manufacturers of mining machinery where technical feasibility of coal extraction from thin seams with use of shearers manufactured and offered by those companies was subjected to thorough analysis. Despite the pending efforts intended to design a machine dedicated to efficient cutting of coal from thin seams no reasonable solution seems to be available in the market that would be able to compete against plow systems in terms of efficiency and safety of coal extraction from seams with thickness ranging from 1.0 m to 1.5 m. Having considered the above, the Managing Board of JSW S.A. came up with the decision to apply the plow technology to cut coal from thin seams. The plow was first introduced to the Hard Coal Mine "Zofiówka", then to the Hard Coal Mine "Jas-Mos" and in future it shall appear at the Hard Coal Mine "Pniówek".

The plow system is an extraction method where the entire coal cutting process is managed by the preprogrammed algorithm executed by a computer. The control system for a plow – based longwall offers a number of benefits, including but not limited to the following:
– monitoring of essential parameters that depict operation of the longwall,
– identification of the plow position and measurement of its traveling distance down the longwall,
– predefinition of the plow immersion and web of the plow head during operation,
– automatic reversion of the cutting direction and control of cutting speed for the plow head,
– automatic advance of the roof support units,
– sequential spraying of the cutting head with water,
– manual control of the powered roof support units in case of equipment failures,
– monitoring of hydraulic pressures across supplying and drainage lines,
– on-line monitoring of the longwall front with possibility of necessary adjustments.

The plow systems, as compared to shearer ones, feature common applications of roadway roof support units and the need to construct large-sized tail and head gates due to large dimensions of driving units for both the plow and the AFC located in longwall headings.

3 OPERATION OF THE PLOW SYSTEM AT THE HARD COAL MINE "BORYNIA-ZOFIÓWKA"

The first, brand new set of plow equipment was applied for coal cutting at the B-1 longwall within the 406/1 coal bed. Intense efforts were made to appropriately prepare the personnel for correct installation, operation and maintenance of the plow system, both in theoretical and practical aspects. Trainings for the staff were commenced in December 2008 during the compatibility test of the automated set of plow machinery that was carried out at the premises of the plow manufacturer. The subsequent phase of trainings took place at the Specialized Training Centre in Czech Republic as well as DSK coal mines in Germany. In addition, to make the maximum possible number of employees familiar with the plow technology, a part of plow-based longwall equipment was assembled on the surface of the "Zofiówka" coal mine and one cycle of personnel training was carried out therein.

The actual operation of the longwall lasted from August 2009 to April 2010.

The B-1 longwall within the 406/1 coal bed was extracted with the use of longitudinal longwall system with roof fall.

Basic technical parameters of the longwall were the following:

- average length of the longwall – 244.6 m
- average thickness of the longwall /maximum operation gate – 1.17 m / 1.44 m
- average volumetric density of coal – 1.31 t/m^3
- total longwall panel length – 806 m
- crosswise slope – $-4 \div +6^{o}$
- longitudinal slope – $0 \div 8^{o}$
- average advance per day – 5.97 m
- average daily production between the start-up period and preparation of the longwall to have it closed down – 2500 t/day
- maximum output achieved – 5100 t/day

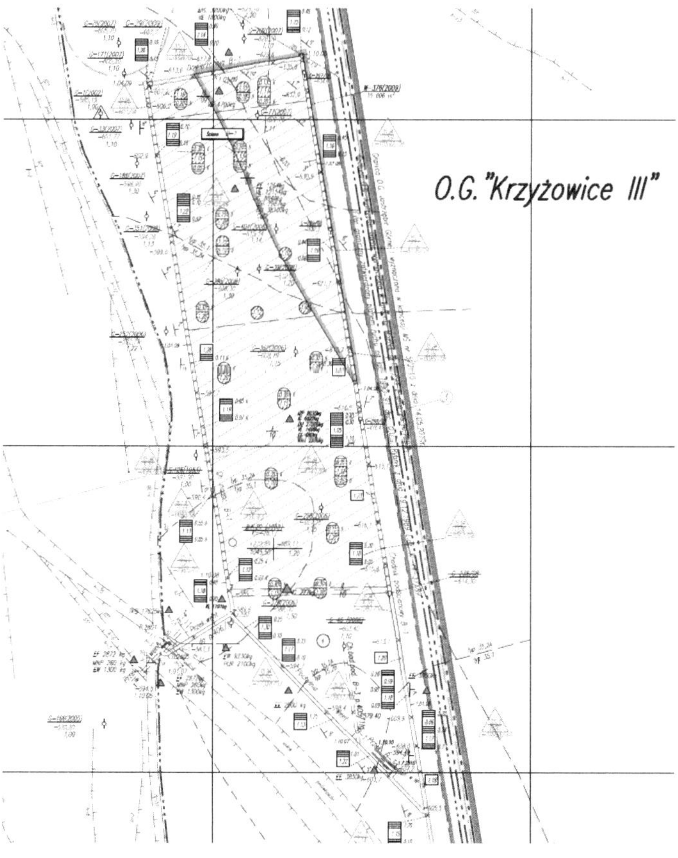

Figure 1. Extract from the map of the 406/1 bed with the B-1 longwall

The experience acquired during operation of the plow-based system clearly shows that application of the plow technology for coal extraction from the B-1 longwall within the 406/1 coal bed was the most beneficial solution.

The use of plow technology made it possible to reduce contamination of the coal output with gangue, improved occupational comfort and personnel safety (lower noise level, increased automation of the coal cutting process) and allowed to reduce number of miners directly involved in operation of longwall equipment. The overall production from the 406/1 coal bed amounted to 299,400 tonnes of coal. None of that coal could be mined with use of a longwall shearer.

It must be mentioned here that coal extraction from the B-1 longwall within the 406/1 coal bed was carried out in difficult mining and geological conditions. The longwall featured with substantial methane hazard (the average methane emission was 13.3 m^3/min) and climatic hindrances (the initial temperature of virgin rock was 37 to 42ºC). In addition, the roof was made up of rock with low cohesion and vulnerable to spontaneous falls of rock lumps.

Extraction of coal from the B-1 longwall made it possible to gain necessary experience related to application of the plow system to coal cutting from thin beds.

4 OPERATION OF THE PLOW SYSTEM AT THE HARD COAL MINE "JAS-MOS"

In order to maintain high utilization of the plow system and to minimize the duration of downtime periods the decision was made that after completion of coal cutting at the B-1 longwall within the 406/1 coal bed at the Hard Coal Mine "Zofiówka" the plow system shall be transferred to the Hard Coal Mine "Jas-Mos" to extract coal from the longwall no. 22 within the 503/1-2 coal bed.

Initially, the "Jas-Mos" coal mine intended to use the plow system exclusively for coal extraction from the longwall no. 22 but due to receiving the system for long-term use and fabulous opportunity to benefit from the equipment application at the Hard Coal Mine "Jas-Mos" the coal mine has developed the program of coal extraction from thin beds with use of the plow technology. The program assumes that not only the 22-C3 longwall within the 503/1-2 coal bed can be mined that way but subsequent longwalls will also become available.

The 22-C3 longwall within the 503/1-2 coal bed was extracted from November 2010 to April 2011.

Basic technical parameters of the longwall were the following:

- average length of the longwall – 188 m
- average thickness of the longwall /maximum operation gate – 1.3 m / 1.45 m
- average volumetric density of coal – 1.37 t/m^3
- total longwall panel length – 570 m
- crosswise slope – 0÷10°
- longitudinal slope – 0÷7°
- average advance per day – 6.9 m
- average daily production between the start-up period and preparation of the longwall to have it closed down – 2500 t/day
- maximum output achieved – 4200 t/day

The overall output from the longwall no. 22 amounted to 217.6 thousand tonnes of coal that could never be mined with use of the shearer technology.

It must be mentioned that during the initial phase of the longwall extraction, the longwall no. 22 within the 503/1-2 coal bed exercised a number of faults with drops about 0.5 m as well as the coal bed was disturbed by extraction from the 505/2 coal bed deposited some 55–60 m below (the longwall no. 29).

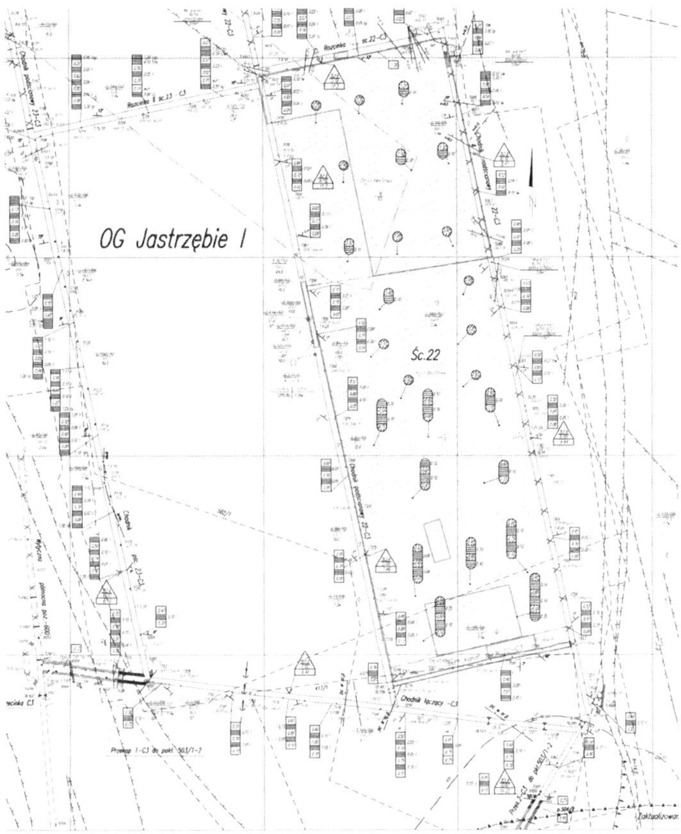

Figure 2. Extract from the map of the 503/1-2 with the longwalls no. 22 and 23

4 PROSPECTS FOR FUTURE APPLICATION OF THE PLOW TECHNOLOGY AT COAL MINES OF JSW S.A.

The JSW S.A. follows the policy of reasonable coal extraction from its reserves and therefore it is going to continue coal extraction from thin deposits with use of the plow technology.

Hard Coal Mine "Jas-Mos"

After completion of coal extraction from the longwall no. 22 within the 503/1-2 coal bed the plow-based equipment was transferred to the longwall no. 23 within the same coal bed, where coal cutting was commenced in August this year. Subsequently, the equipment is anticipated to be applied for the longwall no. 53 within the 505/1 coal bed and then the longwall no. 4 within the 510/2łd coal bed.

Table 1. Parameters of longwalls to be extracted with use of the plow technology

Parameter	Unit	longwall no. 23 coal bed 503/1-2	longwall no. 53 coal bed 505/1d	longwall no. 4 coal bed 510/2łd
Longwall length	m	240	150	90–245
Total longwall panel length	m	620	450	900
Longwall height	m	1.4	1.4	1.4
Reserves	kt	270.8	125.6	374.7

Hard Coal Mine "Borynia-Zofiówka", the Zofiówka colliery

After having completed coal extraction from the B-1 longwall within the 406/1 coal bed the colliery intends to keep cutting coal from thin deposits with the use of plow technology. The reserves to be extracted that way include the "F" part of the 406/1 and 408/2 coal beds, the "B" part of the 407/1 coal bed, the "H" part of the 409/3 coal bed and the "D" part of the 410 coal bed.

Since the existing set of plow-based machinery was handed over for further operation at the Hard Coal Mine "Jas-Mos", the "Borynia-Zofiówka" purchased a new plow system that comprises a plow-dedicated AFC, the plow itself, the power supply and control system, powered roof support units together with auxiliary equipment (the BSL equipment, floor loaders, etc.). The system shall be applied to extraction of coal from the F-1 longwall within the 406/1 coal bed and the longwall operation is anticipated to commence September this year.

According to the adopted schedule, from the year 2011 to 2018 the coal mine intends to extract coal from seven longwalls with use of the plow technology.

Table 2. Parameters of longwalls to be extracted with use of the plow technology

Parameter	Unit	F-1 coal bed 406/1	F-3 coal bed 406/1	F-5 coal bed 406/1	D-2 coal bed 410	B-3 coal bed 407/1	B-3a coal bed 407/1	H-3 coal bed 409/3
Length	m	245	250	245	246	170	180	195
Height	m	1.20	1.20	1.20	1.30	1.70	1.70	1.60
Total panel length	m	1,080	760	520	685	1,070	1,280	600
Reserves	kt	412.8	296.4	198.7	284.8	402.0	509.2	243.4

Hard Coal Mine "Pniówek"

It is anticipated that after completion of the plow system use for coal extraction in longwalls of the "Jas-Mos" coal mine the machinery in near future shall be transferred to the Hard Coal Mine "Pniówek" where it can be operated from 2014 for coal cutting from longwalls within the coal beds no. 356/1, 357/1 and 358/1 in the eastern part of the mining area.

It order to extend the accessible reserves JSW S.A. has commenced activities that are necessary to develop the coal seams directly adjacent to "Pniówek" coal mine and "Zofiówka" colliery in order to make them available for extraction. These seams include deposits called "Pawłowice 1", "Bzie-Dębina 2-Zachód" and "Bzie-Dębina 1-Zachód". The structure of reserves in deposits that are to be extracted demonstrated a substantial portion of coal beds with thickness below 1.5 m that can be preferably extracted with the use of plow technology. The anticipated method of the seam partitioning and development assumes that the longwalls shall have lengths from 200 to 250 m and the panel lengths significantly above 1000 m, which shall make it possible to use the machinery and equipment in much more efficient way.

Due to substantial contribution of thin coal seams in the future reserves of the company, further operation of JSW S.A. great lyrelies on successful application of the plow technology. That technology will enable to:

1. Increase the degree of coal bed extraction. Pretty high number of longwalls is totally unmineable without application of the plow technology, and such reserves would have to be irrecoverably abandoned.
2. Reduce contamination of coal output with gangue.
3. Improve occupational comfort and safety of the personnel involved in exploitation.
4. Extract coal from thin beds in seams called "Bzie-Dębina 2-Zachód", "Bzie-Dębina 1-Zachód" and "Pawłowice 1" with substantial contribution of coal beds with thickness from 1.0 to 1.5 m.

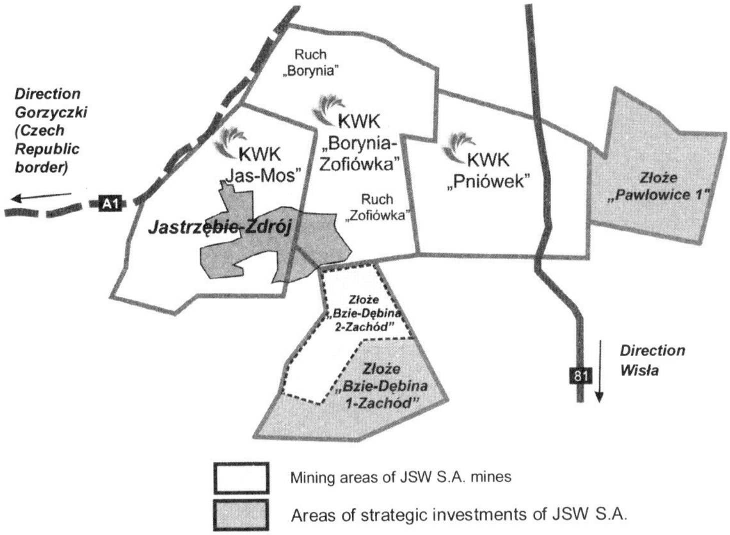

Figure 3. Location of coal seams called "Bzie-Dębina 2-Zachód", "Bzie-Dębina 1-Zachód" and "Pawłowice 1"
Notations: złoże – coal seam, Ruch – colliery

6 EXPERIENCE OF COAL MINES OPERATED BY JSW S.A. WITH REGARD TO APPLICATION OF THE PLOW TECHNOLOGY

To date only two longwalls in coal mines of JSW S.A. have been extracted with the use of innovative plow technology. However, it enabled to acquire much experience and to formulate some specific suggestions how to improve operation of plow systems under conditions of coal mines operated by JSW S.A.

Due to methane and climatic hazards that are typical for these coal mines it seems reasonable to apply the "Y" style of ventilation with discharge of used air down goafs in order to minimize effect of these hazards on efficiency of coal extraction from the longwalls.

In addition, to mitigate methane hazard, it is also sensible to apply advancing methane extraction that consists of serial drilling of methane-draining boreholes and suction of methane much before coal extraction begins.

Due to low thickness of coal beds, most of roadway and gate excavations are drilled in floor and roof rock. In order to achieve maximum possible drilling efficiency it is essential to appropriately design dimensional parameters of such excavations with consideration to their future use, in particular concerning dimensions of machinery and equipment that must be installed in head and tail gates. These purposes may entail the need to apply non-standard types and sizes of roof support facilities.

Provided that appropriate drilling and support technologies are applied, relatively low extraction height makes it possible to maintain the already drilled head or tail gates in working conditions and reuse them for subsequent longwalls.

In order to improve occupational safety and speed up the closure of plow-based longwalls with evacuation of equipment it is reasonable to drill an excavation (a roadway) in advance down the line where the longwall is to be finished. Such excavation shall be used as an evacuation passage when the longwall extraction is completed. Such solution was successfully applied to the longwall no. 22 of the 503/1-2 coal bed in the "Jas-Mos" coal mine.

During operation of the plow system in the first longwall (the B-1 longwall within the 406/1 coal bed in "Zofiówka" colliery) some problems were encountered with durability of some structural nodes of the snaking system for the beam stage loader (BSL). The experience gathered in that scope enabled to introduce some design alterations that allowed to avoid similar problems when the subsequent longwall was mined.

The entire experience acquired from operation of plow systems under geological and mining conditions of coal mines operated by JSW S.A. makes it possible to conclude that the most hindrances in application of the plow technology may be associated with the following factors:

- occurrence of faults, in particular in regions of cohesive roof and floor rock,
- sudden and unexpected waving or interleaves of coal seams with intruded minerals,
- occurrence of spontaneous falls of rock lumps from roof and the associated need to protect the excavation roof within a low longwall.

Currently JSW S.A. assumes that the plow technology can be applied exclusively under geological conditions that can be assessed as favorable or with medium difficulties.

7 CONCLUSIONS

Jastrzębska Spółka Węglowa S.A. would like to maintain its market position among leading worldwide producers of coking coal. This is why the company undertakes new measures with the aim to expand its accessible resources and extends the existing coal mines (development of new extraction levels) as well as makes new seams available. The analysis of reserves and their structure demonstrates that a substantial portion of coal is deposited in thin beds with the thickness below 1.5 m. In order to take the maximum advantage of available deposits and prolong lifetime of coal mines JSW S.A. has commenced efforts to efficiently extract coal from such thin deposits. Since no shearer-based solutions proved to ensure the required level of productivity, safety and cost-effectiveness the decision was made to purchase a plow system. For the first time the system was used to extract coal from the B-1 longwall within the 406/1 coal bed in the Hard Coal Mine "Zofiówka" in August 2009. Successful results from operation of the plow system confirmed that the decision was right and further beds could be mined with its use as well.

Currently the plow technology is used in two coal mines of the Company: the former "Zofiówka" coal mine, nowadays called the "Zofiówka" colliery within the "Borynia-Zofiówka" coal mine as well as the "Jas-Mos" coal mine. In near future the technology shall be applied to "Pniówek" coal mine. However, the technology shall become really important when the new coal seams called "Bzie-Dębina 2-Zachód", "Bzie-Dębina 1-Zachód" and "Pawłowice-1" are ready for extraction. Therefore it is essential to acquire necessary experience and good practices for safe and efficient coal extraction with the use of plow systems. The expertise, gained sufficiently in advance, makes it possible to train the mining personnel and adapt the necessary technical solutions, develop the concept of coal bed partitioning, calculate dimensions of head and tail gates according to results and past experience from coal mining with use of plow systems.

REFERENCES

Tor A., Czernecki Z., Kubaczka Cz., Celmer M., Pluta Z., Lubryka M. 2011. Dotychczasowe doświadczenia i dalsze plany w zakresie zastosowania techniki strugowej w Jastrzębskiej Spółce Węglowej S.A. – Szkoła Eksploatacji Podziemnej 2011. (*Past experience and further plans related to application of the plow technology at coal mines operated by the Jastrzębska Coal Company S.A. – The School of Underground Extraction, 2011)*

International Mining Forum 2011, Dyczko et al. (eds) © 2011 Taylor & Francis Group, London, UK. ISBN 978-0-415-68938-0

Efficiency Optimization of Plow Systems Through the Precise Planning of New and Comprehensive Enhancement of Operating Longwalls

M. Myszkowski
Caterpillar Global Mining, Luenen, Germany

ABSTRACT: Plow technology reemerged and increased in importance in recent years, as many operators worldwide searched for an effective technology for extraction of thin and medium thickness coal seams. Plow systems were significantly improved in recent years. This resulted in higher production from plow faces. Nevertheless, there is an appreciable potential for further performance increase by the means of optimal technical, procedural and organizational improvements. Very important for the achieving of planned production rates is a correct design of plow faces. On the other hand, in many cases a performance improvement is possible without significant changes of the longwall system, simply requiring consistent analysis and effective realization of conclusions.

KEYWORDS: plow longwall operation, proper design, operational procedural and organizational factors, performance improvement and efficiency increase

1 INTRODUCTION

Plow longwall systems have been working for more than seventy years. In that time thousands and thousands of plow faces were in operation worldwide. The technical layout of plows has changed dramatically since the first years. The rated power increased by 10 to 12 times, the equipment became decisively more powerful and first of all: computer based systems were implemented for controlling plows allowing their full automation. There is a clear correlation between above named factors and performance of plow faces, clearly identifiable by analyzing the respective numbers from the last four decades. On the other hand the total running time of plow systems was moving within certain range over the years, proving to be difficult to increase. This paper concentrates on technical, operational, procedural and organizational factors aiming at maximizing the performance, as in most cases there are numerous possibilities to increase production significantly without changing the overall system configuration.

2 PLANNING OF A PLOW SYSTEM

2.1 *Technical layout of a projected face*

The technical layout of a longwall has a crucial influence on its future performance. In this phase, an in-depth analysis of all data, and information, influencing factors for all the important areas is strongly recommended, indeed critical.

In general terms, the equipment selection for a planned face depends on the required production in connection with the all important mining, geological, logistical factors depending on the disposable financial means. It is a vital subject to select the face equipment in a coordinated way in order to reach the highest production and to avoid bottle-necks, which could hamper the production. The choice of the right equipment has to be carried out most thoroughly, as potential mistakes made during the design phase either cannot be eliminated at all or if so, only with immense efforts. In the case of plow faces the most important equipment features that count:

- Installed power and speeds of plow and AFC,
- Plow body type and height range,
- AFC width and spill plates height,
- Discharge type,
- Height range and support density of shield support,
- Type of automation and control system,
- Power supply,
- Conveyance method for the extracted mineral.

Modern plow systems use gearboxes with overload protection for both the plow and the AFC. The overload protection secures the chains against forces resulting from rotational energy stored in turning motor rotors. For that purpose multi-disk clutches are used to limit the carried torque.

Another important element of modern plow systems are variable frequency drives. These drives contribute the following properties:

- Protection against electric network overloads,
- Unlimited number of start-ups of asynchronous motors,
- Infinite speed setting in the range 20-120% of nominal speed,
- Equal load shearing between two drives.

2.2 *Economic efficiency of the planned face*

Frequently while planning new equipment the lowest capital costs present for operators the pivotal criterion. While designing a new plow face the cost-effectiveness of applied equipment should be an object of in-depth analysis. In this phase an analysis of all initial capital expenses for different options should be carried out. Furthermore, for the eligible options, all the following expenses; i.e. operational, repair, maintenance and overhauling costs for the complete service life of the equipment should be identified and calculated. Last but not least, the overall production rate and total tonnage considered over the whole operating life for the options in question should be thoroughly examined and collected.

Having the above listed numbers for reckoned options, the lowest costs of one ton saleable coal should be considered as the determining criterion for the best option selection.

2.3 *Safety aspects*

While planning a new face, all necessary safety aspects have to be consciously considered. In order to optimize the future operation, the technical and procedural design aspects should be harmonized with safety related conditions. Frequently restrictions related to strata pressure, rock burst, fire, methane and water hazards can have a strong negative influence on the face performance.

3 ENHANCEMENT OF RUNNING FACES – OPERATING FACTORS

3.1 *Plow face shape*

Plow faces are most commonly operated in a bow shape, where the face in the middle is more advanced than at the drives by 1–2% of the face length, as a rule of thumb. This advance in mid-face has following reasons:

- Allows an effective horizon control along the face,
- Increases avoidance of roof fall-outs,
- Enables easy adjustment of differences in distance between gate-roads along a panel.

The bow should not be bigger than 2% face length, because in case of a more bent AFC, higher than normal forces of chains cause an increase of power consumption for the plow and AFC and as a consequence higher wear on all moveable parts.

The AFC should be kept untwisted along a face. A horizontal seesaw AFC run increases frictional resistance of the plow's chain or AFC's chain assembly.

Vertical undulations should be followed, although in case of very sharp synclines and anticlines, if possible a smoothing through the floor and/or roof cut could help to increase radiuses of AFC vertical bends. Sharp AFC bends close to drives, especially to the main drive, should be possibly avoided, as they strongly increase the power consumption and thereby wear.

3.2 Pre-tension of chain driven face equipment

For the proper functionality of a plow and AFC, the chains of both devices need a correct pre-tension. Too low a pre-tension can lead to too big chain slack and result in breakages. On the contrary, too high a pretension causes high power consumption as well as high wear, provoking in worst case a chain breakage.

The manufacturers of plows and armored face conveyors deliver for their products values of optimal chain pre-tension, calculated for given conditions as rated power, chain speed, face length and type of chain(s). Nevertheless, as a rule of thumb it can be stated; that the chain has its optimal pre-tension when during the highest load the biggest allowable slack chain occurs.

3.3 Operation of overload protection devices

Every comprehensive face system has its overload protection. Thus plow and AFC have over-current protection for their electric motors and protection against too high torques on drives. Hydraulically driven devices as shields are protected against too high a pressure. The settings for those protection systems are vital on one hand for a proper operation and on the other hand for sufficient safeguarding of the equipment. As the working conditions can change in the course of a panel the named setting have to be frequently checked and if necessary adjusted accordingly.

The periodic check of overload protection settings is the best pre-condition for long operating life of the equipment. For that reason the most important settings should be registered by control systems for a troubleshooting analysis.

3.4 Height of the plow body

Contemporary Caterpillar plow bodies have an adjustable height. The main height adjustment will be carried out by adding or removing intermediate extension blocks of the plow body. A height of a single intermediate extension block amounts to 250 mm. Intermediate blocks can easily be stacked, allowing the height adjustment in a wide range. Additionally, for the fine setting of the plow body height the continuously adjustable middle bit column can be used. The adjustable bit column can be extended by 300 mm.

Plow body height in most cases amounts to 60 to 80% of the face height. The top layer of coal seam left at the roof falls down as a rule after a number of plow cuts. This method of plow operation delivers most advantages like:

- The highest attainable webs,
- Low wear of plow body guidance parts because of small lever arm,
- No cutting of the roof by top bits,
- No danger of potential collisions with canopies.

When the coal seam is getting harder, cutting of full face height can prove necessary, as with growing coal hardness the parting of the seam at the roof deteriorates in many cases. In this situation the plow height is the same as the face height, with all disadvantages connected to that type of operation.

The following statement can be made as the general rule: the height of a plow body should be kept as low as the circumstances in the face allow it to be.

3.5 Bits configuration

Bits used on today's plow bodies are divided into:
- Carving bits short,
- Carving bits long,
- Bottom bits,
- Top bits.

The carving bits are narrower than the bottom or top bits. For hard seams or for seams with dirt bands, the plow body is normally equipped with all short carving bits, apart from the lowest and highest bit holders where bottom and top bits are used accordingly. In case of weaker seams, short and long carving bits are staggered on a plow body. Such a configuration delivers a higher stability in the vertical run of the plow body and lumpier coal.

3.6 Horizon control

The correct navigation of a plow system along the coal seam is one of most relevant activities in plow faces. There are two complementary mechanisms for the height control:
- Vertical position of the bottom bit on the plow body. There are four different positions on every plow body: two for diving, one neutral and one for climbing. The setting of the bottom bit position has an influence on the complete face length.
- Position and angle of plow guidance and AFC pans along the face by means of horizon control (called also outrigger steering) mechanism. Hydraulic cylinders placed on the gob side of pans enable setting of the guidance inclination towards coal face and thus diving or climbing of the plow system.

Under conditions of hard floor the horizon control actually requires no action. Once a neutral position of bottom bits and outrigger steering cylinders has been found, the plow system will glide automatically along the seam-floor layer.

The situation is more complicated in case of a weak floor. An efficient horizon control in weak floor conditions demands skills from the workforce. While commencing a face in new seam miners need to first learn the behavior of the AFC under new conditions and find out by trial and error the optimum settings. The horizon control is one of most difficult operations in plow faces, as the horizon control using outrigger steering is very sensitive and the changes of settings can be observed first on next shifts. Thus the workers who did the changes of settings do not have the possibility to observe the results of those changes, which complicates the learning process for the miners. For that reason, a moderate adjustment of outrigger steering settings in case of inexperienced workforce is recommended.

3.7 Avoidance of bottle necks

In order to optimize the use the longwall equipment, all elements of the face and coal clearance have to be harmonized. Otherwise, less capable elements of a longwall system will present a limiting link of the production chain. As a result the overall performance of a longwall decreases.
It is important in the design phase to consider all influencing factors resulting from:
- Geology – properties of the seam and adjoining rock,
- Strata pressure – roof support in gate-roads and in the face,

– Ventilation – air supply, methane occurrence,
– Energy supply – electric energy, compressed air, hydraulic emulsion, water,
– Logistics – conveying of coal, material transportation, personnel movement,
– Control system efficiency and functionality.

4 ENHANCEMENT OF RUNNING FACES – PROCEDURAL FACTORS

4.1 Optimal plow and AFC speed selection

Speeds of plow and AFC have to be considered very precisely, as they are crucial for the performance of a plow face. While selecting the speeds, a number of influencing factors have to be analyzed such as:

– Achievable plow cutting performance and according coal stream on AFC,
– Available cross section of all conveyors,
– Admissible torques of gearboxes,
– Maximum permissible chain forces,
– Maximum utilization degrees.

Every longwall is in a certain way individual and there is no universal procedure of ensuring the best performance. The most common procedure used in plow faces in recent years is the so-called "overtaking procedure" with 3 m/s speed of the plow and 1 m/s speed of the AFC. At a proportion of both speeds equal three, the coal stream coming out from the face is approximately constant, while plowing with long passes. The AFC is covered partially by threefold layers of load. If the required cross section of the conveyors is smaller, then in the case of speed proportion there is a difference, i.e. smaller than three.

In case of very weak coal and high faces the so-called "combination procedure" is used. At this procedure, the plow is moving on the pass to the tail drive with a speed comparable to the AFC speed, and on the pass to the head drive, i.e. while traveling in the same direction as the AFC, is moving with half speed. There are no multiple layers of load on the AFC at this procedure and thereby the coal stream is relatively low, allowing usage of conveyors with a small cross-section or operation with very high webs at wide AFCs.

4.1 Long passes or sectional plowing

In the first decades of plow face operation extraction was carried out using so-called "sectional plowing". In this procedure, reciprocating plow movements in a certain section of the face were piloted by miners distributed in that area. After achieving the intended face advance the roof support was advanced in this face section and the piloting team moved to next one. Typically, a face was divided in several sections. This type of operation was also used in face areas with roof fall-outs or faults, in order to have easier control the face.

In recent years, plowing from one drive to another was systematically implemented. This type of operation was enabled first of all due to implementation of effective automation. With improving level of control systems in plow longwalls, plowing in long passes is becoming more and more often applied, as it provides a decisively higher performance of the face.

4.3 Double web or double cut

There are two methods of cutting face-end areas, before moving back to the other drive:

– To cut the face-end twice with single web,
– To cut the face-end only once with double web.

Cutting the face-end twice with single web is used in faces with hard seams, where a double cut is difficult or not possible because of too high power consumption of plow motors and, accordingly, high chain forces. Cutting a face-end twice requires moving in low speed over long distances and more frequent changes of direction and thus more frequent switching of plow motors. These effects provoke a reduction of the procedural utilization degree and thereby cause a reduction in face performance.

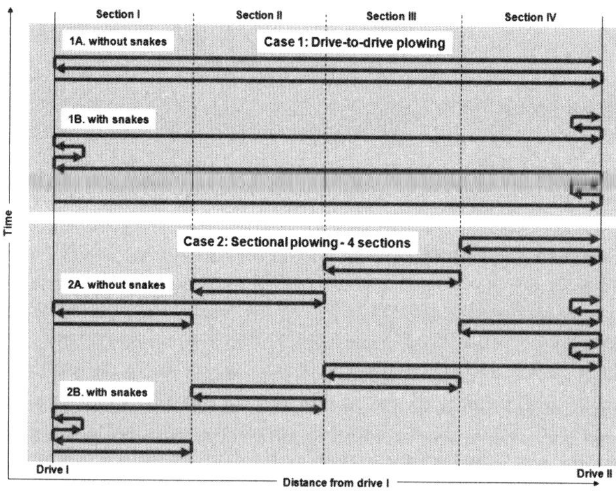

Figure 1. Plowing in long passes vs. sectional plowing

If the seam hardness allows cutting with double web at face-ends the procedural utilization factor and its related face performance increases by 10 to 20%.

Summarizing: the double web cutting at face-ends should be preferred, if technical circumstances allow the usage of this procedure.

4.4 *Procedural utilization degree and face performance*

The procedural utilization degree (PUD) describes a production diminishment caused by time and space losses related to an uninterrupted plow extraction with full web within the running time of a plow system. Using other words: PUD describes the efficiency of plowing. PUD is directly proportional to daily face production and thus if for example the PUD will be increased from 40 to 80% the daily gross production of a face doubles.

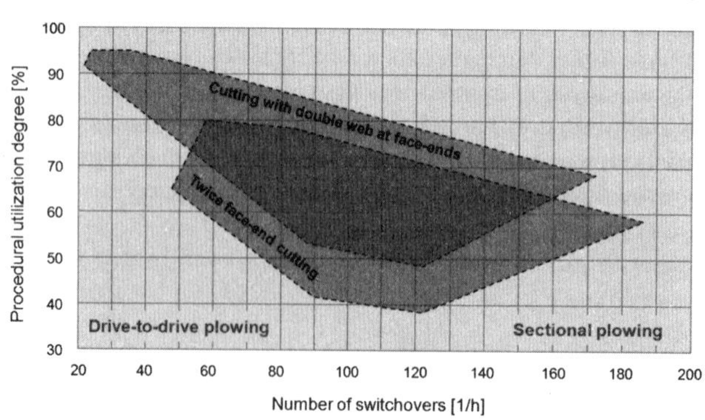

Figure 2. PUD as a function of switchover frequency

The procedural utilization degree depends on the number and duration of plow moves with speeds lower than the nominal one, on short production breaks as well as on the length of face sections plowed with a web smaller than the nominal one. In order to reach PUD as high as possible, the following recommendations should be implemented:

- Plowing exclusively in long passes from one drive to the other,
- Plowing with double web at face-ends, if circumstances allow it,
- Keeping the breaks between direction changes as short as possible,
- Accelerating the plow body in the shortest technically feasible time,
- Switching to low speed while approaching the opposite side of the face as late as possible, with consideration of the necessary safety distance.

One of the most important and measurable values describing the magnitude of PUD is the number of direction changes, i.e. switchovers in a time unit during the plow running time. Figure 2 shows the range of PUD as a function of switchover frequency in an hour.

Generally it can be said: the longer a face and the lower the nominal plow speed is, the higher the PUD results. The highest PUD can be reached while plowing in long passes with double web at face-ends, while using short switching times and while decreasing the plow speed in a short distance while approaching a drive.

5 ENHANCEMENT OF RUNNING FACES – ORGANIZATIONAL FACTORS

5.1 Optimization of works

Good organization of works and discipline in plow faces is crucial in order to reach satisfying performance. While watching the operation of longwalls, frequent interruption of works can be observed. The most common reasons for those breaks are:

- Load congestion at conveyors,
- Switching-on the emergency break,
- Interruptions in power supply,
- Overloads of plow or AFC,
- Material transportation,
- Works at face-ends.

Many breaks can be avoided by having a good organization of work in and around the face and employing a disciplined work force. In order to optimize operational sequences, a wide knowledge of the activities in the working time is necessary. That is why it is vital to observe the operation and analyze the breaks in order to identify the most frequent and time consuming interruption of works.

5.2 Education, training and motivation

A good level of education of the workforce is crucial for achieving higher production levels. Money and time used for vocational education and intermediate training proves to be a very good investment. The complexity of contemporary plow systems requires in-depth knowledge of equipment control. The majority of devices and machines working in plow systems are controlled by microprocessor based units with a wide variety of functions, which have to be well known to the work force in order to reach sufficient operating efficiency. On the contrary, false operation caused by a lack of necessary knowledge can lead to serious damages and performance deterioration.

Some operators require specific level of education for certain jobs; e.g. plow operators have to get a university degree at one of successful mining companies. This very high requirement will result in reaching high performance.

Another important attribute referring to the work force is its motivation. Experience proves that good motivated workers contribute decisively to more efficient and higher production. In particular, it is important to use positive individual and group motivation, initiating efforts to improve their own personality and achieve better results at work. Adequate motivation, facilitating identification of workers with their workplace and occupation, is a condition "sine qua non" for high performance. To obtain high motivation of the crew first of all highly motivated and appropriately trained management is necessary.

5.3 *Equipment maintenance*

In order to avoid damages resulting in production losses and high repair costs and to keep the face equipment operational, continuous maintenance plays an important role. The equipment working in underground longwalls today is highly sophisticated. This property in connection with dynamic type of operation and very harsh environment underground creates high requirements on regular equipment maintenance. Frequently, a relatively small and uncomplicated action carried out on time can help to avoid serious problems and consequent losses.

Typical examples include gearboxes, where a regular check of oil quality and when necessary change of oil and oil filter allows a proper operation and ensures an avoidance of damages and a prolongation of service life.

Another important subject of maintenance constitutes the hydraulic high pressure supply system. A proper hydraulic medium with sufficient concentration and filtration, without dangerous minerals or organic contaminations is crucial for efficient and faultless functionality of a hydraulic face system. In case of deterioration of hydraulic emulsion quality, extensive damage to components of hydraulic equipment occur and their results hamper the operation, decreasing face performance. As damages are widespread, the necessary repairs are expensive and time consuming. For that reason a frequent and detailed inspection of emulsion quality and check of hydraulic equipment constitutes an important pre-condition for faultless face operation. Correct maintenance means saving a lot of money!

Good maintenance requires high level of "technical culture". This is one of the more difficult topics while implementing a modern technology. In case of sufficient financial resources, a purchase and implementation can be realized in a relatively short period. A much more complicated task includes the change of work force attitude to new technologies. This is what the term "technical culture" means. In order to change that attitude next to afore said education a constant "technical indoctrination" over long time periods is required.

It is important at this point to mention an implementation of condition monitoring at modern longwall systems, as a part of preventive maintenance. Under the term "condition monitoring", it should be understood that online monitoring of the most important longwall machinery. Condition monitoring covers a constant surveillance of all the main parameters, inclusive vibration monitoring. Such a system means slightly higher capital expenses for operators, on the other hand allows much deeper analysis of "machinery health", delivering the best basis for avoidance of surprising damages and significantly elongated life cycle.

5.4 *Plow control systems*

The control system belongs to the most important components, determining the magnitude of plow face performance. Modern plows are controlled either from a cabin located in the energy train underground or in a special control room on the surface. The majority of plow faces worldwide have control cabins underground. In Germany, all plow faces have been controlled from the surface for many years.

As the data transfer distance using modern fiber optics is not important anymore, it can be assumed that in future controlling of plow systems will be increasingly carried out from surface centers. A control center at the surface has a wider functionality than a narrow cabin underground. In contrast to the underground cabin, where the space for control devices is limited, in surface control centers all important information is presented directly on several monitors without the necessity for switching between applications.

Figure 3. Surface control center at a German mine

A control center plays an important role in achieving high production. The dispatcher's duty next to the effective controlling of all plow system devices is to react as quickly as possible in case of any unscheduled breaks. In order to act effectively, the dispatcher has to have a direct access to all relevant monitoring systems like power supply, pump stations for high pressure medium as well as for cooling and spraying water, conveyors with bunkers and shaft, ventilation and methane control, material transport etc. Having all this information at one glance allows a quick reaction in case of a stoppage. Considering that in the case of high performance plow faces, a one minute break means a loss of 15 to 30 tons, the importance of a quick reaction for unusual situations becomes obvious.

5.5 Time utilization degree and face performance

The time utilization degree (TUD) is a proportion of total daily plow running time to daily face operating time, expressed in percent. A certain part of a working day in a face is used for scheduled maintenance or some activities necessary for proper operation but requiring the stopping of production. Further, the increase of production within the running time is difficult and expensive; raising the plow running time in contrast is much easier to reach and does not require significant investments but good organization and work discipline.

The operating time of an average plow face accounts for 18 to 22 hours a day. A part of that time is necessary for process indispensable operations. The residual time can be theoretically used for production. Unfortunately, it can be said based upon available statistics that only less than half of that time is used for active production. The rest is lost for frequently trivial reasons.

The longer the total daily running times of a face the higher its production. Improvement of organization is the easiest way to increase TUD and in that way to raise production.

6 CONCLUSIONS

Plow systems were significantly improved over more than seventy years of operation worldwide. The size and installed power increased dramatically and on the other hand control systems were implemented and improved, allowing in some applications full automation.

A comprehensive planning of a plow face from the all important perspectives is an important pre-condition for reaching the desired production size. Potential mistakes done at this stage can seriously hamper future operation and their subsequent improvement can be costly or even impossible to carry out in practical terms.

While designing a new plow face the cost-effectiveness of applied equipment should be an object of in-depth analysis. The height of initial capital costs is not the correct indicator for the equipment selection, but the rolling costs of one ton of saleable coal calculated over the whole life span of the equipment. Thus it can be said: cheap is expensive in the long run.

An effective enhancement of running longwalls has to be regarded from the operational/technical, procedural and organizational point of view.

Some important operational aspects having an influence on performance of plow faces count the correct shape of a plow longwall, the optimal setting of plow body height and bit set, the proper operation of horizon control and avoidance of bottle-necks.

The combination of AFC and plow speeds describes one of most important procedural factors influencing the plow face performance. Furthermore, the way of cutting the face ends and the length and frequency of plow moves in the face is of high importance for that performance. All those factors affect the procedural utilization degree, used as an indicator of plow face operational efficiency.

The proper organization of a plow face plays an important role while reaching the highest production size. In order to reach that level, a running analysis of production processes is necessary. Good organization requires also very good level of miners' education with periodic training. Effective controlling of the plow face and successive maintenance of the equipment presents a further relevant issue for achieving high performance. The quality of organization in a face is expressed by its time utilization degree.

Concluding: The maximization of TUD and PUD means an attainment of high effective running time. The highest possible daily running time is the most important key to success!

REFERENCES

Myszkowski, M. & Paschedag, U. 2008. Longwall mining in seams of medium thickness – comparison between shearer and plow. Lünen: Bucyrus.

Myszkowski, M. Plum, D. 2000. Verfahrenstechnische und maschinentechnische Auslegung von Hochleistungsstreben. RWTH Aachen.

Myszkowski, M. 1996. Auslegung des optimalen Hobelverfahrens für Hochleistungsbetriebe mit Maximierung der Gewinnungs- und Förderleistung. Essen: DMT.

Myszkowski, M. 2004. Erstellung eines Berechnungsmodells zur Auslegung und Optimierung von Hobelanlagen auf der Grundlage von Betriebsdaten und experimentellen Untersuchungen. TU Berlin.

Myszkowski, M. 2011. Stan techniki strugowej w światowym górnictwie. Kraków: XX SEP. (In Polish)

Paschedag, U. 1988. Hobeln harter Kohle. Glückauf 124, 848-853.

Paschedag, U. & Schwaak, A. 1988. Steigerung des effektiven Flächenverhiebs durch langfronthobeln. Glückauf 124, 1154-1160.

International Mining Forum 2011, Dyczko et al. (eds) © 2011 Taylor & Francis Group, London, UK. ISBN 978-0-415-68938-0

Plough Longwall Operations under Challenging Geological Conditions

H.W. Voss
RAG Anthrazit Ibbenbüren GmbH, Ibbenbüren, Germany

H. Witthaus
RAG Mining Solutions GmbH, Herne, Germany

ABSTRACT: In German coal mining, comprehensive experience is available in the grading of geological faults. The available experience has produced a level of expertise which enables the maintenance of high production figures even under difficult conditions. This is demonstrated by the many successful projects of RAG. The paper describes in the following several successful case examples and illustrates that the collected know-how can be used globally, even for future challenges, by German RAG Mining Solutions GmbH to successfully operate modern high-performance plough companies under difficult conditions.

KEYWORDS: hard coal mining, geological conditions, plough technology, planning of exploitation

1 GEOLOGICAL CONDITIONS

In German coal mining, the deposits are of sedimentary origin, and in the course of the geological past are often characterised by major movement processes.

The seams are therefore embedded in layers of different strength and structure. Following their creation, tectonic influences occurred, which lead to a displacement of previously level deposits. Flexures and ruptures were caused by different stresses on the rock strata. The example of the Ruhr Carbon in Figure 1, left shows these tectonic elements clearly in plan view. Vaults run from north-west to south-east, accompanied by horst and graben structures. Synclines and anticlines which describe flexures run from south-east to north-west (Figure 1, right).

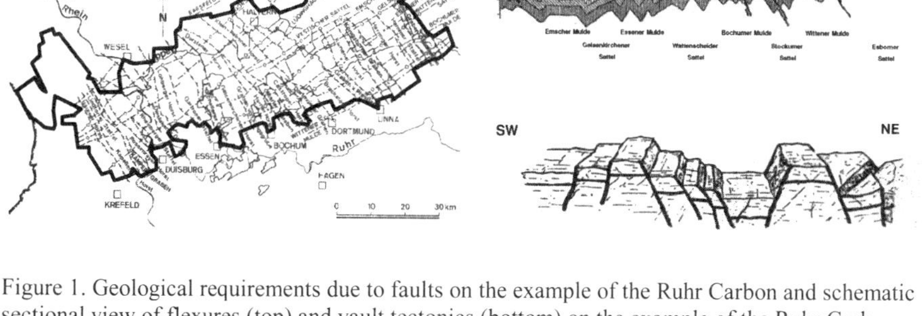

Figure 1. Geological requirements due to faults on the example of the Ruhr Carbon and schematic sectional view of flexures (top) and vault tectonics (bottom) on the example of the Ruhr Carbon

Difficult parts of the deposits differ from easy deposits above all in that the intensity of faults increases. The possibilities of development and planning therefore come up against their technical and economic limits.

It can therefore naturally happen that panels with only thin seams are also mined, and the longwall face comes up against geological anomalies.

In order nevertheless to design these fault gradings efficiently and economically with plough technology, RAG has developed a range of measures which are used at different times. These include:

- technical developments,
- intensive mine surveying developments and exploration,
- planning foresight,
- organisational and preparatory measures.

The relevant know-how of RAG Aktiengesellschaft, which is marketed internationally through its subsidiary company RAG Mining Solutions GmbH, is described below in the form of case examples.

2 HISTORY

In order to demonstrate the potential of plough technology, a brief outline is first given of the rapid development of this technology up to its present status. The history of plough technology is based on developments which began in the year 1936. The practical testing of these initial developments took place at the Ibbenbüren mine in the year 1941. Only one year after its introduction, 14 mines were already using this plough technology. The use of ploughs therefore characterises the early trials of mechanised coal mining.

Figure 2. RAG standard plough system GH 42

The extensive use of plough operations in Germany only came about however in the 1960's. The subsequent development was characterised above all by the optimisation of individual machine components such as the plough body, bits, chain guide, drive technology and control elements. Only with the introduction of automated mining operations was the plough considered in the complete system of the longwall faces, including conveyors and powered supports. It then became possible to work unmanned longwall faces, which can be regarded as the ultimate solution for overcoming the following dangers:

- Methane gas influx,
- Spontaneous combustion of the seam,

- Gas outbursts and
- Rock burst.

The end-point of development by the year 2011 is the GH 42 plough system, which is used today as the standard equipment by RAG (Figure 2).

The technical features of the system can be summarised as follows:
- Plough drive performance: 2 x 800 kW, converter technology,
- Gears: PL45 ÜL; planetary gear with overload protection,
- Plough chain: 42 x 137 mm,
- Plough speed: up to 3.6 m/s,
- Usage range: up to 2.0 m seam thickness.

3 EXAMPLES FOR PLANNING AND IMPLEMENTATION

One of the main requirements on modern planning is to adapt the high-performance technical solutions to the difficult conditions of the deposit. In addition to performance, safety is an equally important objective. In the understanding of German coal mining, this includes the areas of work safety for the miners, the health protection of all person directly or indirectly involved and sustainable environmental protection.

In deep coal mining, and in addition to the purely technical failure risks of operating equipment for mining, transport, drifting, ventilation and logistics, major risks are posed by rock pressure, rock burst and fore damp.

The rock pressure makes it more difficult to maintain the required cross-sections in the roads and also to control the working room and avoid caving ground. These risks must be expected and taken into account in particular with fault gradings.

3.1 Fault grading with a plough face, panel 6/7 West, seam 65/68, Ibbenbüren mine

At the Ibbenbüren mine of RAG Anthrazit Ibbenbüren GmbH, longwall face 6/7 West, seam 65/68 was to be prepared. An original planning provided to set up the panel at a depth of 1.370 m with a working length of 1.200 m up to a fault bordered in the west (Figure 3).

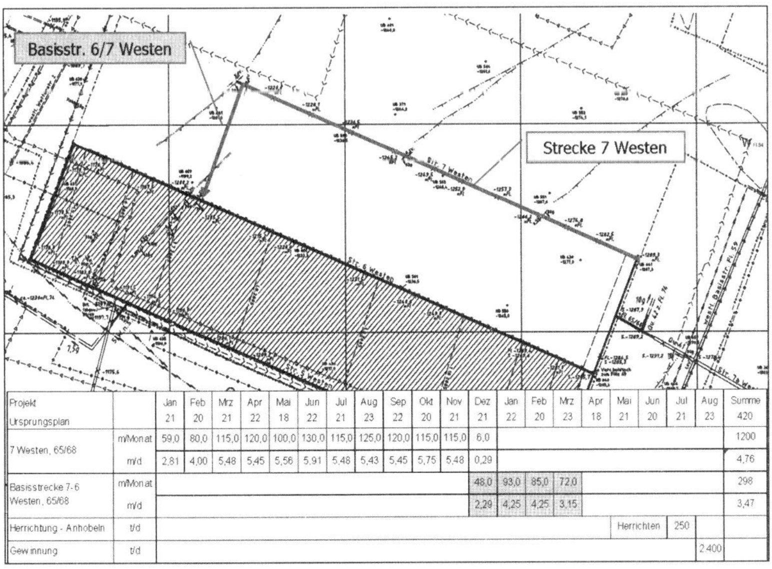

Projekt Ursprungsplan		Jan 21	Feb 20	Mrz 21	Apr 22	Mai 18	Jun 22	Jul 21	Aug 23	Sep 22	Okt 20	Nov 21	Dez 21	Jan 22	Feb 20	Mrz 23	Apr 18	Mai 21	Jun 20	Jul 21	Aug 23	Summe 420
7 Westen, 65/68	m/Monat	59,0	80,0	115,0	120,0	100,0	130,0	115,0	125,0	120,0	115,0	115,0	6,0									1200
	m/d	2,81	4,00	5,48	5,45	5,56	5,91	5,48	5,43	5,45	5,75	5,48	0,29									4,76
Basisstrecke 7-6 Westen, 65/68	m/Monat												48,0	93,0	85,0	72,0						298
	m/d												2,29	4,25	4,25	3,15						3,47
Herrichtung - Anhobeln	t/d																	Herrichten		250		
Gewinnung	t/d																				2400	

Figure 3. Longwall face 6/7 West, seam 65/68, original advance plan

An optimization of the planning according to Figure 4 provided the grading of the fault at roadway meter 1.220 and further drifting of the roadway 7 west to 1.600 m. In this way, with an additional gate road drive of only 400 m, around 120,000 t usable output more of high-quality anthracite coal could be mined. It was an essential requirement however that the preparation date in July could not be postponed. The phase of fault grading also had to be taken into account in the plan in terms of time and quantity.

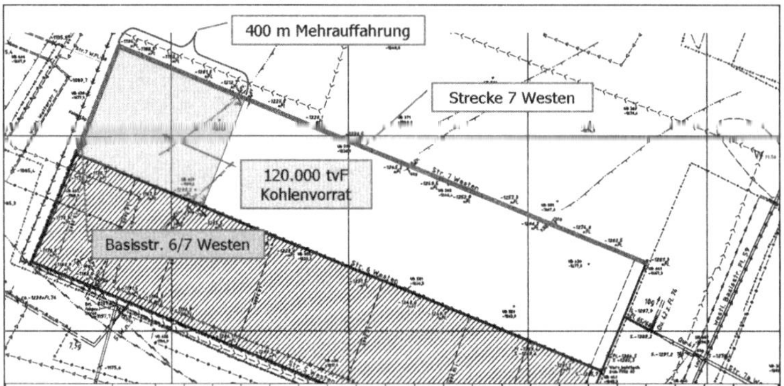

Projekt Neuplanung 2005		Jan 21	Feb 20	Mrz 21	Apr 22	Mai 18	Jun 22	Jul 21	Aug 23	Sep 22	Okt 20	Nov 21	Dez 21	Jan 22	Feb 20	Mrz 23	Apr 18	Mai 21	Jun 20	Jul 21	Aug 23	Summe 420
7 Westen, 65/68	m/Monat	59,0	80,0	115,0	120,0	100,0	130,0	126,0	138,0	132,0	120,0	126,0	126,0	132,0	96,0							1600
	m/d	2,81	4,00	5,48	5,45	5,56	5,91	6,00	6,00	6,00	6,00	6,00	6,00	6,00	4,80							5,44
Basisstrecke 7-6 Westen, 65/68	m/Monat															63,5	76,5	88,0	70,0			298
	m/d															2,76	4,25	4,19	3,50			3,63
Herrichtung - Anhobeln	t/d																	Herrichten		250		
Gewinnung	t/d																				2.400	

Figure 4. Longwall face 6/7 West, seam 65/68: Optimisation

The driving of the base road at 298 m was carried out in 54 days, at an average drift speed of 5.52 m/d. At the same time, 147 longwall face conveyor channels (PF 2/4, 1.032 mm wide) were installed and 108 shield support units (SaarTech IV STS 7/18 L) each consisting of 198 units were transported, swung in and fitted. The remaining installation of 51 longwall face conveyor channels and 90 shield support units, the installation of the auxiliary drive and the complete electrical installation of auxiliary and main drive, shield computers etc. and the installation of all other supply lines was completed in six working days (Figure 5).

Figure 5. Base road 6/7 West on the day of the breakthrough

The ploughing from the base road and the construction of the two explosion-proof closure dams at roadway 7 and roadway 6 West was carried out in eight working days.

As the sum of all results, the longwall face was able to be handed over to the mining, 12 working days before the planned date.

During the early part of the year, a situation had arisen at the mine which not only necessitated the punctual commissioning of longwall face 6/7 West, seam 65/68, but the new longwall face also had to be brought up to full productivity in the shortest possible time, while still observing the high quality requirements. The word was: "8 m mining progress in one week".

As shown in Table 1, this target was also beaten. With the longwall face handed over in June (ploughed in 5 m), the mining team achieved a mining progress in the month of June (5 working days) of 26 m. In the next month, a total of 96,224 t of high-quality anthracite coal was mined with a mining progress of 255.5 m or an average of 12.17 m/d.

Table 1. Performance data of the start-up phase of longwall face 6/7 West, seam 65/68

Month	Mining progress		Production	
	m/month	m/d	t/d	t/month
June 2006	31.0	5.20	2,075	12,373
July 2006	255.5	12.17	4,582	96,224

Following the successful start-up phase, the next known problem had to be solved: the successful grading of the geological fault, which was found in roadway 7 West with a casting of 3.90 m. For this purpose, the personnel from the coalfield and the longwall face hydraulics were invited to a lean workshop. The surveyors had produced for this occasion a detailed investigation of the optimum fault grading. The overlying strata would be easier to plough in comparison to the base, so that the team would have to control the plough system as shown in Figure 6.

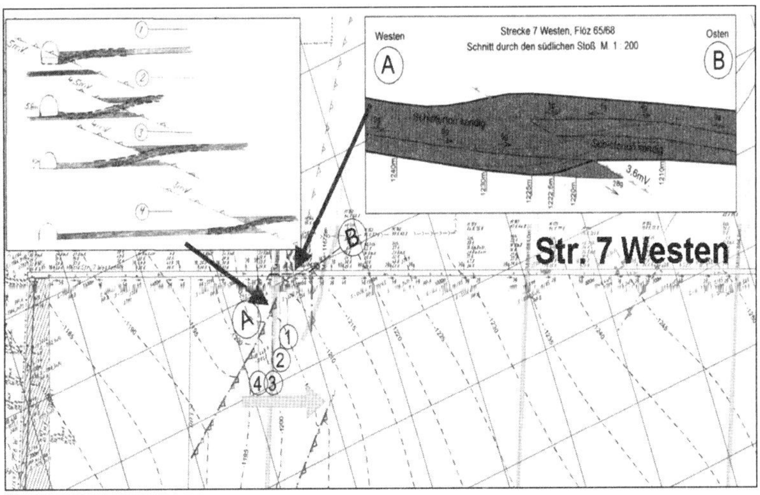

Figure 6. Planned fault grading, sectional drawing of longwall face and roadway 7 West, seam 65/68

Particularly problematic was the fact that after the grading of this fault in roadway 6 West, another fault was found, which should however run out in the course of the longwall face (Figure 7). This fault was already known from the neighbouring panel. The plough system had to be driven into the seam from the base for grading of the fault.

The following measures were carried out as additional preparatory work.

The planned longwall face progression (position of the longwall face conveyor at the main drive) was drawn in roadway 7 West at the support arches. Since roadway 6 West is a gate road with dual use, the area of the second fault grading was extensively injected with cement, in order to fill the fissures created by initial usage. This area was also treated shortly before the longwall face passage with rock reinforcement (polyurethane) (this is referred to in mining as activation of the system anchoring fitted with the road drive and therefore securing of the longwall face-road transition).

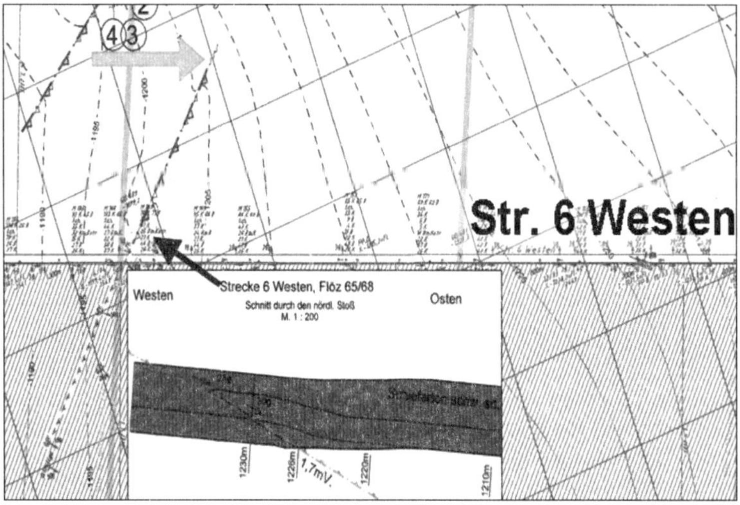

Figure 7. Fault with 1.7 m casting in roadway 6 West, seam 65/68

The detailed mining progress in the phase of fault grading is shown in Figure 8. During the grading, mining progress of 1 to 6 m/d was achieved. In the month of September, the performance was increased to 13.33 m/d mining progress and 5,074 t/d of daily production.

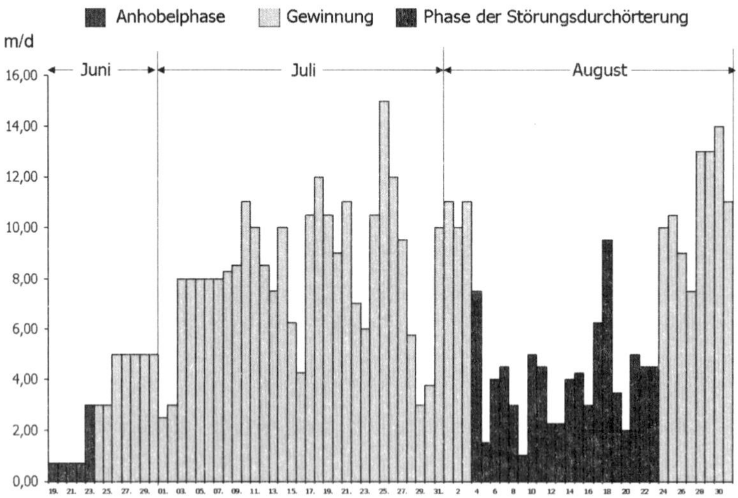

Figure 8. Daily working speed from preparation (red) to fault grading (blue) of longwall face 6/7 West, seam 65/68

Both these measures proved to be successful, and the team was able in August to grade the fault successfully at an average working speed of 8.78 m/d. This gave a daily production of 3,439 t/d (Figure 9).

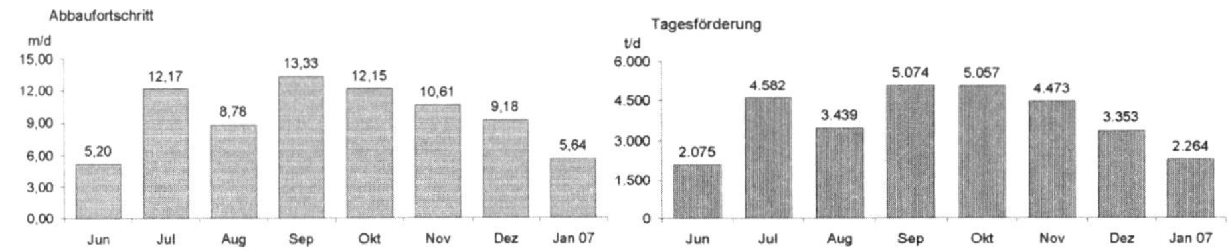

Figure 9. Average working speed (l) and daily production (r) of longwall face 6/7 West, seam 65/68

2.2 *Further case examples from the Prosper-Haniel mine*

The following case examples show that this fault grading is nothing unusual for RAG (Figures 10-14).

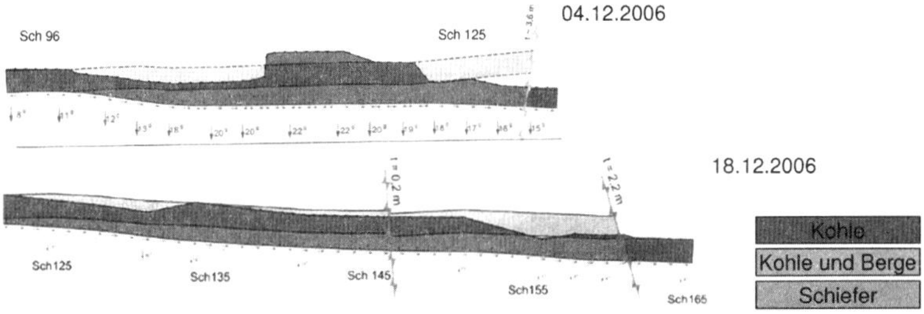

Figure 10. Fault grading panel 432, Prosper-Haniel mine

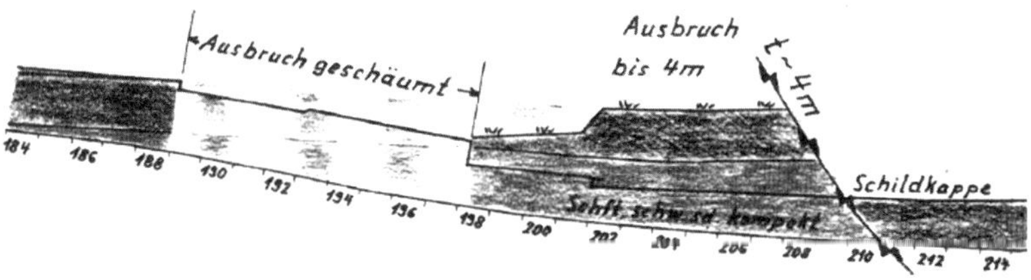

Figure 11. Outburst areas in panel 432, Prosper-Haniel mine

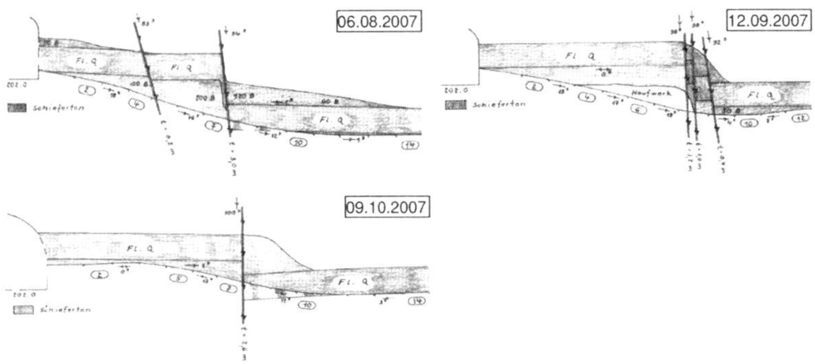

Figure 12. Fault grading in panel 202, Prosper-Haniel mine

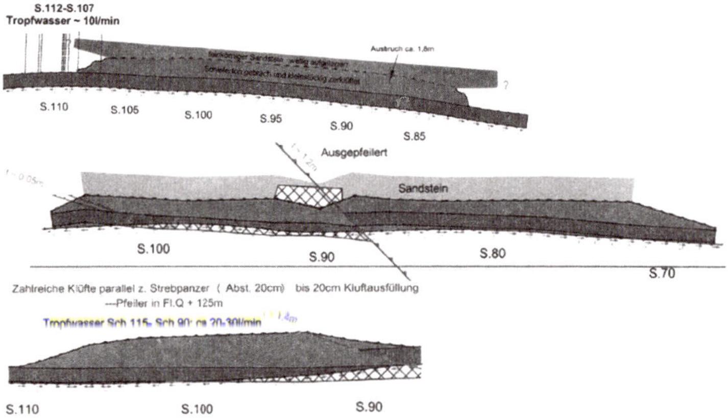

Figure 13. Fault in panel 355, Prosper-Haniel mine

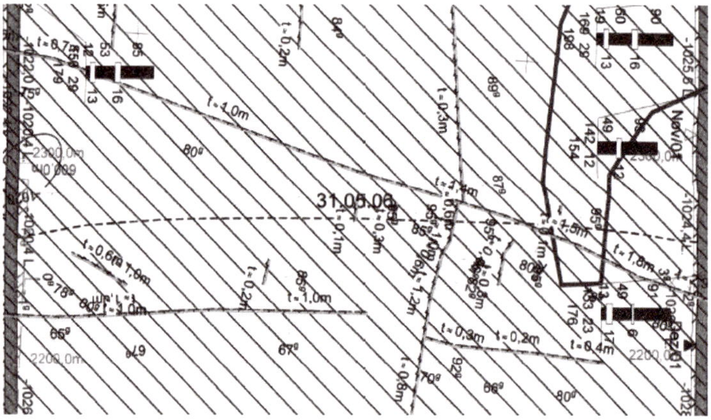

Figure 14. Plan view and fault progression of panel 201.

Table 2. Overview of different fault grading and output data

Panel	Φ Output [t/d]	Φ Output (with fault) [t/d]	Φ Advance [m/d]	Φ Advance (with fault) [m/d]	Type of geologic faults
201	12,379	10,936	8.91	8.4	lateral faults up to 0.4 m; faults up to 1.8 m; up to 1.5 m dirt in the advance front
202	8,963	7,992	6.27	5.5	sandstone deposits up to 0.25 m; caving formation up to 0.5 m; washouts up to 0.6 m; lateral faults up to 1.4 m with up to 1.1 m dirt in the advance front; slough tight pack building material filling and steel bolts in the roof
203	8,965	8,52	6.03	5.7	lateral faults up to 2.1 m; cliffs in floor and sand-cliffs in roof up to 1.8 m; sand stone deposits up to 0.5 m
250	8,248	8,076	7.46	7.2	interbedded strata (t≈0.3 m); drilling and blasting work on AD; caving formation up to 1.1 m
258	11,689	9,279	7.39	6.2	interbedded strata (t up to 2.5 m)
355	6,921	4,6	5.69	3.9	interbedded strata (t between 0.1 m to 2.6 m); distance to seam partially only 4 m
432	8,678	5,868	5.14	3.9	lateral faults up to 2.7 m; dirt with cut on floor up to 1.8 m

The results of the achieved performance are compiled in Table 2.

As shown in Table 2, the average output within the geologic fault makes obvious that inspite of the more difficult circumstances 80% of the average output and advance could be achieved.

4 TRANSFER OF KNOW-HOW TO OTHER MARKETS

The know-how obtained from the various applications in German coal mining can also be made available by RAG for other projects throughout the world.

The above listed case examples illustrate the large amount of experience of RAG in the grading of geologic faults. Success factors for the use of modern high-performance systems under difficult conditions here are mainly:

- Getting huge face advance and continuous rate of travel of powered support.
- Coping with caving formations and roof eruptions with minimum downtime even with low face heights.
- Safe working and detailed organization of special actions, such as drilling and blasting in areas with floor or roof sections with permanent surrounding rock areas according to plan.
- Optimization of the process sequence under individual technical prerequisites through Lean Processing.

To achieve total success, the technology and all processes must be already coordinated with each other in detail during planning and organization.

Founded in the year 2009, RAG Mining Solutions GmbH hereby acts as a linking member to apply this success to other projects. RAG Mining Solutions acts as a consulting and engineering service-provider, and a provider of on-site training and management workshops.

The solution contributions include all project phases from the geological evaluation, the technical planning and the organization, risk management, creation of safe working processes and also process optimization.

4.1 Consulting and Engineering Services

Successful projects begin right at the face development. Even here, as shown by the numbers from the example of the Ibbenbüren mine, maximization of extraction can be already achieved in the start-up phase of the face.

Individually planned solutions for face development out of the base course and fast achievement of maximum output, as shown schematically in Figure 15, generates a corresponding extraction. But if experience must first be gathered, which is typical especially when introducing new technology, then at first small face advance progress and output is to be expected. If RAG's experience is already used in the start-up phase, then higher output can be reached "from the start" and learning the hard way by gathering one's own experience can be avoided. This normally costs a lot of money since along with low output, higher wear usually results from non-optimal use of the equipment, which later can appear in the form of increased maintenance costs.

The hatched areas underneath the "Current output curve" (green) in the time period of the first four weeks describes the excavation for the operator using measures form the know-how transfer, specific training and optimum technical standards. The excavation here can be determined as follows, as an example: Over a time period of four weeks, e.g., 22 production days, more than 1 m more face advance was achieved. With a face length of 250 m and a seam thickness of 1.5 m and a density of 1.6 t/m?, this represents more than 13,000 tons of additional usable output just in this time period.

The planning process required for this, the process knowledge and the technical know-how are tested at RAG successfully and lead to individual and successful solutions. Core competencies are here in all areas of the operating company's knowledge. Along with the corresponding planning and implementation knowledge, also the quality assurance is of critical significance here (Figure 16).

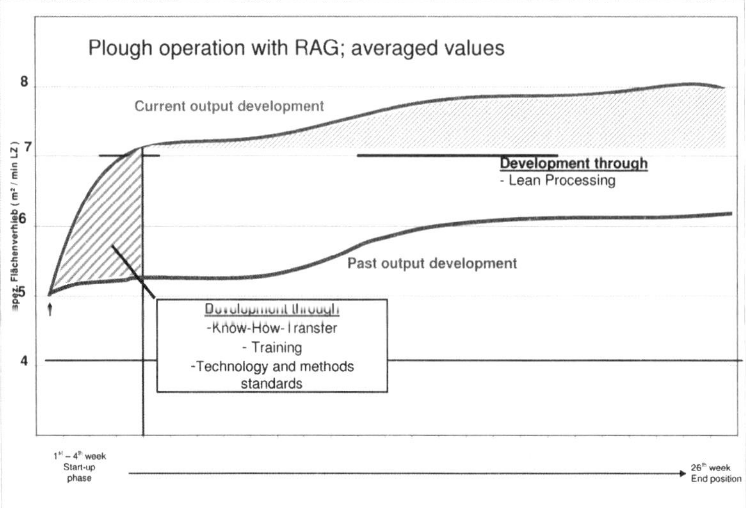

Figure 15. Output during face start on examples from RAG, past and current

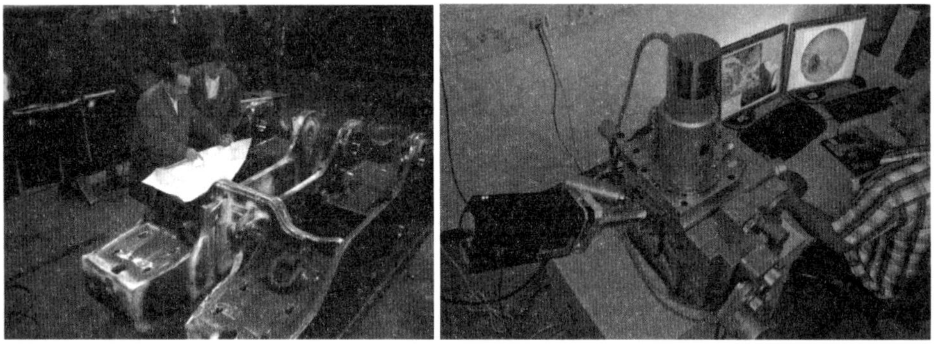

Figure 16. A comprehensive range of tests belong to the product pallet of RAG Mining Solutions

A comprehensive product pallet of RAG Mining Solutions stands out in front of this background:

- Results of the testing method for checking the technology already used during manufacture or maintenance, as well as explanation of events causing damage during operation.
- Planning as total technical planning, taking into account the aspects of geology, advance work, mining, infrastructure, support technology, logistics, ventilation and air conditioning.
- Courses, training and workshops for sharing experience on RAG mines for the expansion of operational experience and competencies under difficult constraining conditions for management and employees.

4.2 Training opportunities

The proven training packages for international project cooperations are designed as modular and flexible units (Figure 17).

Training block 1: Operational management for high-performance ploughing operations

- Target group: Middle management,

- Training by RAG specialists and management with the following content:
  - Operational organisation,
  - Success factors for high-performance ploughing operations,
  - Controlling instruments,
  - Subject-related workings (e.g. ploughing, fault grading).

Training block 2: Process technology for plough operations
- Target group: Supervisors, shift foremen, team,
- Training by experienced trainers and operational engineers with the following content:
  - Structure and organisation of plough operations,
  - Control of plough operations (longwall face location, location of drives, plough procedure, optimum use of existing control elements),
  - Simulations by means of virtual reality,
  - Controlling instruments,
  - Subject-related workings in small groups.

Training block 3: Operational supervision
- Target group: Middle management, supervisors, shift foremen, team
- Operational supervision by experts in strata mechanics and production with the following content:
  - Operating consultation for the improvement of technical availability and performance in the run-up phase,
  - User-orientated training for production optimisation of production under existing conditions and operating situations,
  - Support in the application of the training contents learnt in Blocks 1 and 2.

Figure 17. Participants at a training session at RAG Mining Solutions

In cooperation with a plough manufacturer Caterpillar, formerly Bucyrus, plough technology has already been successfully optimised at many mines. In Mexico in the mine Mimosa, in Poland in the mine Zofiofka, in Russia in the mines Perwomaiskaja und Severnaja and the Czech Republic by OKD on the basis of these concepts by RAG Mining Solutions the production increased in some mines up to 100%. All training is carried out according to individual customer requirements with the motto: "Practice-related exchange of experience from the operator to the operator".

4.3 Operational support on site

Operational support on site by experienced experts from RAG Mining Solutions with planning, engineering and operating competency was shown to be especially successful in several projects (Figure 18).

Figure 18. Safe working conditions in face and face/coarse transition

Finally, the procurement of a modern plough system represents a significant investment for every mining company. With this the reliable mastery of the plough used, as well as the optimization of the extraction processes, are of important significance. It especially applies through the use of a new technology, always even to increase the production numbers and therefore the efficiency of the entire system.

4.4 Lean Processing

"Lean Processing" has also proved to be a further important building block for the lasting increase of output in the companies (Figure 19). RAG Mining Solutions can also include these processes in external projects through planning and training opportunities. With this, this can be transferred to other company aids, to reach a permanent increase in output and positive productivity development, and thus an excavation maximization for the customer.

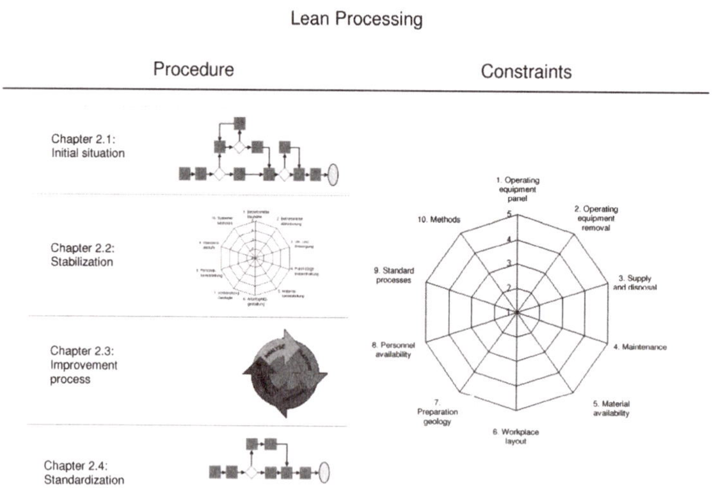

Figure 19. Lean Processing for mining companies, procedure (left) and constraints of optimization (right)

"Lean Processing" was developed to achieve process improvement, both in production and in administration. It is aimed not only at punctual improvements, but rather the processes are systematically improved in a total outlook, using a variety of methods and instruments, and losses in the processes are detected and avoided.

The involvement of the employees is here the most important factor for success. The starting point is the analysis of the processes and the values to achieve from them based on process specific key figures. Starting from the analysis results, improvements can be planned and implemented. Finally the implemented actions undergo controlling. The development of the key figures and their analysis indicate the success. Building on this, improvements in technology, methods and processes are defined as standard. In the implementation of "Lean Processing", not only new approaches are researched in the individual companies, but many times "Best Practice Solutions" of other companies are also fallen back on. The intensive involvement of employees during the formation of the processes, leads to a permanent improvement in work safety, health protection and protection of the environment.

International Mining Forum 2011, Dyczko et al. (eds) © 2011 Taylor & Francis Group, London, UK. ISBN 978-0-415-68938-0

Experience with Plow Equipment at OKD

J. Matula
OKD, a.s., Ostrava, Czech Republic

ABSTRACT: OKD has rich experience with plowing as the first mechanized shields were deployed already in 1973. New or completely refurbished plow sets with shields have been acquired in the scope of POP 2010 project between 2008 and 2011. OKD had high expectations when investing in new plow sets with a planned daily production exceeding 1,000 tons. However, these goals have not been achieved due to adverse geological conditions combined with equipment issues. It is expected that share of plow production will grow in future at OKD so the company needs to do its best to fix all the problems currently incurred at the operations.

KEYWORDS: hard coal mining, plow technology, OKD experience

1 KEY DATA

OKD is the only hard coal producer in the Czech Republic with the annual production of ca. eleven million tons of high-quality, low sulfur and low additive hard coal. The company operates four deep mines, of which three in the Karviná region and one in the Frýdek-Místek region.

Over the past three years, OKD invested almost CZK 10 billion within its POP 2010 project in new mining equipment acquisition and another half a billion within its SAFETY 2010 project in state-of-the-art personal protective equipment and tools for miners.

In longwall equipment, there were 6 mid-seam set (DBT 1300/3100 shields with Eickhoff SL300 shearers), 4 high-seam sets (DBT 2600/5500 shields or DBT 2800/6000 shields with SL500 shearers) and three plow sets (DBT 600/1400 shields with an RHH-42 plow, refurbished KG 291 – 5/15 shields with an RHH800 plow and refurbished MHW 650/2005 shields with a GH 9-38 plow) purchased.

Plow equipment is currently deployed at two sites only – at Lazy Site, Karviná Mine and at Paskov Mine (where, besides shields, single props are also used).

2 PLOW PRODUCTION AT OKD

Shields in combination with plow equipment were first deployed at OKD in 1973 at František Mine. The longwall face was equipped with Westfalia K1.1 shields and a PL-7 plow. Between 2008 and 2011, there were 3 full new or refurbished sets delivered to OKD within the POP 2010 project. At the moment, there are usually two to three plow shield faces operated at the same time at OKD and the same number of single prop faces.

The share of plow production in the total LW production is at the moment ca. 9.5% and, except for 2009, it is a relatively stable trend since 2007. It is expected that the plow production share will grow as the Ostrava seams will be mined at greater depths. Due to insufficient seam thickness development, these seams cannot be operated with shearers. Figure 1 shows the share of plow production at OKD since 2005 (the outstanding quantity up to 100% is represented by shearer production; roadheader production is not included in the statistics).

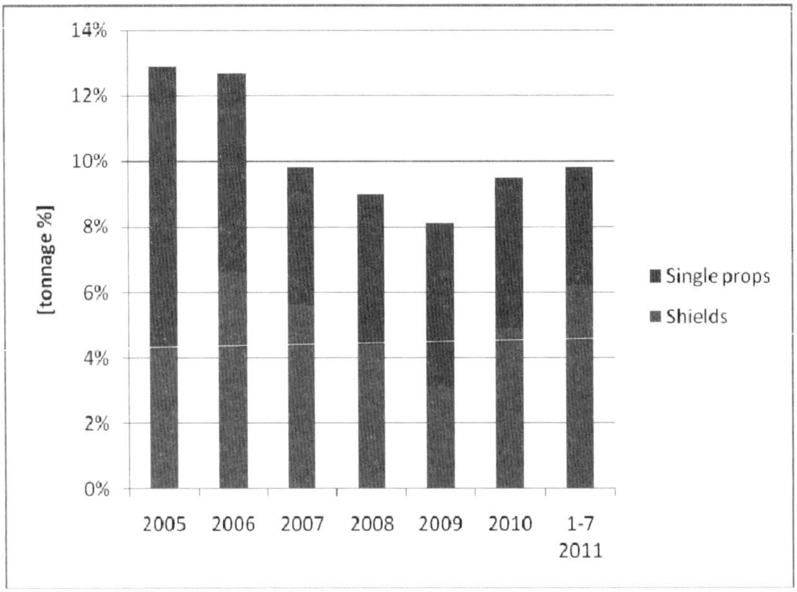

Figure 1. Plow production in percentages

Since the new or fully refurbished plow sets were acquired, OKD expected major increase in production efficiency assuming the average daily production from such longwalls of ca. 1.000 t/day.

As shown in Figure 2, this major increase in productivity of plow longwalls did not materialize though.

The figure shows a more or less steady average daily production from plow longwall face. It is also important to know that the difference in productivity of single prop faces and of shield faces is very small. Compared to other plow faces operated in the world, the performance is low. In view of the fact that OKD plow sets are not significantly different as concerns the technological procedure, the causes of low production need to be seen in particular on the side of adverse geological conditions, crews with less experience with new equipment or an unfavorable combination of all the three factors.

The most important fact is the significantly higher LW performance of shield faces. When looking at Figure 3, it is clear that the LW performance of shield faces has almost doubled in 2010 compared to single prop faces. In view of the fact that labor costs in plow faces represent up to 50% of all production costs, it is clearly advantageous to use mechanized shields even though the difference in the average daily production is not so high.

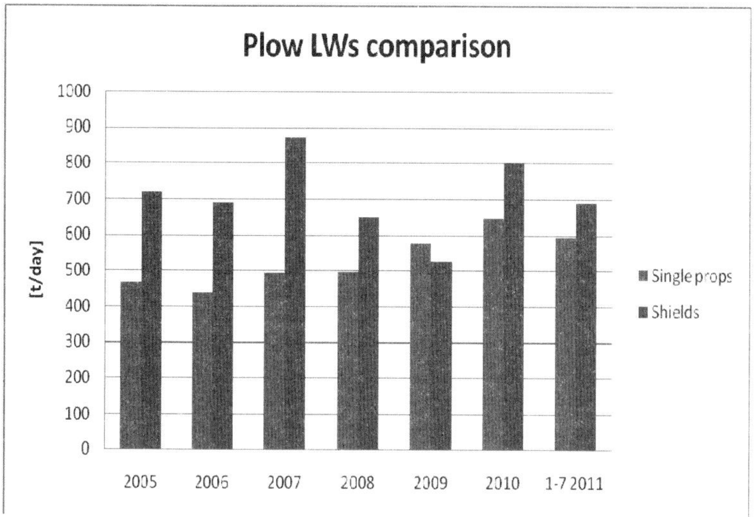

Figure 2. Plow performance

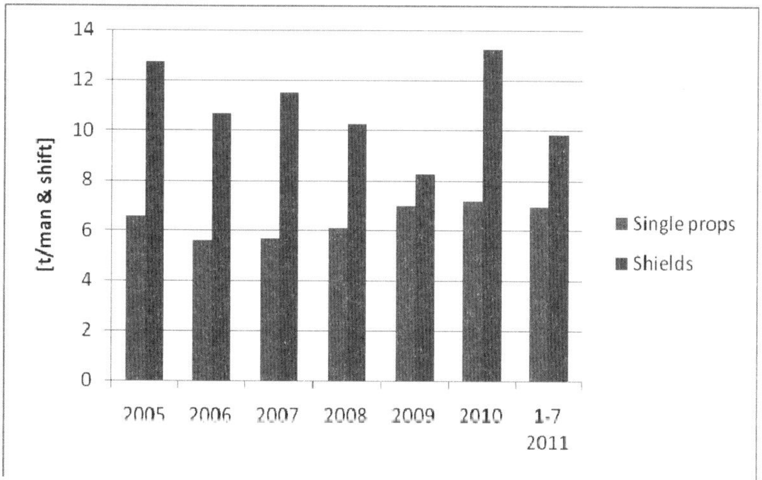

Figure 3. Plow LW performance

3 REASONS OF LOW PERFORMANCE OF PLOW SETS AT OKD

3.1 Geological conditions

At Paskov Mine, the seams are typically featuring squeezes with sandstone lentils in seams and in roofs. They are not coherent with the surrounding rock (there are planes of inconsistency in contact with the surrounding roof mass). The occurring sandstone lentils are irregular in quantity, shape and size and are overlapping in some cases. In case of too many planes of inconsistency and cleavage, spontaneous sloughing may be produced in the cut-out space or the inconsistent roof is sloughing off with the coal seam in the forefield up to a few meters in front of the LW face edge. The sloughing is also facilitated by inconsistent coal seam bedding.

The low performance of the plow set at Lazy Site, Karviná Mine was caused mostly by insufficient coal seam thickness. The average net seam thickness of the seam is only ca. 0.85 m.

Mechanized shields can be operated in seams with the minimum seam thickness of 0.6 m but, in view of maintenance and shield control, it is a purely theoretical number. Based on the experience, a realistic mined seam thickness is at least 1 meter causing the necessity to cut roof or floor which has a major impact on LW performance. In many cases, it is needed to start blasting which is also influencing the performance.

3.2 Equipment

Plow equipment used at OKD mines is not significantly different from plow equipment used elsewhere in the world. The newly purchased (or refurbished) mechanical and electrical equipment is provided exclusively by top world manufacturers such as Bucyrus/Caterpillar or Becker. The difference to plow longwall faces operated elsewhere in the world is the ratio of plow/AFC speed. In general, it is recommended to use the 3:1 ratio (based on simulations and hands-on experience). At OKD, a significantly lower ratio (2:1) was applied based on the manufacturer's recommendations and the subsequent gearbox standardization which caused problems in particular with the even load on the belt haulage system and with insufficient utilization of the plow's kinetic energy in case of the necessary rock or floor cutting. We have to note here that this objective difference cannot cause such a high production drop.

As described above, Paskov Mine faces the worst geological problems with inconsistent roof falling in between shields and causing cavities if the tip-to-face distance is too high. As the currently mined seam thickness is almost at the upper edge of the shield's operational range, the open area between the canopy tip and the coal face is rather high causing major problems (considering the above mentioned geological conditions).

When purchasing/refurbishing the equipment, extendable canopies, that could eliminate this problem to a major extent, were also considered but this idea was skipped due to negative experience from abroad (the extension mechanism is very sensitive and gets easily irreversibly damaged in case of higher roof stress). It is true that Paskov Mine is not facing major problems with high roof stress; just on the contrary, the roof is not consistent enough and shows no high stress. So the question remains open if the canopy extension mechanism would incur any damage.

Another technical (or technical/personnel) issue is the so-called over-steering, in particular at Lazy Site, Karviná Mine. At a seam height of 85 cm and an operational height of a little less than 1 m, it is difficult to optimize OR to best utilize the steering capability located on the gob side of the AFC. A hydraulic stroke adjustment of just 7 mm in this cylinder causes a vertical change in the AFC at the coal face side of 220 mm after 1 m of advancement of the face. Initially, it occurred quite often as the inexperienced operators made excessive steering adjustments in either direction trying to make the plow "follow" the coal seam. This over-steering actually did the opposite by having the plow "climb" or "dive" out of the seam resulting in either too much rock being cut or the shields and AFC becoming ironbound (physically wedged). Both of these occurrences significantly reduced the available plow cutting time during the shift and adversely affected production results.

4 POSSIBLE PERFORMANCE OPTIMIZATION OPTIONS

Most of the optimization methods are based on technical actions. However, we should not forget a high-quality regular education and training of the crews. There are some technical actions here below to improve the performance of plow faces.

4.1 Paskov Mine

Canopy replacement - the tip-to-face distance can be reduced if the existing canopies are replaced with longer ones. This solution is not convenient though as one set would have to be idled for the time of the repair; there is also a risk of collision between the shield canopies and the plow body when mining seams with very low seam thickness or with significantly variable horizontal development (rolling seams).

Reducing the stroke - this is a short-term solution as a shorter stroke means also a lower performance due to more frequent push. Frequent shield setting is also impairing the already inconsistent roof as well as the mechanical status of the shields and their control units.

Adjustment of block anchorage system - the supplier delivering the refurbished shields must adjust the design of their block anchorage in order to provide a re-bolting option to push the shields 150 to 300 mm closer to the face using the bolting system. Currently, due to a faulty design of anchoring cylinders on the shield pontons and base frames on 75 shields it is necessary to extend the distance from the face by 150 to 300 mm.

Acquisition of new/refurbished shields with a longer canopy - considering the option of extendable canopies - as there is no major roof stress incurred at Paskov Mine (just on the contrary, the roof is very inconsistent and sloughing), it can be assumed that the system of extendable canopies would not be subject to major stress and would therefore not suffer any damage. Using extendable canopies could help to keep the optimum constant tip-to-face distance depending on the seam thickness mined, seam conditions and to prevent any possible collision with the plow.

Optimization of the speed ratio between the plow and the AFC at 3:1 - based on the experience, this ratio can ensure a more even spread of the belt haulage load and a more optimum utilization of the plow's kinetic energy.

4.2 Lazy Site, Karviná Mine

Optimization of the speed ratio between the plow and the AFC at 3:1 - see above. Considering the implementation of automated steering cylinder control depending on the current seam development (using inclinometers checking the difference in the incline of canopies and the AFC) - the problem of over-steering was described in detail here above. The automated steering could probably help the equipment to "follow" the coal seam with more accuracy.

5 CONCLUSION

The article describes key characteristics of OKD, share of plow production on the total longwall production and, in particular, problems with low performance of plow longwall faces, of which we can quote, for example, deteriorated geological conditions, high tip-to-face distance and "over-steering". Some options that are currently under discussion were also outlined. As there are some expectations to increase the plow share in future as mining activities will focus on the Ostrava seams with lower seam thickness, OKD must find clearly a way to fix these problems and to increase plow face performance to a level comparable with other countries.

International Mining Forum 2011, Dyczko et al. (eds) © 2011 Taylor & Francis Group, London, UK. ISBN 978-0-415-68938-0

Application of Plow Mining in Ukraine

Dychkovsky R.O., Tymoshenko Ie.V.
National Mining University, Dnipropetrovsk, Ukraine

V.S. Tenia
PJSC "Shahtoupravlinnia Pokrovske"

Dedicated to the scientists of the National Mining University that took part in creating the first plow units in Ukraine and in the Soviet Union

ABSTRACT: The paper discusses application of plow mining in Ukraine. Both historical and present aspects of plowing technique development have been described. Particular attention has been paid to the conditions in which application of plowing is the most advantageous. This concerns both external and internal (e.g. mining and geological) conditions. Advantages and problems occurring during plow exploitation are depicted in detail on the basis of theoretical and practical studies conducted in National Mining University.

KEYWORDS: hard coal mining, thin seams, history of plow system, Ukraine

1 INTRODUCTION

At present time, one of the basic trends of technical improvement of Ukraine's coal industry is a complex mechanization of operations that foresees full mechanization of basic and auxiliary production processes in the stopes.

At the coal industry the works directed to technical re-equipment of coal mines are being conducted. These changes have to speed up coal industry development and provide exit to new higher technical borders, and make a crucial turn to the production intensification. Plow coal extraction is one of the basic methods of this trend during mechanical breakage of coal.

Available power of the plow units makes up 440–800 kW. Plow movement speed is 0.6–1.8 m/sec, scraper conveyor chain – 0.6–1.2 m/sec (Bondarenko et al. 2003).

Since plow units are designed to work under stable wall rocks then in order to avoid secondary subsidence of the roof, the full pneumatic backfill into the worked-out area is used.

2 HISTORIC ASPECTS OF PLOW COMPLEXES USED IN UKRAINE

2.1. Origin and basic stages of plow extraction

Basic works focused on creation and the first industrial tests of plow units in the Soviet Union has begun on the territory of Ukraine in Central region of Donbass several months before the end of WW2. In February 1945 the Lugansk construction department of Giprouglemash began projection works of experimental coal plow of static action US-1 (plow unit – 1), that was built in the beginning of 1946. Further plow models were constructed until the beginning of the 50s. In the years 1948–1950 over 20 plow complexes were working in Donbass mines.

Basic disadvantages of contemporary plow units were the unreliable scheme of rock pressure management and impossibility of use under conditions of solid rock interlayers inclusions that has led to low rates of faces advance (25–30 m/day) and hence to unsatisfactory performance. Thus, in the beginning of the 50's of the last century the decision to completely dismantle new models of US-4 plow.

Average daily output from stopes at mines of Donbass equipped with plow units at that time exceeded 600 t/day. It was quite high indices compared to other means of mechanization.

All decade of the 70's was followed by stable increase of number of mechanized stopes equipped with plow units. Eventually, single prop support use decreases which was basic restraining factor of production growth.

This time is joined with maximal indicators of coal production in Ukraine. During 10 years since 1970 the annual production had been exceeding 200 mln tons. Maximal level of production was reached in 1976 and made up 218.2 mln tons.

It is worth mentioning that production growth was not occurring by intensive method implying production growth of stopes but by extensive – stopes number increase. At that, level of plow extraction did not exceed 10%. With that, overall number of plow units reached 150 pieces.

The same situation remained in the 80's of the last century. The attempts to introduce technological improvements in terms of length increase up to 300 m and to widen field of application onto more strong and viscous coal were not successful. During this century about 120 plow units were working at mines of Ukraine. With that, their use was factually limited by extraction of high metamorphosed coal, basically, anthracite.

Seeming catastrophic situation in the 90's that led to mass dismantling of plow complexes and stopping of researches in this field by many scientific schools, has had a line of positive moments. Rethinking of the ideology of plow and shearer extraction technologies has occurred. Privatization of the mines has led to search for the internal reserves, intensification of processes and production concentration.

Gradual decrease of stopes number leads to output growth on them. Today, average productivity of mining stopes has risen by 3 times, and at the leading enterprises – by 10 times compared to 1991. Starting from 2000 the number of plow stopes has been constantly rising. Although this index (20–30 plow mechanized complexes) today does not correspond to our country's mines requirements. Being in a complex economic condition, the single specialized Lugansk state design and engineering institute of coal plow extraction continuous developments in the field of technical and technological improvement of coal extraction. Together with OJSC HMZ "Svet shahtera", Russian and foreign scientific-research and production enterprises have developed new models of plow extraction equipment. Last developments (SO75M, USK30, SN75, UST2M) provide an opportunity to increase longwall length to 200 m and field of application based on qualitative indices of coal (Figure 1) (http://coal.in.ua).

Figure 1. Appearance of plow units: SO75M, UST2M and SN75

All development and experimental works carried out by the researchers of National Mining University in Dniepropietrovs'k in the field of plow, scraper-plow and wire-saw extraction were followed by deep theoretical studies, scientific substantiations and practical check under conditions of coal mines. Given works find their reflection also today. New generation of scientific workers of our university together with the representatives of other research institutions continue to lead studies in this field.

2.2. Field of application

Seam thickness. Under conditions of Ukrainian mines, coal extraction by plow units is applied on coal seams with relatively steady roof allowing its exposure with width equal to 1.1–1.3 m during extraction process. Plow units work both with mechanized power supports and with single prop supports. Rational seam thickness suitable for extraction varies within 0.55–1.25 m. This value can insignificantly change to greater value depending on types of plow units.

Application of foreign production plow complexes gives a possibility to perform effective coal extraction with thickness being 0.7–1.8 m. Application of plow mining for thicker seams is factually not rational. It is connected with a row of negative factors, the basic of which is discrepancy of separate elements productivity integrated in the complex. First of all, it concerns transport link of reserves movement within the mine. When seam thickness varies at separate areas of the extraction panel the special retractable adapters for more complete coal extraction are used.

In order to manage rock pressure, the sections of mechanized plow complex of stepping type are used which are connected with to other and advance block-like in a staggered order. This method is quite effective under conditions of weak bottom rocks. During work of a complete support the sections advance begins after longwall advance for one advance increment. The sections advance increment depends on the width of extracted coal layer. Also the scraper conveyor advance is carried out. This advance is implemented with help of hydraulic jacks which rest against the face support props.

The scraper conveyor advance scheme is consecutive. For example, in the complexes of Bucyrus Company these values are defined depending on geologic-mining conditions and used in the software of the control system. The advance process itself occurs automatically after the plow run.

This company has proposed quite an effective solution focused on coal complete extraction. Holding of the plow in the coal seam plane is implemented by special laser devices. Technologically it is carried out by hydraulic jacks (rams) and leverage system of the complex holding it in a vertical plane. At horizontal advance of the longwall the conveyor rams move perpendicular to the face plane. At this, the leverage system of the plow holding functions in normal mode. In order to change extraction direction up or down the leverage system turns the conveyor line through a definite angle. Further, during plow movement along the conveyor line, coal extraction is occurring within set limits with its keeping within the coal seam plane.

For more qualitative coal breakage off the massif and its mechanical break, various types of teeth are used. At all three levels of the plow body the teeth of multidirectional action on the coal seam are applied. It allows implementing various mechanical efforts along the face plane for more effective coal breakage (http://ru.bucyrus).

Steering of the plow is performed by a remote control located at the drift. Special software gives an opportunity to receive full visualization of mining-geological situation in a stope. Management of the whole equipment is carried out from the remote control that is contained in delivery set of plow mechanized complex.

If necessary, the work of a plow unit can be coordinated with other technological processes of a mine and included in mine information system of control. Using world Internet it is possible to receive current information about the occurred exploitation problems from informational and maintenance services of the manufacturing company.

Seam dip angle

Plow units are designed for coal extraction in seams with the following dip angles:

- along the strike up to 18° (25°),
- to the rise up to 8° (12°),
- down-dip up to 5° (10°).

Application of scraper-plow units is possible to use with dip angles up to 90°.

Qualitative characteristics of coal seams

As the plow exploitation experience under conditions of Ukrainian mines shows, their application is the most expedient under the following qualitative characteristics of coal:

- resistance of a seam to cutting must be not more than 250 kN/m in stable zone and 110–125 kN/m in the zone of a cutting instrument work,
- cleavage (natural fracture) of coal – with angle of 5–40° to the face line and effective coal spall under influence of rock pressure,
- metamorphism level – preference is given to gelitolitam with the highest degree of metamorphism including minimal amount of mineral admixtures (coking coal and anthracite),
- "viscosity – brittleness" – with increase of brittle characteristics of coal the expediency of plow application for extraction of coal increases,
- hypsometry – stable, low-wavy,
- presence of tectonic violations – extremely negative influence on the plow coal extraction, substitution of coal within the stope is possible only by weak rocks,
- longwall length – up to 200 m, for full seam thickness or with self-caving upper pack,
- ash content of rocks, sulphur content, phosphorus and other elements – given qualitative indices of coal depend on the metamorphism level, hence, when using plow mining these indices must be minimal,
- seam bottom – not lower than average strength, not inclined to heaving,
- immediate seam roof – reasonable for the following types – B_2, B_3, B_4, with various methods of roof control,
- main roof – reasonable for a middle-caving main roof A_2.

For comparison, coal resistance to cutting is factually not a limiting factor for German producers of plow technique because of powerful drives used in the equipment. Hydraulic unit for the pulling chain tension allows to create necessary force at the plow chain and regulate its preliminary tension. This provides a possibility to effectively conduct extraction at the areas of rock inclusions and replacement of coal seam. Longwall length makes up about 300 m using such technique.

Technical solution: wallface – preparatory workings

Longwall supporting technology, first of all, depends on the openings support methods and possibility of the repeat use of drifts. In case of weak surrounding rocks, it is necessary to conceptually resolve the task of abutment stability only in the zone of the stope influence.

Specialized face-end support is included in the plow mechanized complexes for management of the massif at the abutment of longwall with drift. Given support is installed in the drift at the level of a stope plane. It gives a possibility to take off necessary number of the arch support legs for the mechanized complex line advance.

Given support solves the questions concerning the drift protection in the worked-out zone for the drifts reuse.

Various types of the artificial protection are used under the conditions of Ukrainian mines to solve this problem: rock lanes, chock-type supports, overhead logs and others). Used for the development workings protection, these supports do not prevent big roof dislocations due their high yieldability and low bearing capacity. Their installation requires a definite hand labor capacity that leads to decrease of technological processes safety.

One of the progressive methods of drifts protection is cast rigid strips providing expenditures decrease for the workings maintenance and their repeated use under the conditions of coal seams intensive development by highly productive longwalls. In order to make a strip it is possible to implement full or partial mechanization of technological processes.

Applications of cast strips of a various structure have shown themselves as a quite effective method of rock massif management at the abutment of longwall with drifts.

Big volume of researches focused on chemical compound and physical-mechanical properties of the artificial strips has been conducted at the Institute of geotechnical mechanics (Dnipropetrovs'k city, Ukraine). Successful industrial approbation at many mines of Ukraine has shown their effectiveness under various mining-geological conditions and a row of advantages compared to other artificial protection elements:

- possibility of the operations mechanization directed to protection means installation with help of serial equipment,
- decrease of the roof lowering and loads on the support,
- provision of roof caving along the edge of the strip,
- expenditures decrease for the workings maintenance,
- goaf isolation provision, the prevents air loss and coal spontaneous combustion,
- provision of a possibility to use forward-flow airing of the longwalls,
- effective massif management at high stope advance rate and increase of the stope output.

Materials for the strips mounting must satisfy the following requirements:

- quickness of setting – 24 hrs after the material stacking into a strip, the material strength must make up not less than 10 MPa; material should acquire ultimate strength at monoaxial compression of standard samples equal to not less than 20 MPa 28 days after its stacking into the strip,
- steadiness to external sources – material must not lose strength under influence of mine waters, must not be toxic and combustible,
- transportability – initial material must be suitable for transportation in the pipes in dry or solution-like state.

Concretes of various compounds for provision of a required rigidity of a strip are used as solidifying mixtures.

To transport the mixtures to the place of the strips erection, pneumatic and hydro-mechanical methods are used depending on their compound.

Problem of support and steadiness provision of the workings behind the stope is actual not only from the view of its repeated use but also for forward-flow airing since reverse-flow airing does not allow to achieve high output of a plow longwall because of the restrictions by airing factor.

Above described developments are a part of technological schemes of mine workings protection, developed by the Institute of geotechnical mechanics and approbated under the conditions of Ukrainian mines. Based on the conducted researches – "Temporary technological regulations for development openings protection of coal mines with cast strips made of solidifying materials have been designed" (Bulat et al. 2004).

For effective management of rock pressure in a plow stope it is expedient to divide a longwall into technological schemes. Behavior of the massif at these areas is subordinated to the identical laws of rock mechanics. Hence, conducted measures for support and protection of mine workings do not vary significantly within the technological zone.

Practical implementation of plow mechanized complexes under the conditions of Ukrainian mines has shown that with given type of extraction the rate of wall rocks convergence is 4–6 times less than when using shearer method of coal extraction. This, obviously, contributes to the massif integrity conservation and stability of host rocks in a face space of longwall. Hence, it can be concluded that longwall advance rate increase leads to the stresses concentration in the rock massif.

Analysis of a seismic-acoustic activity of the seams shows that there are zones of strongly marked concentration of the stresses. This favors coal breakage and face support since the seam is located under the influence of increased stresses. Under the conditions of developing dangerous by sudden outbursts and pressure bursts coal seams, the plow mining application is inadmissible as it leads to an artificial relaxation of accumulated stresses.

Analysis of foreign plow technique suitability for coal extraction by a comprehensive analysis of mining-geological conditions of coal seams bedding.

Quite detailed evaluation of a plow complex support suitability to the area stresses from the side of a seam roof. Many producers have factually lifted any restrictions concerning coal resistance to cutting, as the drives power allows to mine seams with A_p = 380–400 kN/m and also quite strong rock inclusions in a coal seam.

When designing plow longwall, the works are conducted in the following consequence:

- coal resistance to cutting determination during a plow mining,
- determination of a required and possible output of the longwall (t/hour),
- plow and conveyor movement rate calculation, and also cutting width of a working member (depth of cutting),
- calculation of a required set power of drives,
- establishment of a technological scheme of a plow longwall.

During plow mining of a seam, to some extent, all methods of the roof management are acceptable but the most expedient one is "gradual lowering". Given method implies curving strains of rocks and gradual contact of roof and bottom rocks at a definite distance from the face plane. Such dependence can be received by help of variation of longwall advance rate.

Bottom heaving

As the longwall advance rate and mining depth of coal seams increase under the conditions of weak wall rocks, the bottom heave occurs both in the development and stope workings. Out of the mining stoping influence zone, heaving of the bottom presented by clayish and sandy shales, occurs as a result of elastic-viscous-plastic strain and increase of rocks volume during its breakage. This causes formation of inelastic strains and breakage in the bottom the sizes of which increase with time and lead to significant heaving of bottom rocks.

Practical experience of coal extraction under the conditions of weak wall rocks with stope advance rates up to 10 m/day was being followed by abrupt bottom heaving. At one of the mines of the Western Donbass during mining a seam with thickness of 1 m, maximal value of a bottom heave was seen at a distance of 5–30 m from abutment of a longwall with a drift. This value has reached 30% of a seam thickness. At this, bottom heave process has stopped factually after the shift end. As a result, 10 sections of a mechanized support got out of order.

Quite often technical structure of a mechanized complex is carried out based on analytical calculations of geomechanical situation of the planned-for-exploitation area and existing practical skills of a mining technique functioning under these mining-geological conditions.

Basic advantages of a plow mining

Plow mining of coal has a row of advantages before other methods of coal mechanical breakage. It is especially shown while mining thin seams where the shearers work with "undercutting" of wall rocks but the plow provides delivery of factually "clean marketable" coal.

The basic advantages are:

- relative simplicity and reliability of the equipment,
- construction, work principle and assembly of a plow unit provide possibility of an unmanned coal extraction. Control system located at a drift allows to perform automation of the basic technological processes in a longwall. This provides a possibility to significantly increase safety measures during thin and very thin coal seams mining,

- maximal productivity compared to other existing mechanical methods of thin and very thin coal seams extraction (productivity of plow units is 1.5–2 times bigger than during shearer mining),
- if necessary, with strong wall rocks the cutting member of a plow unit affects the face along not full height of a seam but undercuts its lower part, and the higher part falls on the conveyor under its own gravity,
- to provide temporary stability of a roof, under the conditions of strong coal and weak wall rocks mining ("Bogdanka" mine conditions) it is possible to leave coal packs as a protective layer,
- the plow drives are located at frames of the conveyor drives located at drifts or at the face ends that simplifies their technical maintenance. When drives of plow and conveyor are located at drifts, the necessity to preliminary mine niches is excluded,
- cutting member shears off a coal layer by large split method providing dustiness decrease in a face space (coal dust emission is many times less than during shearer mining), ash-content decrease and coal lumpiness increase,
- rational energy consumption for coal breakage because of a possibility to direct breaking ability of a plow along the strata bedding, to change plow advance speed and pressure on the face depending on coal spall degree,
- provision of a highest mining quality with minimal rocks undercuts, ash content decrease and improvement of extracted coal and anthracites grade,
- application reasonability at seams with high methane content and seams dangerous by sudden coal and gas bursts. At plow mining there is a possibility to conduct seam degassing where methane gradually comes out and then let go to the surface through ventilation system. At narrow cutting depth – up to 10 cm, the seam has enough time to get degassed,
- simplicity of a plow working member structure presenting by itself solid body with installed on it teeth. There are no electro- or hydro-engines mounted on the moving plow, transmissions and there is no energy transmission to the moving plow,
- economic efficiency. Cost of a plow mechanized complex is several times lower than that of a shearer.

Problems concerning plow units exploitation

Under the conditions of Ukrainian mines the problems concerning plow mechanized complexes have both mine and area (local) character.

Problems of mine character. Application of plow mechanized complexes for the same conditions in which shearer complexes were exploited together with the stope output increase, increased gas emission and water inflow, hence, there is a necessity to check suitability of the entire technological link of a mine for the formed conditions of the *mine separate links capacity*.

Direct calculation of the mine separate technological links capacity is based on the job schedule of the mining operations development and diagrams of the longwalls exploitation. At this, temporary production units with maximal capacity are taken as a basis. Based on the detailed analysis of the mining-technical conditions (possibility of a simultaneous development of seams and so on) the variants of the necessary number of simultaneously working stopes are considered during the period of plow mechanized complex enter into exploitation.

Stoping. Productivity of each stope is defined based on the technical means of mechanization considering natural conditions. It is accepted according to the operative regulations and technological schemes of the reserves mining by the plow complexes. Maximal possible indices of the longwall working are accepted for calculations.

Underground transport. Capacity of all links of the underground transport depends on each separate link technical productivity. When underground transport system consists of various interconnected types of transport (conveyor, electric locomotive, rope and so on) the calculations are made based on the complex system.

Hoisting. Productivity of hoisting units depends on their technical parameters, number of served horizons, distribution and structure of the loads movement. Considering this, it is recommended to check suitability of the hoisting units productivity based on the preliminary regulated time of the loads lifting and without this regulation as well. Each hoisting unit productivity calculation is accepted to each served horizon by determination of common, development or common and development operations (coal hoisting, lowering, lifting of people, rocks etc.).

Mine ventilation. When using plow extraction of coal (especially it concerns powerful plow units of foreign producers) the process of gas emission flows by other dependences than during shearer mining. When breaking coal by method of full spall, sufficient amount of methane releases during coal transportation along production and main mine workings. In order to check production capacity of a mine by airing factor it is determined by factual air distribution along the stopes considering satisfaction of all consumers (development faces, chambers, stores and others). Then coal extraction volume from each stope is defined. Test calculations are made by factual methane releases, gas amount from blasting works, maximal number of people and others.

Mine water pumping. Process of water inflow into the mine workings at plow and shearer mining is factually the same but as the face advance rate increases, the water inflow into the mine workings also grows. Thus, check of mine water pumping suitability must be conducted when mining operations cross water-saturated rocks.

Surface technological facility. Suitability of technological facility productive capacity of the mine is performed by evaluation of technological interconnection of each technological link and correspondence of the bunkers for extraction coal storage. Possible production capacity of surface technological facility, in general, is defined as a result of simultaneous consideration of its all components.

Problems of production unit (local) character:

- discrepancy of separate processes interaction in a stope. Such situation is possible when using mechanized complexes, components of which are made by various producers,
- instant relaxation of the massif energy at high longwall advance rates. Given phenomenon occurs during dangerous by dynamic and gas-dynamic phenomena seams mining. Negative effect of the sudden outbursts and rock bursts is much lower than that during shearer extraction of coal due to absence of workers in the stope,
- imperfection of coal breakage by a plow substantiated by its instability in a stope under influence of internal forces of various types and directions. So, normal weight of a working member is not enough to hold the plow in the face and stabilize its working mode. Increase of the working member weight did not contribute to the task solving since it was joined with the dimensions growth and, hence, with zone increase for plow free movement. Stope angle of tip increase caused insuperable difficulties in controlling hanging massif of caving wall rocks and led to tipping of a working member on the face, and to seam bottom breakage. Also it caused break of host rocks by the working member and decrease of productivity. Given problem is successfully solved in plow complexes of foreign producers where needed weight of plows are used and the complexes drives provide necessary pressure of the plow against the face. Leverage holding system of the plow in the vertical plane provides a possibility to regulate extraction movement direction,
- irrational extraction scheme in terms of cleavage and rock fracture often leads to intensive delamination of roof in the upper part of the longwall, decrease of a working space and violation of ventilation mode at the production unit. Disbalance of applied forces on the working member leads to uneven breakage of coal massif and also to host rocks breakage. When thickness of broken rocks is more than seam thickness, the natural slope does not form and slacking occurs distributing to face zone preventing normal movement of plow,
- plow extraction improvement by introduction of dynamic working members (plow-ram, dynamic plows with pneumatic percussion tools, plows with rotating cutting heads and so on) leads to complication of the construction, labor-intensiveness of service and control. Hence, today the preference is given to static plows with various principle of their interaction with coal massif,

- practical work and industrial tests of new pressure and support-guiding elements (in a shape of ropes, chains, rails, box launders and so on) have shown that it is quite hard to create even pressure loads along the whole stope length. As a result, change of a stope shape occurs (from linear to arc-like). It leads to disproportionate growth of all stresses along the stope. Finally, as the stope line advances the correspondence between these stresses and technical possibilities of movement means and plow feed is violated. At uneven hypsometry and variable dip angle of a seam, the guiding element was violating steadiness of the working member to even greater extent and was making it uncontrollable that made exploitation of the whole mining equipment more complicated. Set shape of a stope is maintained in the technological schemes of the plow complexes. It is provided by quite a rigid support-guiding element and uniform forced fixation of the equipment within the face space. The complex construction foresees reliable fixation of the equipment within the stope space. In order to manage rock pressure and protect face space of a longwall from caving rocks, as a rule, the chock-shield supports are used,
- constructional imperfection of the national plow complexes are as follows: failure of the plow frames mainly because of the support brackets break; exit of the plow chain out of the cover due to its holding bolts deformation; intensive wearing of the tool holders bodies for lower and upper teeth; provision of safety reliability of conveyor pulling member from scrapers falling-out from a pan; cutting knives work efficiency increase and reliability of pulling member; plow unit bearing elements wearing decrease. Application of component elements of increased wearing resistance and also corresponding rigid base together with the drives allowing to significantly decrease necessity of repairing works,
- small life until capital repair, high labor-intensiveness and complexity of repair works conduction under confined space conditions,
- when roof rocks hang – necessity of a goaf full backfill in order to escape repeat hanging,
- when extracting strong coal it is necessary to conduct activities focused on preliminary loosening and weakening of the massif. Given problem is effectively solved in the new plow units structural performance of both Ukrainian and foreign production,
- inconsistent geometrical dimensions of the development mine workings. The necessity to mine niches for the complex drive heads.

Problems in "designing-introduction" system»:

- discrepancy of plow complexes productivity and capacity of separate technological links of a mine: transport, productivity of the shafts, surface facility, ventilation and water drainage. As a result, plow complexes cannot work at full power it terms of "available power",
- discrepancy of the assembly-dismantling system. Regulatory schemes of assembly-dismantling of complexes at separate mining enterprises are adapted for concrete mining-geological conditions and based on practical experience of this process realization. As a result, significant problems of wall rocks increased stresses overcoming are possible with correspondence of set-up and dismantling entries,
- complex hypsometry of a seam. Great complexities emerge when seam hypsometry changes and it is impossible to hold plow longwall in the coal seam plane. Specific difficulties occur when stopping intersects various-amplitude geologic faults and solid rock intrusions. Full or partial complex dismantling should be done quite often under such conditions,
- seam bottom heave. Under conditions of weak rocks bedding in the seam bottom when longwall advance rate increases the rate of bottom heave increases as well that negatively influences on the face set of equipment,
- roof dynamic lowering. When roof rocks hang, dynamic lowering of the rocks is possible. This leads to deformation of mining equipment and transferring of sections on the "rigid base",
- dynamic and gas-dynamic manifestations in a stope. Sudden coal, gas and rock bursts and also pressure outbursts have less negative consequence than during shearer mechanized mining since coal mining is carried out without presence of people in the stope. These phenomena lead to, as a rule, deformation of an equipment and stacking of the longwall working space,

– economical discrepancy of the equipment. Structural and technical performance of the complexes is always done with a definite safety margin. This leads to significant rise in price of the equipment and, finally, negatively influences end cost of extracted coal.

CONCLUSION

Coal plow mining is one of the prioritized and perspective trends of coal extraction using mechanical breakage of a massif. Today in Ukraine there are branch programs of plow mining introduction into production are developed; new models of coal extraction are being developed, produced and approbated: plow complexes, plow-ram units, complexes based on frontal aggregates and other technique of unmanned coal extraction. It will give a possibility to provide longwall output at a level of 1500–6000 t/day depending on mining-geological conditions and extracted seam thickness. Besides, corresponding drifting and transport equipment is being introduced that will give an opportunity to realize such output at perspective mines. With planned capacity of a mine being 2000–12000 t/day such technique has to provide stable operation of mining enterprises by 2–3 longwalls.

More than 70% of quality, low ash-content coal in Donbass is located in seams with thickness being less than 1 meter. Practical application of plow technique has shown that its biggest efficiency is provided during anthracites extraction.

Plow extraction success creates prerequisites for further development of constructions and improvement of the technologies. Volume of extracted coal from plow longwalls is planned to be raised up to 20–25 mln tons during the nearest 10–15 years.

REFERENCES

Bondarenko V.I., Kuz'menko A.M., Grjaduwij Ju.B. 2003. Tehnologija podzemnoj razrabotki mestorozhdenij plastovyh poleznyh iskopaemyh. Dnepropetrovsk: Poligrafist, 707 s.

Bulat A.F., Il'jashov M.A., Usachenko B.M., Bajsarov L.V. 2004. Vremennyj tehnologicheskij reglament po ohrane podgotovitel'nyh vyrabotok ugol'nyh shaht litymi polosami iz tverdejuwih materialov – Dnepropetrovsk: RIA Dnepr-VAT, 33 s.

http://coal.in.ua/vosstanovlenie_donbassa/vyemka_uglja_strugami/708-proektno-konstruktorskie-raboty-i-promyshlennye.html

http://ru.bucyrus.com/mining-equipment/longwall/plow-systems.aspx

http://www.mining-enc.ru/o/ochistnyj-agrega

Ignat'ev A.D. 1967. Sostojanie i osnovnye napravlenija razvitija strugovoj vyemki uglja. – Doklad. IGD im. A.A. Skochinskogo.

Ignat'ev A.D. 1970. Analiz tehnologii strugovoj vyemki v SSSR i za rubezhom. Kratkij nauchnyj otchet.

Kompleksnaja mehanizacija ochistnyh rabot s primeneniem strugov. Sbornik nauchnyh trudov. – ShahtNIUI im. A.M. Terpigoreva. 1984.

Tehnicheskaja dokumentacija promyshlennogo ispol'zovanija strugovyh ustanovok v uslovijah ukrainskih shaht.

International Mining Forum 2011, Dyczko et al. (eds) © 2011 Taylor & Francis Group, London, UK. ISBN 978-0-415-68938-0

Chains – Boon or Bane for Modern Plow Technology?

G. Philipp
THIELE GmbH &Co. KG, Iserlohn, Germany

ABSTRACT: The coal plow is an essential piece of equipment for winning under thin-seam conditions. Today's plow installations, which deliver drive outputs of 2 x 800 kW, are equipped with chains 42 x 137 mm in size and can operate at speeds of up to 3.6 m/s. This places huge demands on the chains and their connecting links. Small face openings, which require low-height conveyor systems and coal plows with minimal guideway clearance, combined with the restricted amount of space available in the gate roads where the compact drive units have to operate, pose a real challenge for the geometry and quality of the plow chains. Achieving maximum face outputs regularly and without interruption, even when operating with highly-stressed plow chains, calls for real manufacturing and engineering skills on the part of the equipment suppliers and chain-makers combined with mining know-how and an effective preventive maintenance system on the part of the operator.

KEYWORDS: Chain, Clearance, Crack, Connector, Friction, Guideway, Plow, Plowmaster®, Pretension, Preventive Maintenance, Shearer, Sprocket, Wear

1 INTRODUCTION

Two types of longwall system are normally used in the mineral mining industry (coal, lignite, phosphate, bauxite and potash) – cutting winning with shearer loaders and stripping winning using plow systems. These two methods can be distinguished not only by their technology and economic parameters but also, and more specifically, by their technical usability.

The following technical distinction between the two systems, which dates back to 1948, is sure to raise a smile today: 'The cutting machine, of Latin origin, has an individual character and is adapted to the national characteristics of achieving agreeable working conditions by improving the environment with water sprays. ... The coal plow, of Germanic origin, is a team based machine that requires strong discipline.'[LEH-48]

The longwall face system always requires a chain conveyor installation to carry away the extracted mineral. On plow faces the power that the winning machine needs to move along the face, strip away the mineral and load it on to the conveyor is supplied via an additional chain system. On shearer faces this also applies to older designs of machine and to installations operating in thin seams.

The question of whether chains are a blessing or not for modern longwall installations is therefore a particularly relevant one. An engineer would phrase this question differently: do chains improve the performance of the winning system or are they only performing a necessary function? This is examined in greater detail below using as an example the plow chains that are commonly used in the coal industry.

2 PLOW VERSUS SHEARER

Of course seam thickness plays a huge role in any comparison between the coal plow and the drum shearer – with seams less than 0.6 m in thickness being classified as unsuitable for mechanized coal winning.

Cutting winning is practically impossible from a technical point of view in seams that are less than 1.5 m thick. In any comparison of the two methods, however, stripping winning begins to lose out significantly when seams are more than about 2.2 m in thickness [MYS-08], while both systems are equally able to operate in the seam range of 1.5 to 2.2 m. In this transitional area in particular the usability of each respective method will be determined by factors such as the composition of the coal and the local geological characteristics (variations in seam thickness, inclusions, faults, coal strength, etc.). While the dissimilar coal-getting methods and varying geological parameters make it difficult to draw a direct comparison between the winnability of different coal seams using stripping and cutting winning systems, the following rule basically applies: as coal strength increases so the advantages of the cutting winning method become more pronounced.

What is unquestionable, however, is that the initiated power density of the coal plow system, which is defined as the installed drive power related to the seam thickness, is clearly superior to that of the drum shearer up to a seam thickness of 2.2 m, though with a downward trend (Figure 1). At a seam thickness of 2 m, for example, the initiated power density of the plow method is still 33% higher than that of its rival.

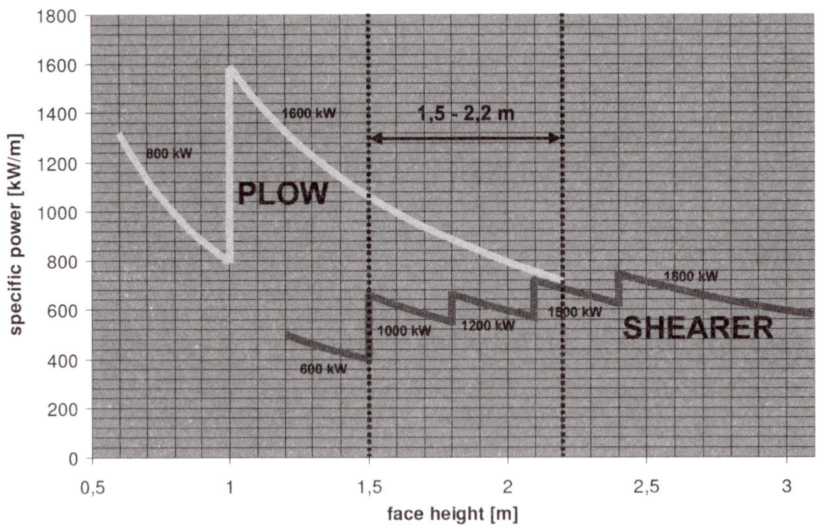

Figure 1. Comparison of power density (installed power/face height) shearer versus plow – subsequently amended [MYS-08]

While there is no doubt that the high initiated power density of the plow installation makes it much easier to win the coal on thin-seam faces, this power density can only be applied to the coal face via the plow chain. This in turn means that the chain and its connecting links will be stressed accordingly.

3 BRIEF REVIEW OF PLOW AND PLOW CHAIN DEVELOPMENT

The first patent for a stripping coal winning system dates back to the year 1912, though it would be another 30 years before Konrad Grebe put his 'coal plow' invention into service at Ibbenbueren colliery in 1941 (Figure 2). This installation was still hauled by ropes, as were the shearer machines of that era, which were the forerunners of today's drum shearer loaders [DRE-05].

Figure 2. The first rope-hauled coal plow at Ibbenbueren mine in 1941 (THIELE archive)

The rope haulage system soon proved unsuited to the task. Large pulley diameters, no capacity for altering the rope length and a lack of repair facilities, plus the added expense of the drive power being provided by locomotives plying back and forwards in the gate roads, were probably just some of the reasons why the switch was made to haulage chains. By 1943 chains, or more precisely anchor stud-link chains, were being used along with their windlasses – as shown in the photograph in Figure 3, which was taken at Westende colliery in Duisburg [DRE-05]. Unlike the modern systems in use today, the plow chain was not part of an endless and continuous chain assembly but was simply deposited into the gate road at the end of the run. This was made possible due to the high packing density that is common to all chains and the presence of the studs, which prevented the chain from kinking as it was being fed back on to the face.

Presumably it was the shortages of the wartime economy rather than technical considerations that led to the introduction of the anchor chain, since these large chain sizes were then to disappear for some time from the mining industry. The first draft standard for mining chains (DIN 22 252), which was drawn up in 1950, only features chains 16 x 64 mm and 18 x 64 mm in size, the first of these deriving from the anchor stud link chain of similar dimensions. Five years later saw the first and only time that DIN 22 252 included the two plow-chain sizes 19 x 75 mm and 22 x 86 mm.

One of the most important steps in the development of plow technology was taken by Wilhelm Loebbe, who was senior engineer at the former Gewerkschaft Eisenhuette Westfalia iron and steel works. In 1947 he designed and built the plow system that was to bear his name. Loebbe was the first to come up with the continuous, endless plow chain that is still in common use today. His mechanized pushing system for advancing the conveyor and plow drives also used chains, which was fortunate for the chain-makers as three chain systems (plow, conveyor and pushing system) were now needed on the coal face [DRE-05].

Figure 3. A 36 x 144 mm anchor stud link chain used as plow chain at Westende mine in Duisburg 1943 (THIELE archive)

As plow drive performance increased over the following years so too did the thickness of the chains (Figure 4). This was initially accompanied by a reduction in the operating tension, which was a positive development as far as plow chains were concerned:

$$\sigma_C = \frac{P}{v \times A_C} \tag{1}$$

where: σ_c = chain operating tension; P = total installed plow drive power; v = chain speed; and A_C = nominal chain cross section (note: 2 circular areas).

This trend was reversed, however, with the introduction of the latest generation of GH 42 plow installations (GH = Gleithobel). The operating tension of the 42 x 137 mm plow chain is about 36% higher than that of the previous generation of 38 x 137 mm chains. The geometrical restrictions imposed on the plow chain guideways, which are a result of the lack of space available on thin-seam faces, simply do not permit the use of larger-sized chains. And yet comparable – and in some cases much higher – operating tensions have been safely applied to plow chains in years gone by (Figure 4).

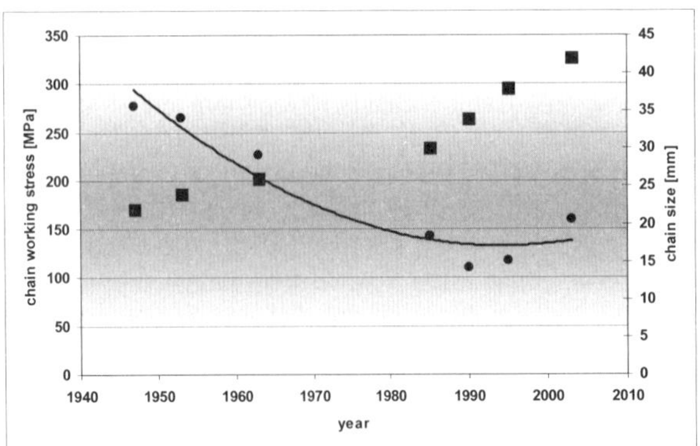

Figure 4. Developments in chain tension and plow chain sizes (source: [DRE-05])

Furthermore, since 2003 numerous successful deployments of plow installations fitted with 42 x 137 mm chains – most recently at Bogdanka colliery (Figure 5) – have shown that chains of the highest quality can indeed cope with operating tensions of this level.

Figure 5. First plow to run a 42 x 137 mm chain at Bogdanka colliery - nicknamed 'Hobel Hans' and seen here with his godfather 'Hobelhans; March 2010 (source: Bogdanka colliery)

4 TECHNICAL REQUIREMTS FOR PLOW CHAINS

4.1 Geometrical requirements

Aside from chain diameter, which – as shown above – determines the chain operating tension and hence the amount of power that can be transmitted, the most decisive factors for a successful plow operation are the pitch and the external width of the chain unit.

For a given number of teeth the pitch will determine the diameter of the drive sprocket and its pitch circle. Operators usually opt for the maximum number of teeth that is technically feasible in order to minimize the amount of bending between the chain links as they pass over the drive sprocket. The latter should have at least six teeth. In simplified terms the pitch circle can be calculated as follows:

$$PCD \approx \frac{p}{\sin\left(\pi/_{2n}\right)} \tag{2}$$

where: PCD = pitch circle diameter; p = chain pitch; and n = number of sprocket teeth.

This technical relationship between chain pitch and sprocket size, together with the space restrictions imposed by the gate roads where the plow drives are positioned, has meant that the pitch remained at 137 mm as the chain diameter increased from 38 mm to 42 mm.

This in turn has had a major influence on the design of the mechanical connecting links (Figure 6). These connecting links are vital for transporting, installing and removing the plow chain and for shortening and lengthening the chain assembly; they are also needed for renewing sections of chain, where short lengths have to be connected together, and serve as replacements in the event of a chain-link fracture. Maximizing the outer dimensions of the connecting links allows the load-bearing profile of the connector to be enlarged, which increases the breaking strength and fatigue limit of the component. However, the shortness of the pitch spacing and the restrictions imposed on external width by the dimensions of the plow chain guideways are diametrically opposed to such a measure.

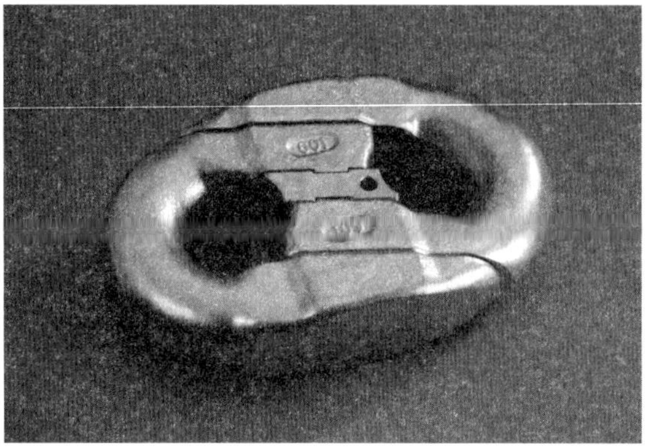

Figure 6. Newly developed 42 x 137 mm connecting link – the Plowmaster®

In order to minimize power losses and avoid additional strain on the chain due to friction between the chain and the guideway the chain should essentially be able to pass through the guide profile without making contact. Modern plow pans and plow guides do in fact permit a mutual play of 6° in the horizontal direction and 1° in the vertical at a line-pan length of 1750 mm [DRE-05]. A simple equation can be used to calculate the required play between the chain and the guideway:

$$C = l_T \frac{\alpha}{2} \tag{3}$$

where: C = clearance chain/guideway; l_T = length of trough; and α = angle between troughs (radian).

For the given values, and at an angular deflection of 6°, there would theoretically be a required play of 91.6 mm, which would in fact be more than doubled given repeated bending in the same direction – as is common on most longwall faces. As the GH 42 plow guideway only has an internal diameter of 160 mm this would imply that the plow would have to run on either air or chains made from anti-matter, as the internal space would be more than taken up by the amount of clearance needed. Such an observation is of course nonsensical from an operational perspective. The plow chain and, more particularly, the chain connector with its larger enveloping circle have to be designed to operate with much less play. At a given connector width of 151 mm, as provided for in technical standard DIN 22258, the clearance in the plow-chain guideway is merely 9 mm. The result is frictional contact at the legs of the connector, and at the chain links too, which inevitably leads to a wasteful loss of performance and increased rates of wear.

This situation led to the first plow installations fitted with special chain assemblies, in which every third link was designed as a thicker-profile forged link. These forged links were intended to provide protective cover for the connectors and, by distributing a series of 'support points' along the entire plow chain, would also reduce the running noise coming from within the guideways. The forged links also featured thickened legs in order to prevent kinking of the chain. [VOS-04] After just a few operational trials it was discovered that this was not an appropriate remedy, with the result that all chains are now fitted exclusively with 'normal' round links – it being accepted as 'operationally unavoidable' that the 'unprotected' broad connectors should act as wear-prone components.

4.2 Wear protection requirements

Leg wear directly weakens the connector because it already has a narrow connecting profile within its leg section. For a normal chain link, leg wear of as much as 20% will have no impact on the component's mechanical properties, the forces generated in the crown area with its load-bearing profile being distributed along both legs of the link. However, the friction stresses generated as a result of the normally high plowing speeds will invariably lead to friction cracks, which – if the chain or connector material is not sufficiently tough enough to be exhibit the required crack arresting capacity – will result in chain failure in the form of a fatigue fracture (Figure 7).

Figure 7. Failed plow chain link with fatigue break induced by friction cracks

Providing wear protection by maximizing material strength, on one hand, and introducing a high crack-arrest capacity in the wear-prone legs by using as tough a material as possible, on the other, are diametrically opposed objectives since the latter can only be achieved at low material strengths. As explained above, this is not a critical factor for the chain itself, while for the connector it can only be achieved by ensuring that this component runs as far as possible inside the profile traced by the chain.

This concept, which was to lead to the development of the special chains used on the first GH 42 plow, was taken up by THIELE in designing the new Plowmaster® plow chain connector, which also meant having to sacrifice some of the connector cross-section because of the narrower 'standard plow chains' being used. Even though the connector had a breaking strength that was some 10% less than that of the chain, initial trials showed that the system was fit for purpose. During the test run the new connector easily outperformed all previous designs, achieving an operating life of nearly 10 000 minutes without failure as compared with the 3 500 minutes normally expected from more conventional connector types.

Chain durability is determined by the wear zone that is located in the shoulders of the chain links. As the chain passes over the drive sprocket material is lost from the shoulder area as a result of the relative movement that must necessarily take place at the shoulders of the links and this action lengthens the pitch spacing. This increase in pitch impairs the way in which the chain interacts with the sprocket, a situation that is further aggravated by the fact that the chain pockets also become worn. A maximum increase in pitch of about 3.5% is technically acceptable –

anything above this and the chain should be replaced. Here it is extremely important to note that the chain strands that connect directly to the plow, and the strand immediately facing them on the opposite run, reach their wear limits twice as fast as the rest of the chain assembly (Figure 8). This can be attributed to the fact that the plow has to 'sump in' at both drive ends in order to start each new plowing run.

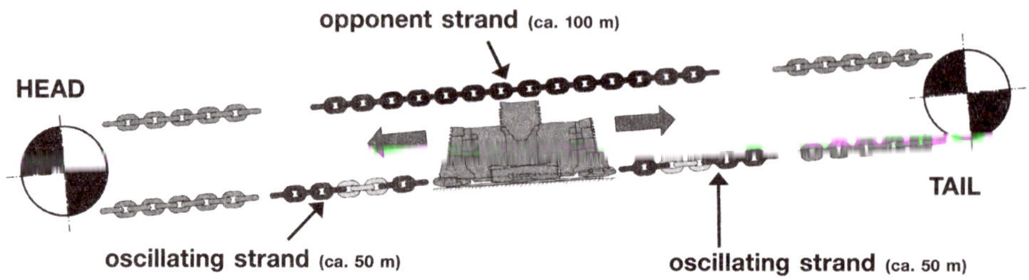

Figure 8. Plow chain schematic showing different chain wear characteristics

4.3 Durability requirements

While the lifetime expectancy of plow chains and chain connectors (see above) is certainly determined by the quality of the products and their geometrical design, the type of operating conditions encountered will also have an influence on the durability of these components.

If the plow installation becomes jammed solid close to the drive unit, resulting in load peaks developing in the millisecond range, the result will almost always be a forced rupture of the chain, as the overload protection devices will not have sufficient reaction time to separate the blocked chain from the pulling drive unit. When undertaking repairs on chain fractures of this kind it is important not only to insert a connector in place of the fractured link itself but also to replace that section of chain located ahead of the fracture in the direction of pull, as this may have been damaged by the load peak generated when the installation jammed solid. Where a chain has become elongated due to wear its opposing strand also has to be replaced in order to prevent any shift in power delivery due to the opposing drives being traversed by unequal lengths of chain.

Fatigue fractures of chains and connectors are almost always attributable to the formation of friction cracks (see above). This problem can effectively be prevented by using a 'proper' quality of chain and by ensuring that when manufacturing the links the material strength of the legs is reduced accordingly (Figure 9).

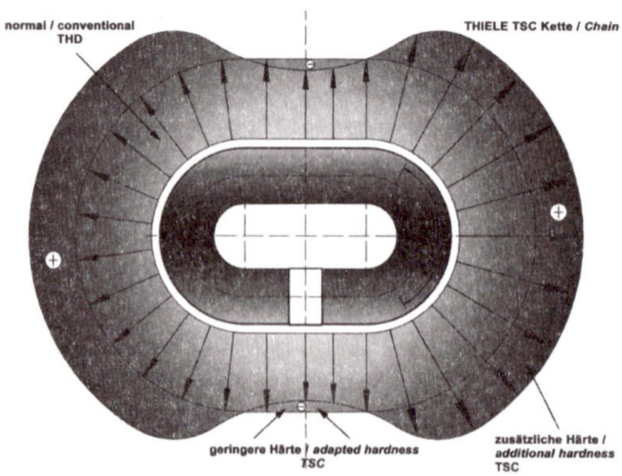

Figure 9. Material strength profile of a plow chain link

The overriding factor for chain replacement, however, is wear induced elongation. The rate of wear can be slowed down by increasing the strength of the material in the shoulders (Figure 9). The shoulders of the chain links are however only part of the tribological system that makes up the 'chain joint', for it is really what keeps the shoulders apart, namely the product being conveyed (and in the best-case scenario a lubricating agent), that essentially determines the rate of chain wear. In fact coal itself has lubricating properties, while stone and shale are abrasive and only accelerate the rate of wear. In the Ruhr Carboniferous the main run of a 42 x 137 mm chain will normally achieve an operating lifetime of about 60 000 minutes, while the sections of chain that attach to the plow, and the strand that lies opposite, will usually make about half of this (Figure 8).

4.4 Pretension requirements

The most important thing that the operator can do to maximize chain and connector operating life is to ensure that chain pretension is regularly monitored and maintained. Constant changes in the position and length of the plow installation, and therefore of the chain too, will also affect the amount of pretension. An overstretched chain not only wears at a faster rate but also ages due to fatigue caused by the excessive operating forces involved. Unlike coal conveyors, plow installations offer sufficient room for a slack chain. The rule to be followed is therefore quite simple: a plow chain should ring, not creak.

OUTLOOK

In spite of the restrictions and requirements imposed by thin-seam faces equipped with modern plow installations, today's plow chains and their connecting links can operate very successfully provided they are professionally designed and manufactured to the highest technical standards. Chain and connector operating life will continue to be the focus of attention in the years to come. Automatic chain lubrication and regular and careful preventive maintenance, including regular chain inspections, will have an important role to play in improving the life expectancy of these components – as will the development of new geometrical designs and new materials for future generations of plow chain.

REFERENCES

Dreher, H. 2005. The coal plow. From its invention to today's state of the art. Lecture THIELE internal technical training – unpublished. Iserlohn.

Lehman, G. 1948. Dust suppression and temperature control, winning and loading machines, prospects for the development of winning technology. Glueckauf 1948 (81/84): 464-466.

Myszkowski, M. and U. Paschedag 2008. Longwall mining in seams of medium thickness – comparison between shearer and plow. Lecture at World Mining Congress. Katowice.

Voß, H.-W. et al 2004. Initial experiences with a new high-performance plow installation operating in hard coal conditions. Glueckauf 2004 (140):510-518.

International Mining Forum 2011, Dyczko et al. (eds) © 2011 Taylor & Francis Group, London, UK. ISBN 978-0-415-68938-0

Power Supply, Control and Automation Systems of the High-Performance Plough Systems Based on the Example of LW Bogdanka S.A.

M. Chagowski
Lubelski Węgiel Bogdanka S.A., Poland

T. Jędruś, J. Lubryka, D. Macierzyński
ELGÓR+HANSEN Sp. z o.o., Poland

ABSTRACT: This paper presents the experience gained during development, implementation and commissioning of the high-performance plough system of new generation at LW Bogdanka S.A. and describes in detail the power and control system taking into account the automation process of plough systems.

KEYWORDS: hard coal mining, thin seams, plough system, control system, automation process

1 INTRODUCTION

Analysis of the situation on the fuel and power supply market indicates that demand for coal, both power and coking one, will systematically grow (Tor et al. 2008). Even in situation of the present economic crisis and decreasing world's resources of oil and natural gas, it is clear that coal is relatively cheap and reliable energy carrier, providing in particular the energy security of Poland and improving the energy security of other countries of the European Union. The condition of proper use of balance resources is rational and efficient exploitation of owned coal deposits. Rational exploitation, i.e. the one that limits to the necessary minimum loss of coal, such as excavating thin seams, and efficient exploitation, i.e. the one that uses modern, safe, efficient systems reducing human intervention in the process of extraction (Tor et al. 2008).

So defined needs direct the coal exploitation towards technology not requiring constant, direct supervision over a mining machine. The ploughing technology, known for many years but at present significantly improved, provides as follows (Myszkowski et al. 2008):

a) possibility of effective excavation of thin seams,
b) less fragmentation of excavated material and thus increase of thick assortments sourcing,
c) possibility of excavating with small webs and fast movement of powered roof support in areas with difficult ceiling conditions,
d) improvement of working comfort and safety due to automated production process (Lubryka et al. 2009),
e) ability of precise following the bed.

Control and automation are the most important components of the highly efficient ploughing technique, an essential base of which there is also electrohydraulic control of a powered roof lining. Control of a ploughing system operation is done from a ploughing station, located in the main gate or from the ground surface of a mining plant. An operator at the ploughing station receives current information about important data related to work of the whole system and can continuously adjust the system operational parameters (Stopa 2008). The ploughing technique for exploitation of thin seams is successfully used in Germany, Russia, China, USA, Czech Republic and Kazakhstan.

In 2006–2008, the ploughing technique was particularly thoroughly analysed with regard to its application for exploitation of the newly opened parts of beds in JSW S.A., KWK Zofiówka and LW Bogdanka S.A. coal mines. The result of the above analysis are two high-performance plough systems successfully put into operation – in the Jastrzębska Coal Company and in the LW Bogdanka SA Currently there are works carried out on implementation in LW Bogdanka SA the next, second plough system.

2 THE POWER SUPPLY AND CONTROL INSTRUMENTS FOR THE HIGH-PERFORMANCE PLOUGH SYSTEM

The proposed technique for highly efficient plough systems forced the use of medium voltage for power supply of the system machines. The innovative solutions of explosion-proof electric devices developed by Elgór+Hansen company, used to supply longwall complexes, confirmed their fitness for use in conditions of underground mine plants, and operational experience gained while tracking the latest scientific studies have allowed to offer, in relatively short time, solutions that provide not only high level of safety, but also the expected utility for the highly efficient plough systems.

Increase of machine voltage up to 3.3 kV, besides provision of power supply reliability of these machines, ensures in particular improvement of operational safety and a number of already repeatedly presented advantages (Morawiec et al. 2008). Positive assessment during operation of flameproof contactor breakers, thyristor starters and transforming stations makes the company to update constantly the produced devices with particular attention paid to improvement of handling security, selection of state-of-the-art components included in the highly efficient longwall systems as well as plough systems of new generation.

2.1 Explosion-proof transforming stations

The transforming stations proposed by Elgór+Hansen for power supply of plough systems meet high technological requirements, ensure operational reliability and significantly improve working comfort and safety of technical services in mines. In power supply systems of the plough, sets of units are applied in the transforming stations in fire-proof enclosures, with transformers of 2100kVA and 2600kVA power and 6/3.3 kV transmission (Lubryka et al. 2009).

All of the above mentioned explosion-proof transforming stations are equipped with a dry-type mining transformers insulated with roving resin (Morawiec et al. 2009). The individual interconnected station segments integrate not only the mechanical housings by means of fire-proof flange connections, but also ensure proper and safe operation of the station. The chamber of upper voltage of transformer stations has been equipped with overcurrent protection of the main transformer and small size medium voltage switch insulated with SF_6, what definitely improved the possibility of effective securing the supply network. This chamber – being a single-element medium-voltage distribution field - determines the possibility of supplying power and proper operation of the whole station.

In the station chambers of lower voltage innovative solutions are applied that enable transfer of all transforming station power from completely secured output onto common connection of manoeuvre instruments – parallel operation of two, three or four cables and/or rubber coated cables on outputs.

Applied electro-power protections and adopted system of control and cooperation with technological locks, with use of intrinsically safe circuits allow for supplying receivers installed in explosive rooms. The transforming stations have been equipped with intrinsically safe system of remote switching off of the main switch on the side of upper voltage. All stations are equipped with high-speed earthing switch on the side of lower voltage, providing a particularly high level of operator's safety (Morawiec et al. 2009). Possibility of moving the stations in underground mines is provided by a set of wheels and transport handles.

An exemplary block diagram of transforming stations with 6/3.3kV transmission of the EH-d31-2600/6/3.3/2/01 type applied in the first plough system at LW Bogdanka S.A. and offered to supply the system no. 2 for LW Boganka S.A. is shown in Figure 2.1.

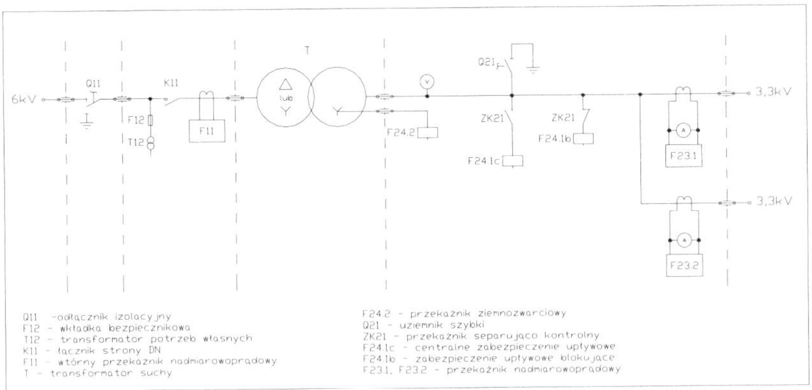

Q11 – isolating switch
F12 – fuse-link
T12 – transformer for internal load
K11 – connector of lower voltage side
F11 – secondary overcurrent relay
T – dry-type transformer

F24.2 – earthing relay
Q21 – fast earthing switch
ZK21 – separating-control relay
F24.1c – central leak protection
F24.1b – interlocking leak protection
F23.1, F23.2 – overcurrent relay

Figure 2.1. The block diagram of the transformer station of EH-d31-2600/6/3,3/2/01 type

2.2 Explosion-proof shunting devices for control of machines in a plough system

The following components included in the company's offer are examples of special constructions for use in highly efficient plough systems: fireproof four-output contactor switch, switch with a thyristor starter and transformer set of 160kVA power and 3.3 kV/1 kV or 3.3 kV/0.5 kV transmissions (Morawiec et al. 2009).

The fireproof contactor switch of EH-d03-W/3,3/I/01.xx type and of EH d03 W/3,3/I/02.xx type (ELGÓR+HANSEN), as well as the switch with thyristor starter of EH-d03-WR/3.3/I/01.01 type (Regulation of the Minister of Economy) are devices that bring completely new quality in the earlier proposed range of shunting equipment for voltage of 3.3 kV. Disconnecting switches on the supply side and earthing switches on the side of supply and output have been used in them. Additionally, the safety switch enables assembly of the devices in every case of shortage of the power supply network in the underground mine. The devices are equipped with a system of mechanical and electrical interlocks, which also causes a significant improvement of occupational safety of technical services at mining plants (Regulation of the Minister of Economy).

The fast earthing switches used in these devices set new standards of safety. The integrated mining systems supplied with 3.3 kV voltage eliminating the need of leading other than 3.3 kV operating voltage to the area of the system, is completed with the fireproof transformer unit of EH-d03-160/3.3/0.5(or 1,0)/0.2/4/01 with 160 kVA power, 3.3kV/525kV or 3.3 kV/ 1050V transmission, besides four three-phase outlets for voltage 500V or 1000V equipped with two completely protected outlets for 230V voltage with 5kVA power, one completely protected outlet for 42V voltage with 500VA power and one fully-protected outlet for 24V voltage with power of 500VA. The unit, like the 3.3 kV circuit breakers of the new generation, is provided with a power supply isolator and fast earthing switch on the supply side.

Block diagrams of the new generation of shunting switches for 3.3 kV voltage and a block diagram of the transformer unit and switch with a thyristor starter are shown in Figure 2.2a and 2.2b. Figure 2.2b presents a block diagram of the power system for machines in the plough system no. 1 in LW Bogdanka S.A.

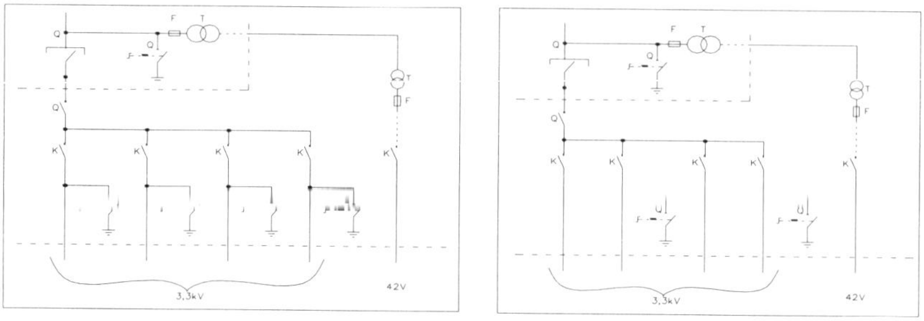

Figure 2.2a. Block diagram of the contactor switch of EH-d03-W/3.3/I/01.xx. type and the switch for reverse power supply of plough main engines EH-d03-W/3.3/I/02.xx

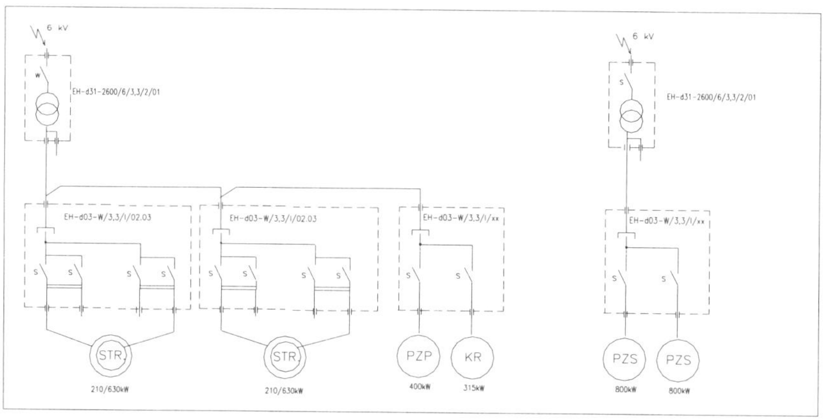

Figure 2.2b. Block diagram of power supply system for the high-performance plough system in LW Bogdanka S.A.

2.3 Control system for the plough system

The condition indispensable for innovation and improvement of operation and safety of the new generation plough system is automation of machine operation with use of general industrial IT and networks of Ethernet type in order to provide complete exchange of information. Degree of advancement, universality and availability of computer networks in mining industry has reached very high level, both on the side of technological processes and devices for their automation. For this reason, producers, constructors, designers, users and operators have been challenged with very high requirements.

The control system for machines in the high-performance plough system is an advanced and very complex microprocessor-based system, that performs many functions such as (Czechowski et al. 2008) control, visualization, registration, archiving, diagnostics.

The spatial distribution of individual elements of the whole system suggests a distributed construction of the control system that uses industrial local networks as transmission media.

The plough system consists of many sub-systems responsible for various control and visualization functions of the system as follows:

1) EH-WallControl – control system,
2) EH-WallView – the system of local visualization and central electric equipment of Elgór+ Hansen company, along with possibility of general parameterization of the system and protections,
3) the Bucyrus PMC-R control system for a powered roof support,
4) the Bucyrus PMC-D control system for a ploughing device,
5) intrinsically safe system of hands-free communication, signalling and interlocking – in LW Bogdanka the APD-1 system has been applied.

Data processing is done hierarchically, starting from the level of individual components control, through parameterization of devices, diagnostics, registration, visualization, local and central data archiving, to the level of management of the entire technological process.

By assumption, the system analyses many dependences to foresee the largest possible number of cases, situations and statuses of both the machine itself and the very difficult surrounding environment (Czechowski et al. 2008). The system is divided into two logical, interrelated parts: control and visualization.

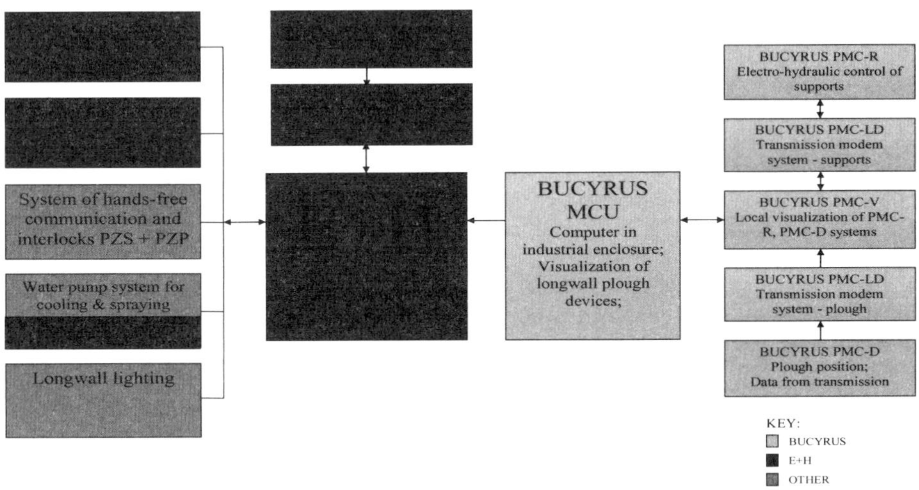

Figure 2.3a. Block diagram of the high-performance longwall systems

2.3.1 Elgór+Hansen control and visualization system for electric devices

The EH-WallControl system was designed as a set of universal functional devices that communicate with each other using the serial transmission bus performing distributed control (Bucyrus DBT Europe):

a) the control device of EH-O/06/02 type (free programmable logic controller – PLC),
b) the control device of EH-O/06/03 type (IPC computer in industrial performance).

Individual controllers (with individually defined configurations) have been arranged at the plough operator's position, at drive of the longwall armored face conveyor, at the auxiliary drive of the plough and at compact stations. At the design stage an emphasis was put on simplicity of handling and control as well as easiness of making changes and modifications. One of the main functions of the system is to ensure reliability and safety in an algorithm of technology process control. Thus, the system has been equipped with an independently operating hardware-analogue system for emergency shutdown. When selecting components for construction of the control system devices a particular emphasis was put on reliability, guarantee of stable maintenance of operating parameters, diagnosability and a high degree of integration. The system software of individual elements in a maximum achievable degree performs the function of testing and self-monitoring of components.

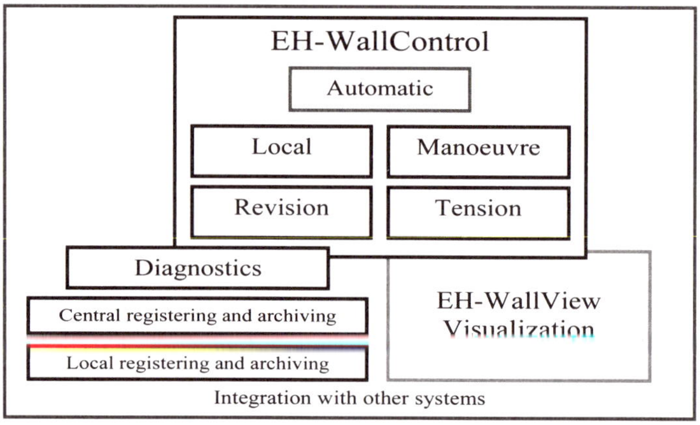

Figure 2.3b. Tasks performed by the control system of the plough system

The individual connecting elements including manoeuvring sets and circuit breakers have been connected with the industrial network operating in master – slave structure, in which the control data pass through a central controller and are visualized and archived in the EH-WallView central visualization system produced by Elgór+Hansen company. All devices and included in them modules are equipped with system software, performing the basic functions of monitoring and control of individual, communication sub-processes with the controller and system resources, addressing of individual modules and communication services enabling remote diagnostics and testing.

The RS485 bus and MODBUS RTU protocol were used for diagnostic communication between the devices. The RS485 bus along with PROFIBUS DP protocol were used for control communication between controllers. At a level of a single device a controller is the master unit, while all functional modules are slave units identified by the assigned address and type. The connection structure of controller network was based on a similar solution, which consists of one master station and several slave units. The system was designed for transmission speed of 9.6 and 19.2 kBits/s.

The next step of the plough control system development is precise defining of control algorithm. Correlations of all input and output signals, e.g. sensor statuses received from the EH-O/06/02 devices, information from the Bucyrus control system, executive elements statuses from manoeuvring sets etc. are taken into account.

On this basis the application software of the central controller and local controllers have been developed. Part of the control process tasks was distributed among individual local controllers, while data between controllers are exchanged in the network protocol frames.

The control device of EH-O/06/02 type, due to use of the industrial PLC controller, allows for implementation of complex input/output logic functions in the control system, in which it was used. The device can be connected with intrinsically safe external sensors and external executive devices (e.g. solenoid valves, signallers, separators, other systems, control circuits of fire-proof connectors, systems of visualization and data transmission, etc.).

While the control unit of the EH-O/06/03 type enables in particular, due to use of the industrial IPC computer, performance of the complex functions supervising the entire control system consisting of many sub-systems by means of various transmission interfaces and various protocols (Bucyrus DBT Europe).

The control unit of the EH-O/06/04 type enables, due to use of the industrial PC computer with monitor, performance of the complex visualization functions of the control system and functions to parameterize of the entire control system by means of various transmission interfaces and various protocols.

The following operation modes of the plough system are foreseen and implemented in the control system:

1. The automatic mode,
2. The local mode,
3. Revision/tension,
4. Transport.

The system controlling can be performed after selection of the operating mode from the plough operator's panel underground, from local control panels at given drives or from the ground surface of the mine.

In the automatic mode the plough system can be activated from the underground or ground surface operator's position. The plough is activated by means of separate buttons. The plough's initial direction of operation is preset by the plough operator. After reaching the end of the longwall during automatic operation, the plough stops and starts movement in the opposite direction automatically, after a time period defined in the system parameters and after obtaining drive permits from Bucyrus system.

In the local mode activation of individual devices of the system is performed from local panels at drives.

Activation hierarchy of individual devices is technologically determined and realized as in the automatic mode, i.e. the crusher → bottom conveyor → longwall conveyor → pump unit → plough. During local mode operation the plough stops after reaching the end of longwall and it restarts in the opposite direction when the operator presses one of the activating buttons.

The plough activated by a fast run button starts running slow and switches into fast run after a period of time preset in the system parameters by an operator and after receiving permission for the fast run from Bucyrus system. After reaching the plough drive on fast run, the plough switches automatically into the slow run in the operator-defined distance from the drive. If the operator decides that the plough has to pass the entire wall on a slow run, he can activate it by the slow run button. In this situation, the control system will not switch the plough automatically into the fast run.

The revision and tension mode allows for individual operation of single drives in the whole system. The revision/tension mode is switched from local panels respectively for individual devices. During the revision/tension of the longwall conveyor the ability of plough activation is inhibited and operation is possible with only a single engine. The revision/tensioning of the plough chain inhibits the possibility of the longwall conveyor activation. Also at the revision of a crusher, the longwall conveyor is blocked, and the revision/tensioning of the longwall conveyor inhibits the crusher operation. Activation of all devices is preceded by an acoustic warning signal.

In the transport mode, it is possible to switch on only the conveyors. Control of conveyors in this mode is done from local panels.

Deactivation of individual devices of the system will be caused by:

a) pushing the switch off button for each device, depending on the operation mode with preservation of deactivation hierarchy of the slave devices,
b) pressing the emergency stop,
c) activation of sensors, depending on the operation mode and impact of a given sensor onto a given device,
d) activation of electro-power protections,
e) turning off the preceding device (in automatic and local mode),
f) no transmission,
g) no control contractors from the system of powered roof support control in local and automatic mode,
h) no confirmation of activation of controlled contactor and similar factors.

The EH-WallView visualization system of electrical devices of plough system was divided into:

a) local visualization,
b) central visualization.

The local visualization covers one unit device such as e.g. compact station, transformer unit. The visualized data have been limited to the given device and respectively grouped in order to simplify local diagnostics for an operator. For example, in circuit breakers and starters of the plough system for local visualization the KSK-1 device has been used, while in the transformer unit one has used the EH-P05/03 LED display.

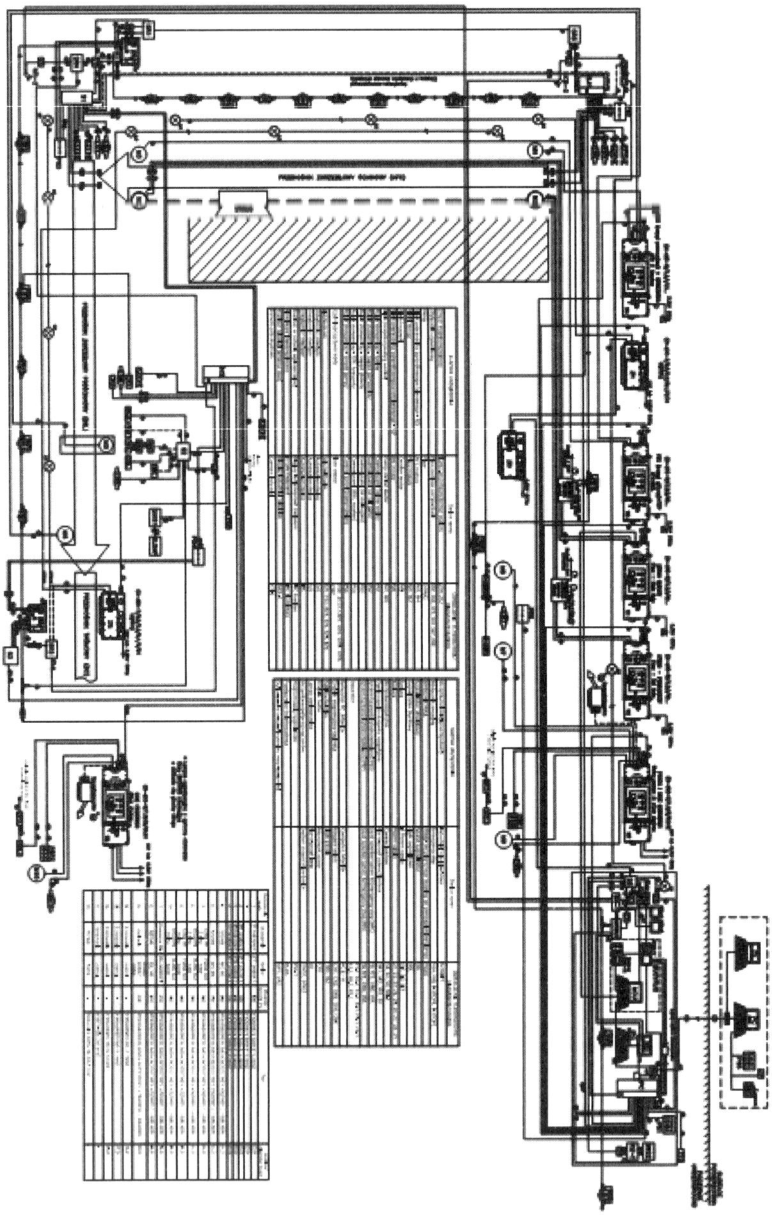

Figure 2.3.1a. Block diagram of transmission bus connection system in the plough system no. 1 in LW Bogdanka S.A.

At some devices, e.g. EH-O/06/02 or EH-O/06/03, the local visualization has been performed by diodes indicating the status of individual components of the device. In addition, the local visualization system based on KSK-1 enables local data archiving. The exemplary local visualization screen is shown below.

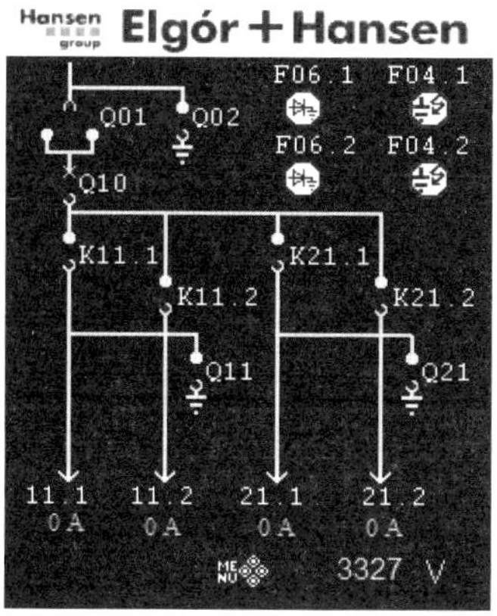

Figure 2.3.1b. The exemplary main screen of local visualization

Figure 2.3.1c. The exemplary main screen of central visualization

The central visualization is based on the device of EH-O/06/04 type, which includes a computer with monitor and intrinsically safe keyboard. It covers all devices included in the plough system. The main performed functions are as follows:

- visualization of the control system status,
- visualization of devices included in the train of electrical instruments,
- visualization of devices included in the control system,
- parameterization of the control system,
- central archiving,
- review of historical data for service analysis.

2.3.2 The Bucyrus PMC-R control system for the powered roof support

The control system for the powered roof support is based on the PMC-R control units, each built in housing section, which respectively configured realize the imposed system operational mode.

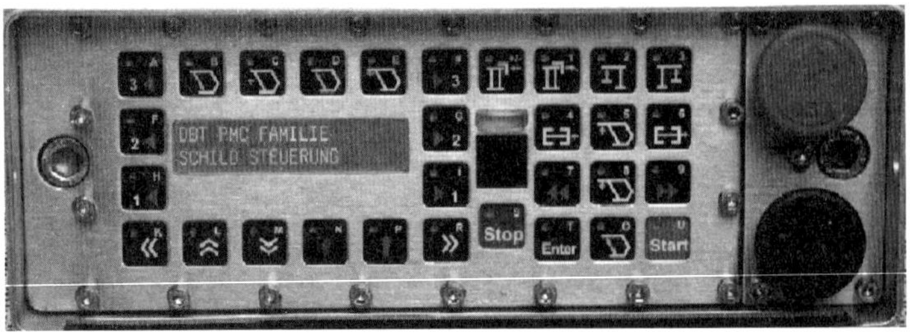

Figure 2.3.2a. View of the PMC-R control unit for a powered roof support

The control unit (Figure 2.3.2a) receives the measured values from elements of the system and, depending on the desired configuration, allows the user both to use the individual selected functions and application of automatic features.

The elements cooperating with the PMC-R controller that allow for the implementation of the required functions are:

a) control strip for PMC-SD1 valves. The strip can handle up to 22 valves, which are connected by special cables perform the desired functions,
b) the SKKT24 encoding plugs, which while connected to the controller allow to encode the desired operation program,
c) pressure sensors with built-in amplifier that generates measuring voltage from 0.5 to 4.5 V for 0–600 bar measurement range,
d) contraction meters for measuring the road of travel, built into the section,
e) system power units with built-in adapters.

Fig. 2.3.2b. View of a double power unit with built-in adapter

2.3.3 Control system for the UEL plough device

The PMC-D control system was used for control of the plough machine. Each drive unit has assigned the PMC-D controller and cooperating visualization device PMC-V.

Both PMC-D devices independently verify the proper operation of the control system and signal the readiness to work of the plough device control system. The PMC-V controllers automatically determine, inter alia, the plough position, and therefore continuously monitor the plough head position sensors (limit and sync) and the sensor of plough drive clutch slip.

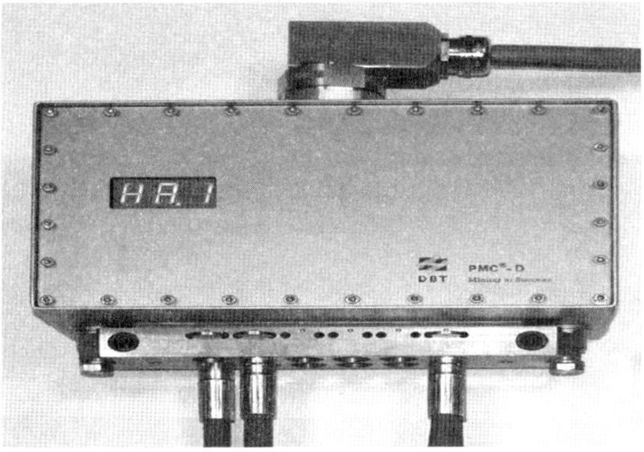

Fig. 2.3.3a. The PMC-D controller

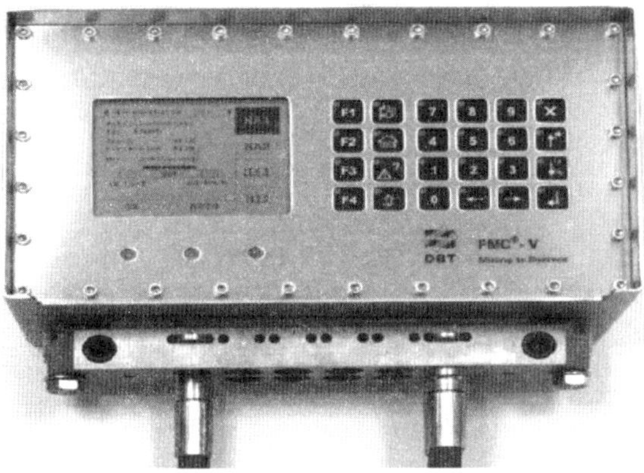

Fig. 2.3.3b. The PMC-V visualization device

Communication of the PMC-D controllers with external systems is done exclusively by the PMC-V visualization device, which sends messages in both directions related to co-operation of individual control systems of the plough system.

2.3.4 The system of central visualization of longwall and plough system

The complete visualization of the plough system operation is provided by the visualization software cooperating with the PMC-R control and automation system:

- the VShield central longwall visualization for the PMC-R system of powered roof support control,
- the VPlow plough visualization for the PMC-D system of plough control.

The software for longwall and plough operation visualization has been implemented at two operator's stations: one at the main gate and the other on the ground surface of the mine. The application program of the central control unit is run under Windows XP operating system, and allows for:

a) monitoring of the basic parameters of the system operational parameters from position of the system operator and from each PMC-R field built into the longwall,
b) identification of location, working direction, speed of the plough head in the longwall,
c) defining a web depth of the plough head while working in range of 0–200 mm,
d) automatic change of speed and cutting direction of the plough head,
e) automatic movement of the support section according to indications of the installed sensors,

f) parameter change for moving the powered roof support,
g) sequential water spray of cutting head from the roof of each section of powered roof supports,
h) deactivation of automatics and manual control of subordinate units,
i) providing current information about operation status and parameters such as: pressure in sectional stands, protrusion of sliding system, activated automatic functions, position of the plough head in relation to all PMC-R controllers built in sections and to the Vshield visualization software,
j) the plough system control from the main gate or from the ground of mine and visualization of operating parameters of the plough system individual elements,
k) diagnostics of the plough system control.

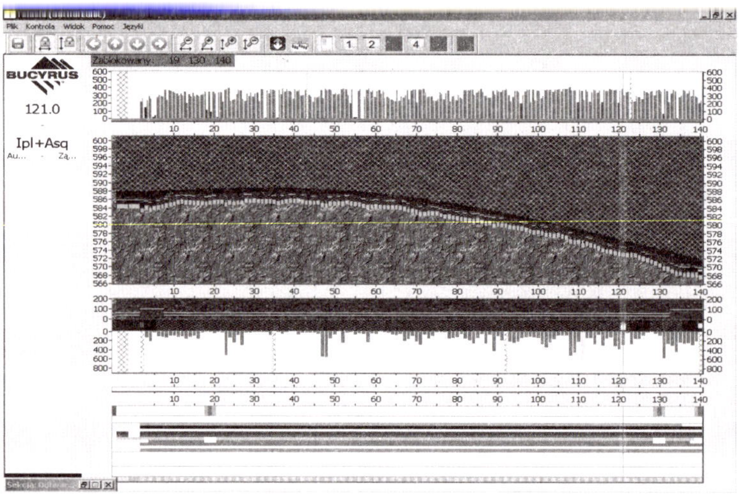

Figure 2.3.4a. The exemplary screen of the VShield longwall visualization software

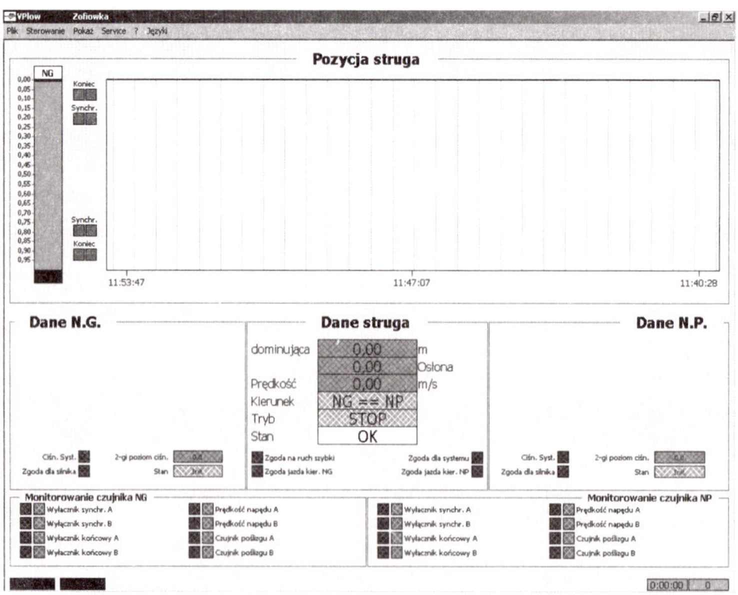

Figure 2.3.4b. The exemplary screen of the VPlow plough visualization software

According to general standards, the software enables complete archiving of the system data: the report is saved in a compressed file at the managing computer disk. Access to the relevant software options is protected by access passwords that protect the system, software and saved data against any adverse actions.

4 CONCLUSIONS

Extensive and detailed analyses of exploitation efficiency of thin coal seams are conducted in many countries. Experience gained from exploitation of low longwalls with use of plough systems indicates that the plough technique is a technically justified manner of low coal seam exploitation. In addition, the plough technique is supported by serious economic arguments that indicate it as a recommendable technique for exploitation of seams with thickness up to 1.5 m (Stopa 2008). Polish coal mining industry exploits – or soon will exploit – more and more seams of low thickness, so the quantity and technical quality of plough systems will develop dynamically. Therefore it becomes necessary to develop power supply and control systems for the high performance plough systems, so as to offer such electrical equipment on ongoing basis, which will completely fulfil the technical and economic tasks. It should be emphasized that the plough technique poses numerous new challenges to be met by each element of the system – only overcoming these challenges guarantees full economic success for mining companies, that translates proportionally to the success of their co-operating parties. The challenges set by the new generation of ploughing technique should result in development of other shearing machines, including shearer loaders not requiring direct handling. Certainly, particular importance will be assigned to non-contact switches enabling flexible adjustment of drive motor revolutions in mining machinery.

REFERENCES

Adamusiński M., Morawiec M., Lubryka J., Jędruś T., Macierzyński D. 2011. State-of-the-art solutions of systems for power supply, control and automation based on example of the high-performance plough system supplied with 3.3 kV voltage at JSW S. A., KWK Zofiówka. (KKEG 2011), Cracow.

Advertisement materials of the Bucyrus International Inc. company.

Bucyrus DBT Europe: Manual Instruction: The electrohydraulic control unit – 10 and 12 functions, the PCM VShield system Central longwall visualization, Control of the longwall support.

Czechowski A., Lubryka J, Lubryka M. 2008. Development of application software for control and visualization system of shearer loaders. The World Mining Congress, Cracow.

ELGÓR+HANSEN Sp. z o. o. Operation and Safe Use Instruction of the contactor switch of EH-d03-W/3,3/I/01.01 type and the EH-d03-W/3,3/I/02.01 type.

Lubryka J., Bugla P., Macierzyński D. 2009. Power supply and control of high-performance plough systems (ATI 2009), Szczyrk.

Morawiec M, Jędruś T., Macierzyński D. 2009. Operational safety of medium voltage network from the perspective of construction and equipment of machinery for underground coal mining plants (EmTech 2009), Ossa.

Morawiec M., Jędruś T., Macierzyński D. 2008. Use of fire-proof 3,3kV medium voltage thyristor starters for power supply of drives of push-plate longwall and bottom conveyors in high-performance longwall systems. The 4th International School of Mechanization and Automation of Mining Industry, Szczyrk.

Myszkowski M., Paschedag U. 2008. Longwall working at medium thickness seams – comparison of cutter and plough .The World Mining Congress – 6th Session – Coal mining – possibilities and challenges, Cracow.

Regulation of the Minister of Economy dated 28.06.2002 on OHS, maintenance of a mining plant and special anti-fire protections of underground mining plants as amended. (Journal of Laws no. 139, item 1169 dated 02.09.2002 as amended) in the scope of ensuring operational safety in hazardous conditions present during operations of a mining plant.
Stopa Z. 2008. Perspectives of thin coal seam exploitation at LW Bogdanka S. A. School of Underground Mining, Szczyrk.
Tor A., Kubaczka C., Olma R., Kapcia J. 2008. Technical conditions of thin coal seam exploitation with use of plough system at Jastrzębska Spółka Węglowa S. A. School of Underground Miting, Szczyrk.

International Mining Forum 2011, Dyczko et al. (eds) © 2011 Taylor & Francis Group, London, UK. ISBN 978-0-415-68938-0

Determination of Cuttability of Coal Seams and the Surrounding Rocks in Underground Conditions by Means of a Device Designed at GIG

St. Prusek, S. Rajwa, A. Wrana, A. Walentek
Central Mining Institute, Poland

ABSTRACT: A device for cuttability measurements of coal seams and the surrounding rocks in underground conditions, designed at GIG has been presented in the paper. Part of the article is dedicated for a review of cuttability of rocks measurement methods, hitherto applied in Poland and abroad. After that, the main elements of the device have been described as well as a procedure for performing measurements in underground conditions. Simultaneously, the examples of the underground measurement results have been given.

KEYWORDS: mining, cuttability, underground measurements

1 INTRODUCTION

Cuttability is defined as a force necessary for loosening a piece of gangue or coal. It influences the selection and stability of mining machines. Thus, this parameter determines the consumption of energy and size of mining machines which is reflected in their cost of development and use. Negative effects of the selection of mine plants with too much power become significant particularly in the process of thin hard coal seams extraction. Thereby, within the international project performed by the Central Mining Institute (Katowice) and the National Mining University (Dniepropietrowsk) titled: "Technical-technological possibilities and economical basis for thin hard coal seams extraction" (the order of the Minister of Higher Education and Science no. 654/N-UKRAINA/2010/0), it was decided to design a device for measurements of coal seams cuttability and surrounding rocks in underground conditions.

In this paper, a review of cuttability measurement methods applied so far, outline of the newly-developed device for cuttability measurements of coal and the surrounding rocks in underground conditions, designed at GIG, as well as a complete measurement cycle have been described.

2 SELECTED METHODS OF COAL AND ROCKS CUTTABILITY DETERMINATION

The importance of rock cuttability evaluation is indirectly reflected by the number of methods developed in the past. Studies have been conducted both in Poland and abroad. The cutability measurements have been carried out in underground mining but also in tunnel building.

The most significant methods of coal-and-rock cuttability evaluation are (Biały 2002):

1. Laboratory methods that indirectly describe the cuttability indicator,

2. Measurements of drilling parameters,
3. Crushing measurements,
4. A cutting tool or a group of tools cutting force measurements.

First methodology includes determination of the Protodiakonov indicator and measurement of the ultimate compression strength (Rc). Measurements during drilling are generally based on the analysis of torque value in drilling rod during the process of rotary drilling in the head. Crushing measurements' reference point is the participation of size grades in coal samples, evaluated after falling from a particular height.

The above mentioned methods are vulnerable to measurement errors because of the lack of cuttability devices work mapping. Therefore, the most suitable investigation corresponding to the real conditions is cutting force measurement. Thus, two groups of methods are available to apply: laboratory methods and in situ methods.

Cutability measurements in laboratory conditions are performed in AGH University of Science and Technology in Krakow (Krauze 2000), CERCHAR centre in France (Bialy 2002), the USA by the use of IN-SEAM TESTER (Sundae 1986), Czech Republic VVUU Ostrava Institute (Sikora 1978) and in Technical University in Istanbul (Balci 2007).

It should be mentioned that research conducted in Technical University in Istanbul has the recommendation of the International Rock Mechanics Association which allows to measure the cuttability in laboratory conditions. In that case, the reference point for cuttability is the *SE* indicator. It is calculated as the ratio of the cutting force to the volume of reduced output on a particular cut length (1) (Eskikaya 2000).

$$SE = \frac{FN}{Q}\left[\frac{MJ}{m^3}\right] \tag{1}$$

where: FN – average force affecting the cutting tool [kN]; Q – volume of the loosened rock at a given length of a cut [m^3/km].

Other methods employ advanced instruments to evaluate the cuttability in underground conditions.

The in situ investigations are predominating as opposed to the laboratory tests because the measurements are taken directly in the coal seam which is in an actual state of stress, corresponding to the particular geological-mining conditions.

In Polish mining industry cutability research has been performed by the use of POS-1 and POU-BM/1 (Bialy, Mizgala 2006) developed by ITG KOMAG as well as devices used in AGH University in Krakow (Andrusiewicz 2008).

In figure 1 a construction of POU-BM/1 device and device used by AGH have been presented.

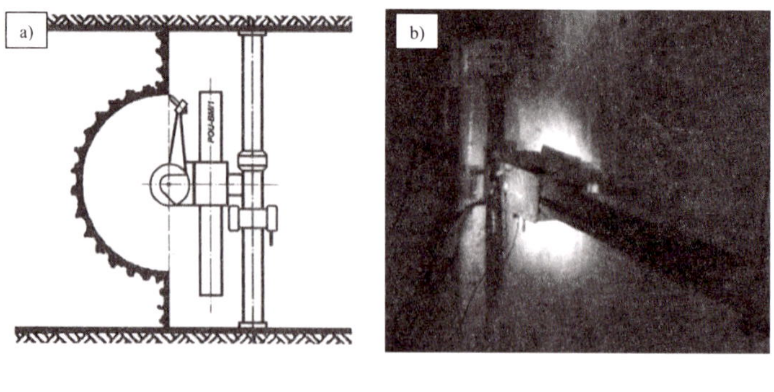

Figure 1. Devices for cuttability measurements: a) POU-BM/1;
b) a device used by AGH (Biały, Mizgała 2006; Andrusiewicz 2008)

In the devices, the measurement of cutting force was based on pressure measurement in the feed system. Additionally, the cut depth has been measured and, as a result, the cuttability indicator *(A)* was calculated:

$$A = \frac{P_{śr}}{g} \left[\frac{kN}{cm}\right] \tag{2}$$

where: $P_{śr}$ – average cutting force [kN]; g – depth of the cut [cm].

For mapping the real shape of the furrow during the measurements, it was described as an angle of lateral disintegration Ψ and reduction lines. In this case the formula for calculating the cuttability indicator (WU_B) was as follows:

$$WU_B = \frac{P_{śr}}{b + 2g\sqrt{1 + tg^2\psi}} \left[\frac{kN}{cm}\right] \tag{3}$$

where: $P_{śr}$ – average cutting force [kN]; g – depth of cut [cm]; b – width of the measurement cutter [cm]; Ψ – angle of lateral disintegration [°].

Cuttability research has also been performed abroad: in Czech Republic (ZP-1 device) and in Germany with the device developed by DMT. DMT's device is widely employed when determining i.e. the conditions of plows' work (i.e. driving motors power) in a particular area. The measurements are taken by registering the cutting force with a model-cutter placed on the measuring head during the process of cutting 2 cm in depth. As opposed to the research performed in Poland, to classify the coal beds only an averaged force (Fs) was applied affecting the measuring cutter with a constant value of cut depth (Myszkowski 2008). The investigations performed according to this method also allowed to classify the coal beds, as follows:

- Easy to plow — Fs < 1.5 kN,
- Normal to plow — 1.5 kN < Fs < 2.0 kN,
- Difficult to plow — 2.0 kN < Fs < 2.5 kN,
- Very difficult to plow — 2.5 kN < Fs

The presented methods were the base for GIG's design of a device to measure the cuttability of coal seams and the surrounding rocks in underground conditions.

3 THE DEVICE FOR CUTTABILITY MEASUREMENTS OF COAL SEAMS AND THE SURROUNDING ROCKS DESIGNED AT GIG

Within the mentioned international research project a device for cuttability measurements of coal and rocks in underground conditions has been designed at GIG.

It has been constructed of the following components:

- Measurement Module,
- Flame-proof Mine Computer,
- Oil Feeder Assembly.

The Measurement module, presented in figures 2 and 3, is the main component of the cuttability measurement device. Its main functions are as follows: adjustment of side wall surface, measurement cut with registered cutting force and scan of the furrow, resulting from the cut.

The above-mentioned functions are realized by the interchangeably installed heads, which are depicted in figures 3 and 4. The scanning head consists of two components: a laser illuminating the furrow and a register. The head is used for measuring the geometry of the furrow which occurred after measurement cut was performed

Figure 2. Measurement Module constructed on SHC props

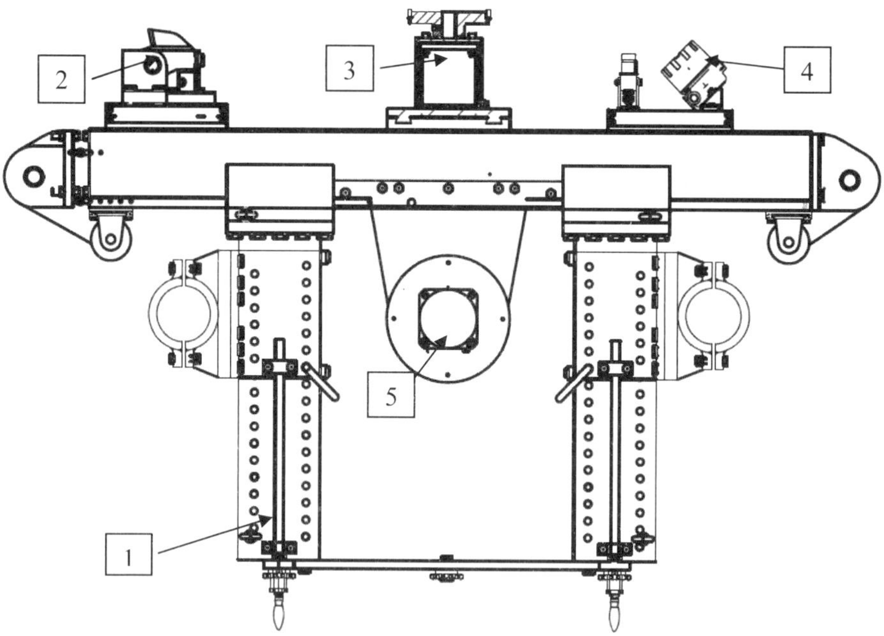

Figure 3. Components of the Measurement Module – schematic picture (1 – traverse gear for side wall device; 2 – measuring head; 3 – face milling cutter; 4 – scanning head; 5 – hydraulic engine)

The measuring (cutting) head consist of a measuring cutter and strain gauge of force. The face milling cutter is equipped with a hydraulic engine, propelling the mill which is used to prepare the side wall before the measurements are taken.

Figure 4. The heads of the Measurement Organ MO: a) measuring head with a cutter; b) scanning head; c) face milling cutter, adjusting the sidewall before the measure cut

Head movements along the side wall are forced by a driving chain which is connected with planetary gear and hydraulic engine. The traverse of the measuring organ is based on a manually propelled mechanism that uses two plug screws.

The measuring and scanning heads are equipped with sensors connected to the Flame–proof Mining Computer OKG1. It allows registering and monitoring the measurement data at the investigation destination.

The measuring organ is fed with the Assembly of Oil Feeder, type ZZO-1. Its function is to generate the flow of oil with a proper pressure for propelling the hydraulic engines build on the measuring organ and face milling cutter.

The pumps draw in the oil from a container, which is a part of the device equipment, guaranteeing constant parameters of the work of the device and the possibility of taking measurements, regardless of the access to water-oil emulsion in the area of investigation.

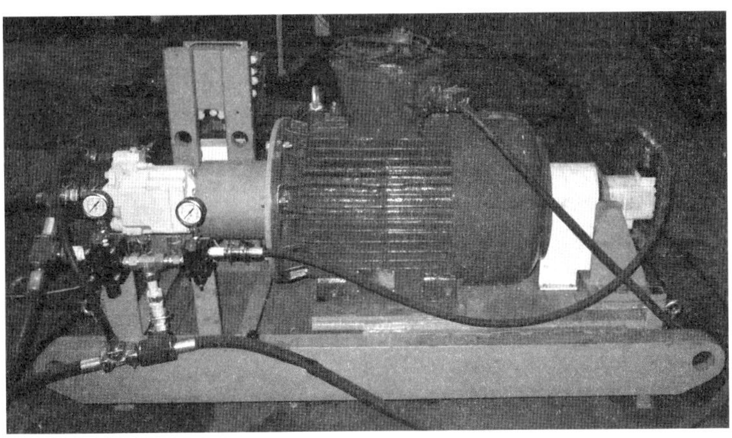

Figure 5. Oil Feeder Assembly, type ZZO-1

The presented construction allowed to prepare a procedure of carrying out research accordingly to the phases depicted in figure 6 (Prusek, Rajwa, Wrana 2011).

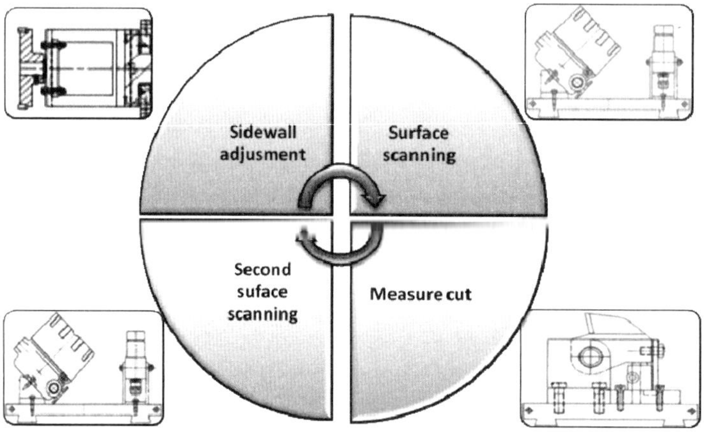

Figure 6. The cycle of taking cutting measurements of coal beds and the surrounding rocks using the device developed at GIG

The procedure presented above is strictly connected with the new methodology of indicating cutting resistance designed at the GIG which describes the cuttability as a ratio of cutting force to the volume of loosened coal mass.

In essence, the method serves for comparison of the registered shapes of undisturbed coal surface before and after making a furrow and calculation of the volume of the mined coal or rock.

The next step is to calculate the coefficient of cuttability from the equation below:

$$A_{Bi} = \frac{F_B}{V_B} \left[\frac{kN}{cm^3} \right] \tag{4}$$

where: F_B – the average of the registered cutting force in the process of cutting, V_B – the value of the mined coal or rock volume.

In order to define the coefficient of cuttability of the coal seam or surrounding rocks in question more precisely, it is necessary to register the actual force value and calculate the actual value of mined coal or rock volume for the elementary furrow segment and then actual values of the cuttability coefficients of coal and rocks. At this point, the value of cuttability coefficient for a single furrow is calculated as a sum of actual values of the cuttability coefficients of coal and rocks which are assigned to this furrow. The equation is presented below:

$$A_B = \frac{\sum_{i=1}^{n} F_i}{\sum_{i-1}^{n} V_i} \left[\frac{kN}{cm^3} \right] \tag{5}$$

where: A_B – cuttability coefficient of coal or rock, F_i – value of cutting resistance force of the i-th elementary measuring segment registered on the cutter, V_i – value of the mined rock volume for the i-th elementary measuring segment.

Main advantages of the device are: the possibility of taking measurements with great precision due to the adjustment of side wall surface before the measuring cut, as well as the use of the innovative technique of laser scanning which allows to map the geometrical parameters of the furrow precisely.

The first practical experiences show that the possibility of the device operation with the access only to the electricity at a destination is essential as well. This is influenced by applying an independent oil feeder.

4 UNDERGROUND RESEARCH

In order to finish the works of the device for cuttability measurements, it was necessary to perform underground research aiming at the examination of the device components. The investigations were performed according to the procedure presented in figure 6.

First phase included adjustment of the sidewall surface by using the face of the milling cutter, then the adjusted surface was scanned by a laser scanner. The scanning process itself has been presented in figure 7 which, simultaneously, depicts the results of the side wall surface adjustment using the face milling cutter.

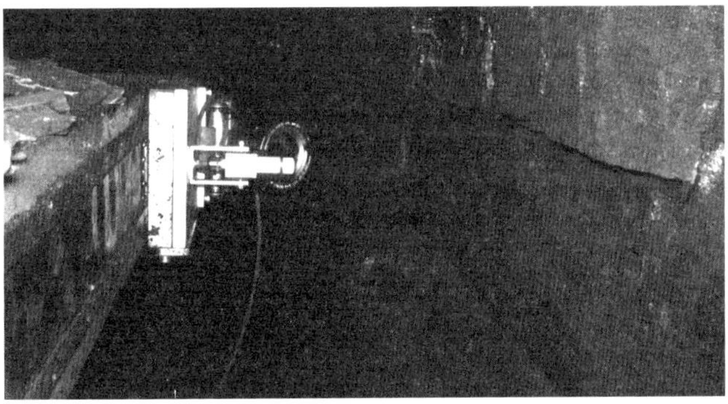

Figure 7. Laser scanning ahead of the measurement cut.

The most important phase of the cut measurement procedure is to proceed with the measurement cut with the simultaneous registration of the cutting force. The picture of the measuring head and the furrow is presented in Figure 8.

Figure 8. Cutting head along with the cutter and the resulting furrow

The last phase of this investigation is to scan the furrow in order to compare its results with the view of the adjusted surface. The scan result of the elementary measurement segment is presented in Figure 9.

Figure 9. Laser scanning result of the elementary measurement segment

The obtained results which include the cutting force and the precise view of the measuring furrow were base for the assessment of coal cuttability. The example of calculations of the earlier described A_B indicator has been presented in Figure 10.

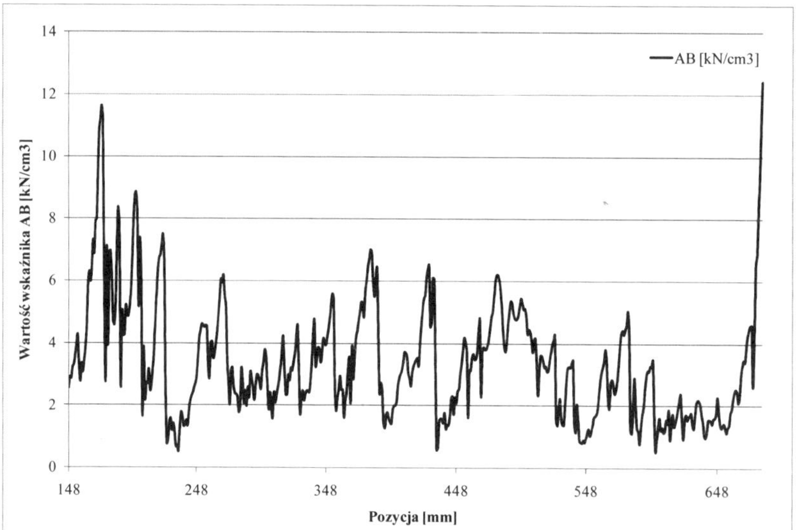

Figure 10. Example results of A_B indicator calculation regarding the position of the cutting tool

By applying scanning for geometrical parameter measurements of the cut furrow, it is possible to precisely indicate the volume of the output being cut as an average and actual value. The recorded parameters, such as force and volume, are essential for the precise and reliable indication of the value of rock cuttability coefficient.

5 SUMMARY

A work to design a new device for the measurements cuttability of coal seams and the surrounding rocks in underground conditions was initiated in GIG. This includes a few phases such as the invention of the concept, the project design, preparation of a prototype, obtaining a compliance certificate of conformity with the ATEX directive and performing underground tests.

Also, a methodology of performing the trials with the use of the new measuring device has been designed.

The device allows to take precise measurements of all magnitudes necessary for the assessment of the cuttability parameter with the methods used so far as well as new method with the use of A_B indicator.

The underground investigations performed in the variable conditions (rock mass parameters, temperature, humidity, slope of working, dustiness) have acknowledged the effectiveness of measurements of the cutting force as well as the laser scanning of the furrow shape.

By applying scanning for geometrical parameter measurements of the cut furrow, it is possible to precisely indicate the volume of the output being cut as an average and actual value.

REFERENCES

Andrusiewicz W. 2008. Możliwość zmiany sposobu urabiania soli kamiennej w KS Kłodawa. Gospodarka Surowcami Mineralnymi. Zeszyt 2. (In Polish).

Balci C., Bilgin N. 2007. Correlative study of linear small and full-scale rock cutting tests to select mechanized excavation machines. International Journal of Rock Mechanics & Mining Sciences nr 44.

Biały W. 2003. Nowa klasyfikacja urabialności pokładów węgla Zagłębia Górnośląskiego. Przegląd Górniczy nr 9. (In Polish).

Biały W., Mizgała J. 2006. Koncepcja przyrządu do wyznaczania oporów urabiania. Mechanizacja i Automatyzacja Górnictwa nr 2. (In Polish).

Biały W. 2002. Analiza metod badania urabialności węgla. Przegląd Górniczy nr 4. (In Polish)

Eskikaya S., Bilgin N., Ozdemir L. 2000. Development of rapid excavation technologies for the Turkish mining and tunneling industries. NATO TU Excavation SfS Program project report. Mining Engineering Department. Istanbul Technical University.

Krauze K. 2000. Urabianie skał kombajnami ścianowymi. Wydawnictwo "Śląsk", Katowice. (In Polish).

Myszkowski M., Paschedag U. 2008. Longwall mining in seams of medium thickness – comparison between shearer and plow. 21st World Mining Congress. Kraków.

Prusek St., Rajwa S., Walentek A., Wrana A. 2011. Wykorzystanie nowoczesnych technik pomiarowych do oceny parametru skrawalności węgla i skał w warunkach in situ. Prace Naukowe GIG. Kwartalnik Górnictwo i Środowisko Nr 1/1/2011. Katowice. (In Polish).

Sikora W., Fels M., Sołtysek K. 1978. Strugalność węgla. Prace GIG Komunikat nr 687. (In Polish)

Sundae L. 1986. Measurement of Coal-cutting Forces Underground with the In-seam Tester. NIOSH Report of Investigations.

International Mining Forum 2011, Dyczko et al. (eds) © 2011 Taylor & Francis Group, London, UK. ISBN 978-0-415-68938-0

Mining of Thin Hard Coal Seams in the Mines Belonging to Katowicki Holding Węglowy Using Shearer Longwall System

Z. Gach, M. Skrabaka, T. Nosal, R. Holewa
Katowicki Holding Węglowy S.A., Katowice, Poland

ABSTRACT: The possibilities for mining thin coal seams in mines belonging to Katowicki Holding Węglowy S.A. are presented in this paper. Amount of coal which can be mined from these seams was analysed as well as its basic parameters i.e. heating value, ash content and sulphur content. Moreover, technical specification of the shearer longwall system for low longwalls, manufactured by the company Famur is presented. This system is used during mining of coal seams in KWK Murcki-Staszic, Staszic colliery.

KEYWORDS: hard coal mining, mining, longwall system, thin coal seams, shearer longwall system

1 INTRODUCTION

One of the fundamental objectives of KHW S.A. achieved through realization of the strategy, for which the basic tasks are among other things:

- Improvement of output effectiveness by the use of new mining technologies,
- Increasing of resource base, through development works in new deposits,
- Increasing of output from thin coal seams.

All above mentioned objectives can be achieved by using longwall systems adapted for mining of thin coal seams.

All pros and cons of the mining technology using longwall plow system and longwall shearer system were analysed during planning output from mentioned beds. Geological-mining conditions decided about selection of the longwall shearer system. In comparison to a plow, a shearer should deal better with the disturbances in the form of washouts, cross-cuts and faults which are expected in the beds intended for mining. The second element, arguing for selection of the longwall shearer system was the purchase costs, which was ca. 74 million PLN (costs for plow system are higher by ca. 100 million PLN). It is also estimated that costs of development of the longwall, in case of shearer system, are considerably lower than in the case of plow technology (smaller dimensions of longwall excavations are connected with smaller dimensions of machines installed in these excavations (drives, turning stations etc.).

In order to realize plans within the range of mining of thin coal seams, Katowicki Holding Węglowy has acquired longwall shearer system type FL 12/18. At the turn of March and April 2011 a pilot mining of wall no. 703 in bed 407/1 in KWK Murcki-Staszic mine, Staszic colliery was started. This system is a typical assembly of mining machines, consisting of longwall shearer, conveyor and roof support, however it cannot be assembled from the devices which are already available on the market. System for low beds manufactured by Famur is the first system executed in the *e-kompleks* standard. It means that it connects operation of set of modern, high-capacity and safe machines with information systems for remote control and diagnostics i.e. components of the e-mine [*e-kopalnia*] system. In the not-too-distant future it is planned that machines will operate in semi-automatic or automatic mode, what will result in minimization of people's participation in works in the most hazardous locations in the longwall.

Currently the mining of longwall no. 703 is coming to a finish. The described system will be then used for mining of longwall no. 702 in the same bed. The positive experience connected with use of the system fitted for mining of low beds caused that KHW S.A. is planning to purchase similar system. Such system will be used for mining of longwalls in bed 364/1 in Murcki-Staszic mine, Boże Dary colliery in the first instance. Beginning from the 2013 all longwalls in the low beds will be mined using mentioned longwall systems.

Current daily output of Boże Dary colliery amounts to approx. 4000 tonnes of stone. It is estimated that use of the system for mining of low longwalls will result in decrease of this amount to approx. 1000 tonnes per day. Average costs of utilization of 1 tonne of stone currently amounts to 12.20 PLN. Annual costs of utilization of stone only for Boże Dary colliery equals to over 13 600 000.00 PLN. Limitation of output and utilization of stones to 1000 tonnes per day will enable to lower the costs of its utilization by over 10 000 000.00 PLN per year.

2 SIZE OF HARD COAL RESOURCES IN THIN COAL SEAMS PREDICTED FOR EXTRACTION IN KATOWICKI HOLDING WĘGLOWY

Thin coal seams are a considerable item in the resources of Katowicki Holding Węglowy. Form this point of view preparation and mining of face in thin coals seams i.e.

- 407/1 in CDF field as well as 402 and 401/1 in S field in Murcki-Staszic mine, Staszic colliery,
- 334/2 in C field, 364/1-2 in F and W fields, 403/1 in W field in Murcki-Staszic, Boże Dary colliery,
- 334 in S field in Mysłowice-Wesoła mine,
- 620 in Wieczorek mine,

belongs to the most important production and business tasks.

The size of resources documented in thin beds in the currently active deposits indicates the justness of taking actions associated with mining of the mentioned beds using longwall shearer system.

Table 1. Operative resources of hard coal deposited in thin coal seams in KHW S.A. in the individual enterprises of the company

Mine		Bed	Bed thickness [m]	Mineable resources [thousands of tonnes]
Wieczorek		620	1.3	2 575
Murcki - Staszic	Staszic colliery	407/1	1.59	2 800
		402	1.61	880
		401/1	1.53	5 370
	Boże Dary colliery	334/2	1.9	1 984
		364/1-2	1.7	6 610
		403/1	1.6	9 677
Mysłowice - Wesoła		334	1.5	23 281
			Total	53 177

The analysis of data shows that the biggest coal resources in thin coal seams likely to be mined using longwall system, are located in Murcki-Staszic (27.3 million tonnes) and Mysłowice-Wesoła (23.2 million tonnes) mines. In the case of their exploitation it is possible to work approx. 37 longwalls in Murcki-Staszic mine and approx. 20 longwalls in Mysłowice-Wesoła mine.

Table 2. Quality parameters of coal from coal seams in KHW S.A.
in the individual enterprises of the company

Mine	Bed / field	Heating value [kJ/kg]	Ash content [%]	Sulphur content [%]
Wieczorek	620/ east	25 240	22.0	0.6
Murcki - Staszic	407/1 / CDF	26 152	7.3	0.96
	402 / S	25 580	9.4	0.96
	401/1 / S	25 106	9.0	0.65
	334/2 / C	26 144	14.48	0.81
	364/1-2 / F	24 706	18.37	0.54
	364/1-2 / W	28 588	10.61	0.54
	403/1 / F	28 045	11.76	0.45
	403/1 / W	28 045	11.76	0.45
Mysłowice - Wesoła	334	27 084	10.0	0.5

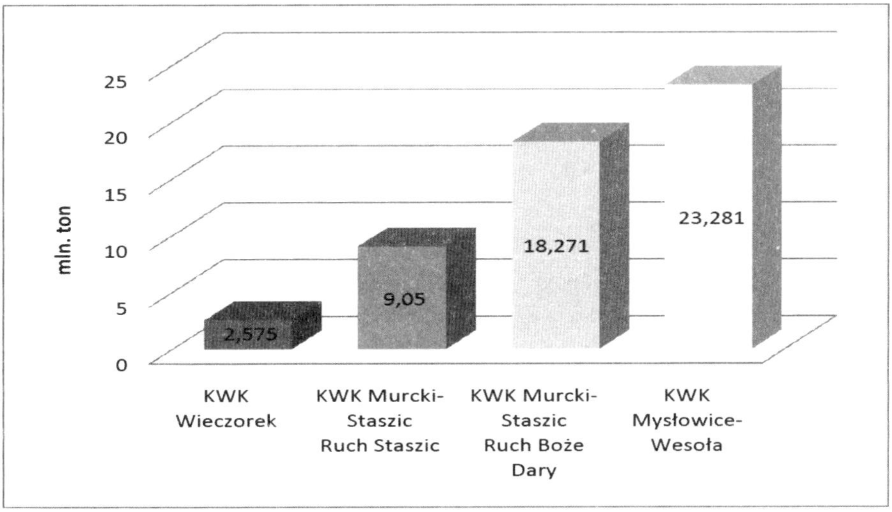

Figure 1. Classification of hard coal resources in coal seams in KHW S.A. mines

2 CHARACTERISTICS OF LONGWALL SYSTEM USED FOR MINING OF THE COAL SEAMS

Technical equipment of the longwall system:

– longwall shearer loader type FS-200 – is a shearer operating "next to" the longwall conveyor equipped with strand-free Eicotrack haulage system. Taking into account construction of the shearer, the attacking element must be lead in bottom in order to extract space for shearer's body.

  Specification:
  - range of cutting – 1.2 –1.8 m
  - diameter of cutting elements – 1200 mm
  - drum web – 900 mm
  - power transmitted on mining elements – 200 kW
  - total installed power – 467.5 kW

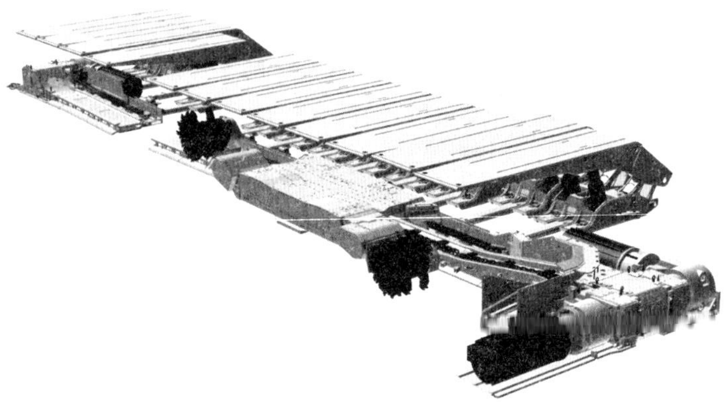

Figure 2. Roof support type FAZOS-10/19-2x2340

- powered roof support type FAZOS-10/19-2x2340 – is a supporting-shielding roof support intended for working of beds endangered and not endangered with crumps, mined using caving longwall system. For the operation in crumping beds, the roof support is fitted because of the release valves. Each section of the roof support is equipped with pressure level indicator, located in the area under prop's piston, which is equipped with own energy source.
  Specification:
  - geometrical operating range – 1.0–1.9 m
  - housing pitch – 1.5 m
  - working support capacity of prop – 2340 kN
  - supply pressure – 25–32 MPa
  - control type – pilot

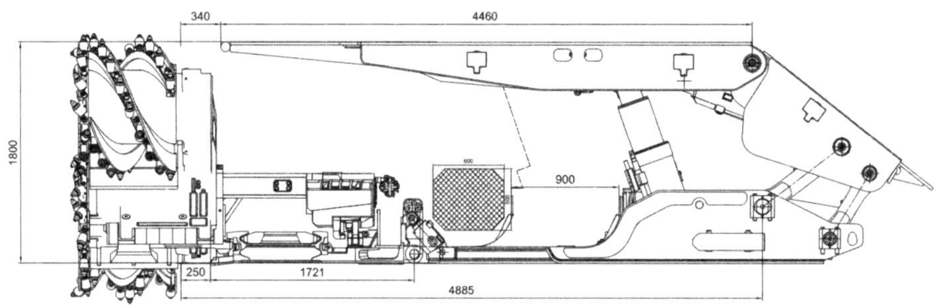

Figure 3. Face conveyor type FFC 750

- face conveyor type FFC 750 – is a conveyor equipped with Eicotrack advance system, fitted for cooperation with FS 200 shearer as well as with FAZOS 10/19-2x2340powered roof support. The conveyor is equipped with three drives using transmission, size 25, type FKPL25 and PL25 conforming the RANG 335000 standard.
  Specification:
  - installed power – 3 x 250/85 kW
  - supply voltage – 1000 V
  - scraper chain – ϕ 30 x 108 flat
  - chain speed – 0.97 m/s
  - capacity – 1060 t/h

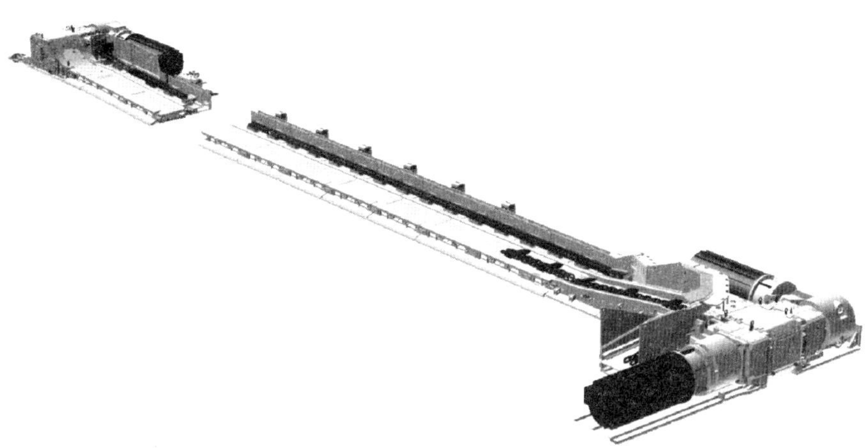

Figure 4. Beam stage conveyor type PPZ-850/1x65/200 Nowomag

- beam stage conveyor – conveyor type PPZ-850/1x65/200 Nowomag with rounds crusher, type KD-1600/1x160 Nowomag, was used as beam stage conveyor. It is fitted for cooperation with the cross drive of FFC 750 face conveyor.

  Specification:

– driving unit power	– 65/200 kW
– chain speed	– 1.8 m/s
– capacity	– 1300 t/h
– scraper chain	– ϕ 30 x 108 flat

The modern construction solutions used in the complex – high support capacity of the powered roof support, lighting and passage for crew members results in the fact that operation, even in a low wall, is comfort and safe.

The FL 12/18 complex was used for mining of wall no. 703 in 407/1 bed in the mine Murcki-Staszic, Staszic colliery for the first time. The wall was opened out in April, 2011.

Before purchase of the above mentioned longwall system, KHW assumed minimum output from the longwall equal to 3000 t/d. The mine in the initial stage of mining of longwall no. 703 achieved average output of 1900 t/d.

The average daily output was 3977 tonnes, with monthly longwall advance of 197 m in August. However, maximum daily output achieved in this longwall was higher than 6200 tonnes.

Technical difficulties and the use of new technologies which the personnel was not familiar with, were the reasons of lower output in the initial period of mining the longwall. After elimination of standstills reasons as well as a result of changes in work organization, the output in September was increased to daily average of 4700 tonnes.

The economic analysis of effectiveness showed that net present value NPV calculated for financial conditions of 2011 (for years 2011-2025) is equal to 34 341 000,00 PLN, while internal rate of return amounts to 16,6% and is higher than limiting rate established at 12%.

The above mentioned indicators of economic efficiency of the investment incurred for preparation and mining of coal seams no. 407/1 in CDF field, 402 and 401/1 in S field, indicate cost-effectiveness of the operation. The average monthly advance of the longwall no. 703 as well as average daily output in six-month period is presented on the Figures no. 5 and 6.

During 8 months of mining, the wall achieved monthly advance and average day output as presented in the graphs below:

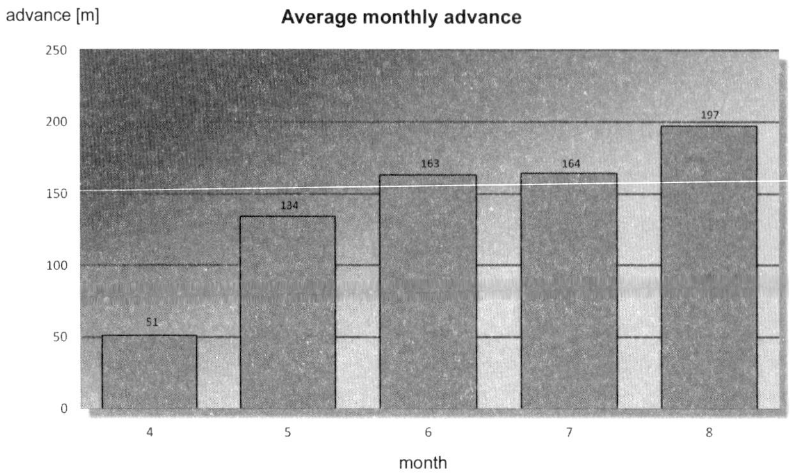

Figure 5. Average monthly advance

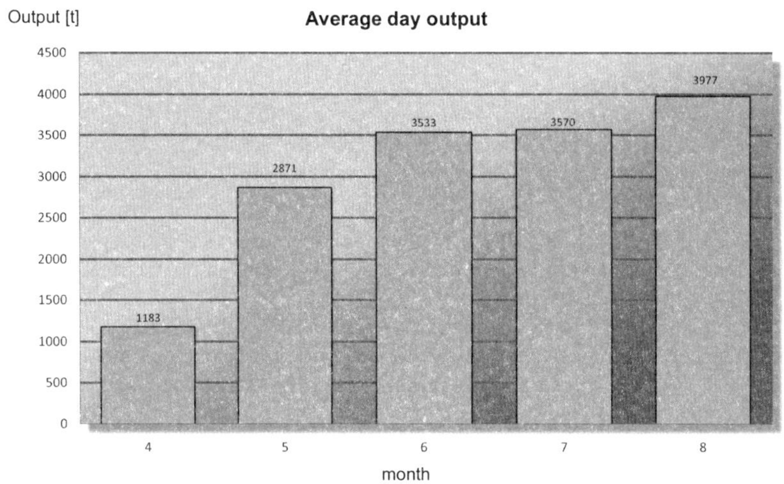

Figure 6. Average day output

4 SUMMARY

Taking into account the presented problems the following conclusions can be presented:

- Mining of thin coal seams will enable KHW S.A. to achieve planned output what should result in increase of share in the hard coal market with simultaneous assurance of high-standard safety in mines which are part of KHW S.A.,
- Longwall system for low walls enables mining of coal beds in the conditions unavailable for a plow.
- Use of longwall system for low walls enables achieving high comfort and safety of operation.
- Use of longwall system for low walls results in considerable reduction of gangue utilization costs.

- Use of longwall system for low walls enabled to incur lower initial expenditures during purchase of the mentioned system as well as to achieve output of 4700 tonnes/day, what is higher than previously assumed.
- Exploitation of thin coal seams using longwall shearer system is cost-effective.

International Mining Forum 2011, Dyczko et al. (eds) © 2011 Taylor & Francis Group, London, UK. ISBN 978-0-415-68938-0

Operational Experiences with Automated Plow Systems at the Pinnacle Mine in the US

M. Myszkowski, S. Bauckmann
Caterpillar Global Mining

ABSTRACT: Coal has been mined at the Pinnacle Mine (former U.S. Steel #50) in West Virginia using plow systems since 1977. Over the decades scores of panels were mined in the Pocahontas coal seam using plow longwalls. In the 1990s the world's first automated plow system was implemented at this coal mine and achieved the world production record of 22,700 tonnes. Pinnacle is currently operating a new modern GH1600 plow system from Caterpillar.

KEYWORDS: Plow systems in the USA, Pinnacle Mine in West Virginia, low seam extraction, longwall, plow technology

1 INTRODUCTION

Longwall mining has been used in the USA since the 1950s. In the early years, many plow systems were used there for mining thin seams. However at that time, plow systems were far inferior to today's technology, with much lower installed power and efficiency, and with no comparable level of automation. Since then, thicker seams have been mined and some early improvements to shearers made them the system of choice of miners throughout the country.

In spite of decreasing number of plow systems in the US, the predecessors of Caterpillar (that is, Westfalia Lünen, DBT and Bucyrus) have supplied a number of plow systems to the Pinnacle Mine near Pineville in West Virginia over the last 22 years. These plow systems performed remarkably, including setting a world record.

2 PINNACLE MINE

The Pinnacle coal mine lies in the southwest of the state West Virginia near the city of Pineville. All exploitation of black coal at this mine is carried out in the seam Pocahontas #3. This seam lies at a depth between 300 to 500 m and has a thickness of 1.1 to 1.4 m (on average 1.25 m). All mine activities are in horizontal deposits, with an inclination of less than 5°. The coal in the Pocahontas seam is classified as 'easily plowable'.

Since 1977 the mine has used roof-fall exploitation with plows and roof supports. All faces were extracted using the retreat mining system. Initially, four parallel entries were prepared on each side of the face, resulting in very high heading costs. The number of entries was later reduced to three to enable timely preparation of the entries for the next panel, an issue because of the high mining speeds of the plow system. For this reason, transport and conveyance were located in a common entry. The entries at Pinnacle have a height of approx. 1.4 to 1.5 m. Only the main transport and conveyance entries have a height of 2 m. Generally, all entries are anchored.

While the face length was fixed to a maximum of 250 m, the panel length was limited to 2200 m for firedamp prevention reasons. To ensure efficient CH_4 control, gas drainage drillings were carried out from the surface as well as from the entries. Because of strong dust development, above-average water spraying is necessary. Initially, coal was transported to the surface by rail carriages, later a central belt conveyor was installed.

Up to 1989, U.S. Steel Mining Company (USM) #50 mine had two longwall systems with a PF 2 600 VS AFC and a type 7-26 plow. Hemscheidt roof supports with the Hetronik electric-hydraulic control system were used. When both these face systems approached the end of their life span, it was decided to purchase a new automated plow system. U.S. Steel Mining Company prepared a tender for what was then a revolutionary technology in close collaboration with German mining engineers from the Westerholt mine. One element of that tender was a modern and highly comprehensive plow system with incremental advance of the AFC at high pressure and automatic plowing with automatic reversal of the plow at the face ends.

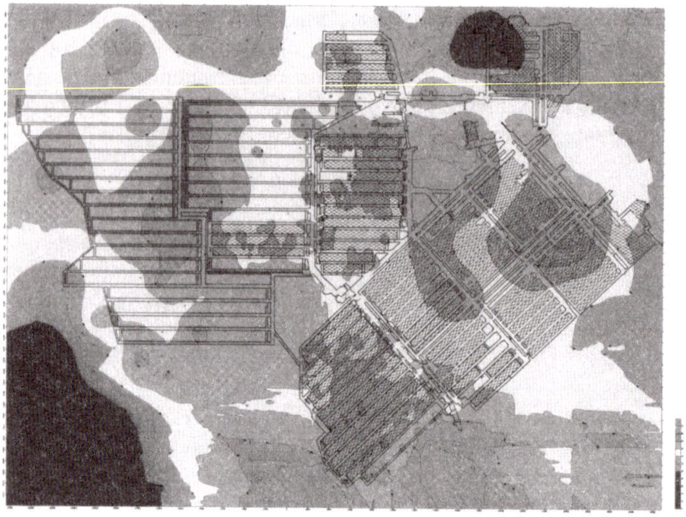

Figure 1. Map of U.S. Steel #50 Mine (now Pinnacle Mine) (date April 21, 000)

This mine map of U. S. Steel #50 shows the sealed area of former mining activities as well as the location of future panels. #50 mine has five shafts, one of which is the inclined main shaft (coal and material transportation). Three Continuous Miners are being used for roadway development.

3 FIRST PLOW SYSTEM FROM 1989

The Pinnacle mine has been successfully extracting coal with plow systems since 1977. Bucking the trend towards shearers, at the end of the 1980s, USM decided to purchase what was then a state-of-the-art plow system.

In 1989, US Steel's Pinnacle mine became the world's first automated longwall face that eliminated the need for an operator at the face. The mine used a complete Gleithobel plow system with full automation built by Westfalia Lünen (now Caterpillar). The two plow longwalls delivered to U.S. Steel #50 Mine in 1989 and 1990 (2 systems) by Westfalia Lünen were the world's first fully automated plows, representing the state-of-the-art at the time.

The world's first fully automated longwall system had following data:
- Face length 250 m,
- 2-leg roof supports with PM 3 controls, 1.75 m centers,
- Installed plow power 2 x 270 kW,
- Plow guide 9-34 ve,
- Plow chain 30 mm (later 34 mm),
- Plow speed 1.54 m/s,
- AFC width 900 mm,
- Installed AFC power 2 x 270 kW.

In 1991 a new plow guide was implemented, allowing the use of 38 x137 mm chain and installation of up to 2 x 400 kW power. In 1994, five years after the start of the system, a new AFC drive system called CST (Controlled Start Transmission) became operational. CST ensured:
- No-load motor start,
- Synchronized heavy-load start-up,
- Accurate load sharing,
- Excellent overload protection.

The initially defective electronic controls were later (in 1996 and 1997) replaced by the reliable PM 4 system built by DBT (a Caterpillar predecessor). In 1995 a new 5.7 plow body was introduced with all parts exchangeable underground. In 1999 a new UEL plow planetary gearbox with overload protection came into use, allowing application of up to 500 kW.

Figure 2. First plow system from Westfalia Lünen (now Caterpillar) used at U.S. Steel #50

Average production has been in excess of 10.889 clean tonnes per day. The peak production record achieved is 18.145 clean tonnes per day.

4 SECOND PLOW SYSTEM FROM 1999

In 1999 DBT (now Caterpillar) delivered to U.S. Steel #50 Mine second fully automated plow system. The technical data of the plow system delivered in 1999 were as follows:
- Face length 319 m,
- Installed plow power 2 x 400 kW with HK 30-2 plow box,

- Plow guide 9-38 ve,
- Plow chain 38 mm,
- Plow speed 1.98 m/s,
- Plow body – triple 2.7,
- AFC PF4/1032 mm with cross-frame,
- Installed AFC power 2 x 600 kW,
- AFC speed 1.86 m/s.

Figure 3. Plow system delivered for the U.S. Steel #50 mine in 1999

The plow at the U.S. Steel #50 Mine operates with a large cutting depth of up to 250 mm. This large web results in large loading streams of extracted coal. Experience shows that a single plow body has difficulty loading such a large amount of coal. For this reason a new triple plow body was developed with bottom bits located near the center of the plow allowing loading of extracted coal at two different points simultaneously. As in the previous plow body, all wear parts were exchangeable underground.

Figure 4. The type 2.7 triple plow body used at the U.S. Steel Mine since 1999

Fig. 5 (below) shows panel 7 O in detail with its three-entry system. The face length was 290 m, the panel length 2,800 m. The seam thickness was 1.4 m. Underground installation took place between February and March 2000, with the panel starting production at the end of March.

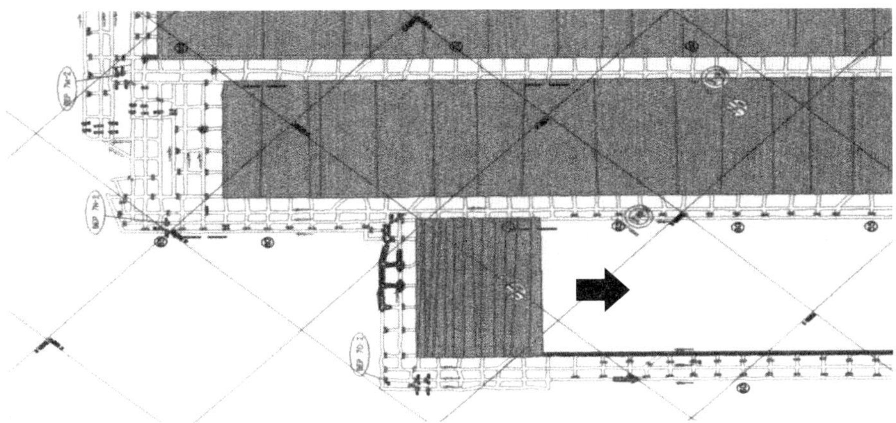

Figure 5. Panel 7 O at USM #50

This plow system reached cutting depths of 240 mm uphill and 220 mm downhill at a seam height of 1.4 m. The plow system achieved production rates of up to 1,400 t/h. Based upon a daily plow run time of 14–16 hours (regularly achieved at #50 mine), daily production of 14,000 to 17,000 t/d of clean coal was achieved. The peak production achieved from that face was 22,700 t/d clean coal, a world record for a plow face.

5 THIRD PLOW SYSTEM FROM 2010

After ten years of very intense mining activity, the Pinnacle Mine decided to purchase a new longwall system for extraction of the Pocahontas seam. A comprehensive study of available technologies was undertaken to select the most reliable and efficient system and ensure the lowest possible production costs. After a very detailed analysis, Pinnacle Mine – which at that time belonged to the Cleveland Cliffs Inc. – decided in favor of a plow system. After comprehensive discussions the company decided to purchase an automated plow system from Bucyrus (now Caterpillar).

The technical data of the plow system delivered to the Pinnacle Mine in 2010 were as follows:

- Face length 300 m,
- Roof support Caterpillar (formerly Bucyrus) 850/1930 2x3181,
- Machine frame MR35-1100/1500,
- Installed plow power 2 x 600 kW / 4,160 V, 60 Hz,
- Plow box HK 40-2,
- Plow chain 42 mm,
- Plow speed 1.98 m/s,
- Plow body GH1600 Pinnacle type (triple body),
- AFC PF4/1132 mm with cross-frame,
- AFC chain assembly 2 x 34x126 mm, spacing 6,
- AFC gearbox KP30 CST,
- Installed AFC power 2 x 600 kW / 4,160 V, 60 Hz,
- AFC speed 2.13 m/s.
- The newest plow system GH1600 for the Pinnacle mine was assembled and extensively tested during the compatibility test, which took place in July 2010 in Houston, Pennsylvania.

A 50-meter long plow longwall system together with roof supports, energy supply and control units was set up, tuned and screened over a period of two weeks. This was necessary as the system incorporated several new technical solutions being implemented for the first time.

Figure 6. The new plow system delivered to the Pinnacle Mine in 2010

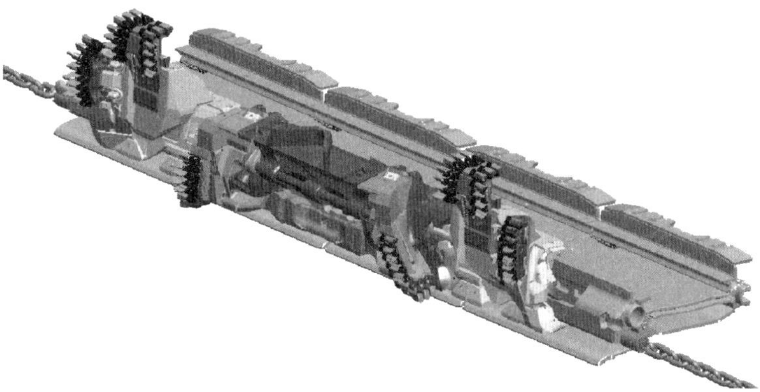

Figure 7. Plow body of the 2010 plow system for the Pinnacle mine

The GH 1600-2 plow body for the new Pinnacle plow face had a similar configuration to the previous one. The plow body consists of two external parts and one intermediate part, with the drive chain attached to the plow body via the pulling sledge. The plow body is designed in this way because heavy coal streams would be difficult to load onto the AFC with a single body. The bottom bits are located near to the middle of the lateral axis to improve the behavior of the plow body when crossing undulations.

The plow system consists of two types of roof supports: face and gate. Face roof supports support the roof where coal is being mined while gate roof supports protect the gate ends of the plow system.

The face roof supports have a retracted height of 0.88 m and an extended height of 1.97 m. A hydraulic cylinder with diameter of 300 mm delivers a yield strength of 715 tons. The transport length, which is critical for a roof support around the mine, is 5.64 m and the roof support weights 16.3 tonnes.

The gate roof supports have a yield strength of 945 tonnes, a retracted height of 1.14 m and an extended height of 3.05 m. The collapsed and extended height ranges are slightly higher to accommodate and protect the equipment in the gate entry. The length and weight increase substantially to 8.18 m inches and 26.1 tonnes.

These roof supports have been designed and validated through rigorous testing. The life expectancy will enhance overall performance and the ultimate return on investment.

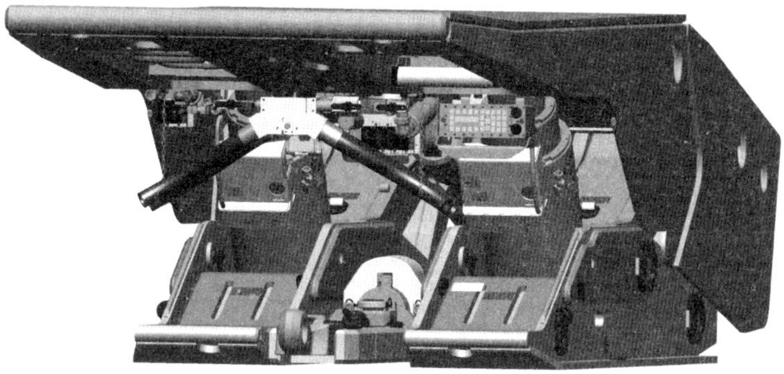

Figure 8. A face roof support for the newest Pinnacle plow system

The plow system commenced service at the Pinnacle mine at the end of 2009 in a relatively short panel. After the move to a new panel, the plow system started again in the spring of 2011. In mid-2011 the Mining Safety and Health Administration (MSHA) closed the Pinnacle mine for several months because of CO occurrence and required countermeasures.

Currently, the Pinnacle Mine and its plow system face have been operational again for some two months and achieving daily production rates of over 10,000 tonnes. The current production restrictions are caused by CH_4 occurrence. The performance of the Pinnacle mine is expected to increase significantly after overcoming these restrictions.

SUMMARY

Based on a performance perception dating back to the 1970s when plow systems were not automated and did not have the cutting and conveying power of today's systems, the US has been reluctant to embrace the plow technology.

Nevertheless, the Pinnacle Mine (former U.S. Steel #50 Mine) in West Virginia has been using plow systems since 1977. In 1989, U.S. Steel #50 Mine purchased world's first automated plow system from the German company Westfalia Lünen (now Caterpillar). This system operated for ten years and achieving excellent performance of almost 11,000 t/d clean coal in Pocahontas #3 with an average seam thickness of 1.25 m.

Ten years later, the next plow system from DBT (now Caterpillar) was implemented in U.S. Steel Mine #50, operating successfully for next 11 years and setting the plow face world record of 22,700 tonnes of clean coal in a day.

In 2010, following exhaustive research into the most efficient mining method for low-seam coal, Cliffs Natural Resources decided to use an automated CAT plow longwall as the safest, most cost-effective and most environmentally friendly way to extract coal from low and medium seams, that is, seams with thickness of less than 1.8 m.

International Mining Forum 2011, Dyczko et al. (eds) © 2011 Taylor & Francis Group, London, UK. ISBN 978-0-415-68938-0

CONSTRUCTION AND MAINTENANCE OF OPENING-OUT HEADINGS FOR PLOW LONGWALLS

International Mining Forum 2011, Dyczko et al. (eds) © 2011 Taylor & Francis Group, London, UK. ISBN 978-0-415-68938-0

Planning of Galleries for High Performance Longwall Productionin Thin Coal Seams

M. Junker
RAG Mining Solutions GmbH, Herne, Germany

M. Growiec
RAG Mining Solutions GmbH, Katowice, Poland

ABSTRACT: Planning work, roadway development and excavation are the main intervals of the lifecycle for a roadway in longwall coal mining. The standard system of planning, which is offered by the German RAG Mining Solutions GmbH follows this phases and is focused on the geological and geotechnical preconditions, the stress distribution influenced by multiple seam working and a wide range of support elements and systems available. Continuous enhancements of monitoring and equipment within the past 25 years give the opportunity of development and use of the roadways with minimum technical risk. The following paper gives an overview to this lifecycle and the know-how of RAG Mining Solutions either for matters planning or of processing under the preconditions of a difficult deposit, a wide range of mine layout variations and various support systems.

KEYWORDS: hard coal mining, thin seam, mine planning, support system planning

1 BASICS

High performance and efficient mining in thin coal seams with high-performance plough systems requires a high working speed per day and an associated high area rate of advance in m²/min.

	BW	Projekt	Gew.-masch.	Förder-tage	Förderung	Abbaufort-schritt	Streb-länge	Mächtigkeit Kohle/Abbau		Laufzeit	Flächenverhieb i.d. Laufz.	*) Schichten/Leistung Streb betreiben	
				n	t/d	m/d	m	m	m	min/d	m²/min	MS/d	kg/MS
						5< # <8				25% < # <50%	3< # <5		
1	PH	571G	Hobel	14	6.257	9,19	330	1,63	2,07	430	7,05	71,4	87.604
2	WS	630G	Hobel	23	5.224	5,14	439	1,67	2,01	282	8,00	49,2	106.095
3	WS	452G	Hobel	23	4.970	8,28	290	1,48	1,52	236	10,17	29,6	168.127
4	AV	494G	Walze	23	4.583	6,80	299	1,87	2,53	456	4,46	48,2	95.125
5	PH	253G	Hobel	23	4.518	6,75	305	1,52	2,11	372	5,53	66,4	68.065
6	SA	557G	Hobel	22	3.762	5,98	353	1,33	1,61	282	7,47	97,5	38.580
7	SA	556G	Hobel	22	2.924	5,45	303	1,32	1,64	208	7,95	80,6	36.285
8	IB	752G	Hobel	23	2.904	5,50	253	1,44	2,00			26,3	110.208
9	IB	453G	Hobel	23	1.631	2,98	300	1,22	1,28			29,8	54.775
10	AV	513G	Hobel	23	1.603	2,43	423	1,40	1,64	117	8,79	38,3	41.902
11	IB	401G	Hobel	22	1.445	3,34	346	0,84	1,48			34,0	42.489
12	WS	572G	Walze	23	1.292	0,95	466	2,17	3,19	236	1,87	41,8	30.899
13	WS	632G	Walze	23	1.224	1,02	450	1,93	3,33	167	2,74	59,6	20.554
14	IB	742G	Hobel	23	1.109	2,26	320	1,06	1,26			16,7	66.585
15	PH	706G	Hobel	23	601	1,36	350	1,08	2,05	144	3,29	69,6	8.636

Figure 1. Key figures of mining operations in German coal mining

The working operations in German coal mining illustrated in Figure 1 clearly show the difference in the area rate of advance between plough and shearing operations.
The output for good ploughing operations is between 7 and 10 m^2/min, but for shearing operations only around 5 m^2/min.

These high outputs of plough operations can however only be guaranteed by long-front ploughing from drive to drive. This in turn requires continual return readiness at the face edges with corresponding formation and holding open of the face roadway transition. Coupled with the geometric requirements placed on the top and bottom roads by a modern high-performance plough system, it is therefore of crucial importance that these roadways have the necessary dimensions at the working passage.

Figure 2 shows the requirements for roadway width and height at the face entrance for high-performance operations in German coal mining. In the case of plough faces the drive and the conveyor transfer are preferably designed for high performance and positioned in the roadway. The space requirement must be suitable for the existing roadway cross-section, and must take into account the roadway convergence. At great depths, height losses frequently occur after driving due to floor lift, as shown by the example in Figure 2, which must be taken into account in the planning of the support and the processes at the longwall face, as well as for the organisation.

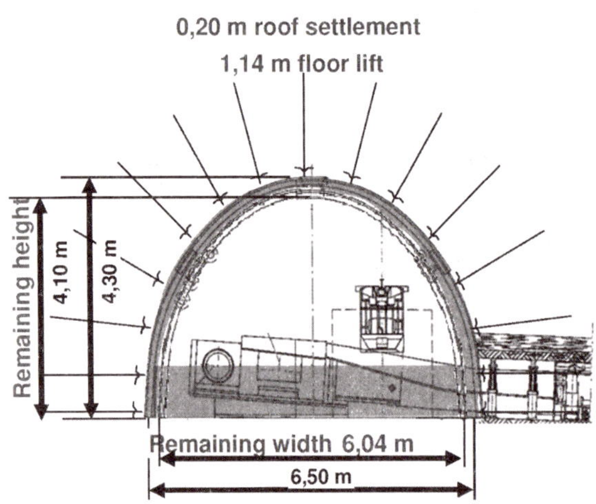

Figure 2. Remaining cross-section at the T-junction

A further key to success for the high face advance of modern plough operations is a stable and prop-free T-junction.

The following section describes how these requirements can be met by comprehensive planning and design of the roadways and corresponding preparation of the T-junction.

2 ROADWAY SELECTION

The key to successful design is the correct roadway selection for the relevant application case. Considerations to be taken into account here include double usage, ventilation, cross-sections for the equipment to be installed and many others. In German coal mining, the support systems shown in Figure 3 have been developed to meet these requirements.

Proven systems are available both for low requirements and for maximum stresses at great depths down to 1.500 m. In the considerations for efficient support even at high rock pressure, the bolting technology plays a decisive role, and ensures high face advancement, iron-free transitions and a high level of work safety.

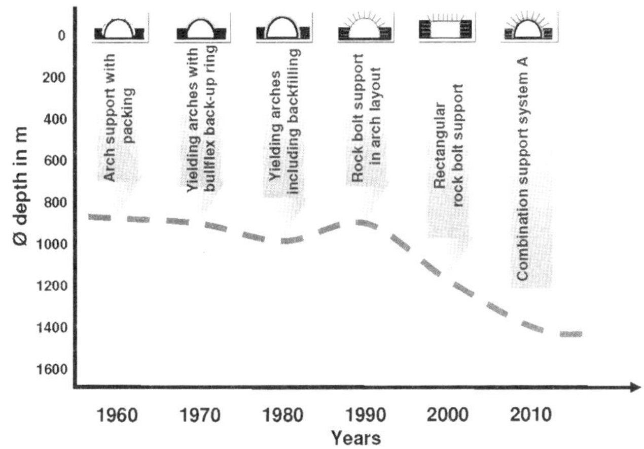

Figure 3. Development of roadway support in gate roads

Figure 4, left, shows an arch profile roadway support with rockbolt support system and on the right a rectangular profile roadway.

Figure 4. Roadway with rockbolt support system in an arched (left) and rectangular roadway (right)

3 ROADWAY PLANNING

The essential requirement for the correct roadway selection is a comprehensive and sustainable planning. RAG Mining Solutions offers for this purpose a modular procedure developed for these requirements over past years which ensures a high level of safety and efficiency. This procedure is described below (Figure 5).

Starting with the geological evaluation and geotechnical analysis the potential loads on the support system and the expected roadway deformation for the whole lifetime is determined.

The planning procedure also includes the panel layout and technical instructions of roadway development and design of roadway development machinery. A synthesis of geotechnical and rock mechanics assessment methods builds the basic platform for an optimized engineering of roadway development and support. Due to this a sense for the rock specification and its stress deformation behaviour is given to the planning engineer. The core of planning result is the recommendation for the roadway support for different parts of roadway development, monitoring system and a transparent documentation of planning parameters which are considered.

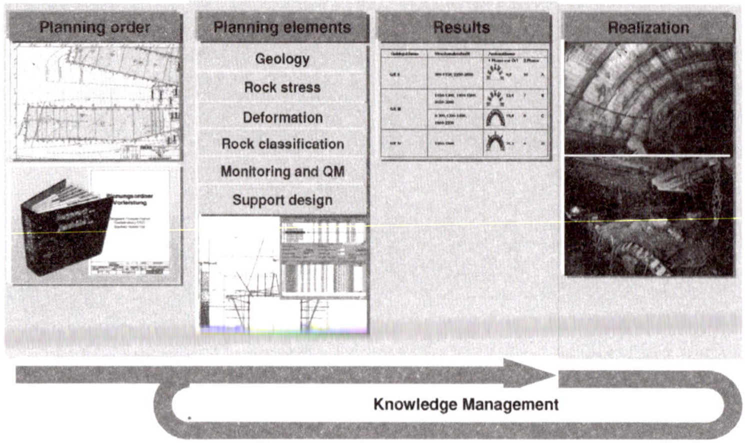

Figure 5. Modular procedure for the planning of roadways

Most important information of the survey are resulting from drill cores. In addition to geometrical data this information is implemented to a consolidated database for German coal mining. For the processing the DUDE IT-system was developed as a platform and data pool.

It gives a complete geographical model of the deposit and all mines – added by operational parameters e.g. ventilation mapping, monitoring results, etc.

The planning work starts with the determination of geological and geotechnical characteristics. All of the survey results are collected within the DUDE geographical data pool as well as in the project-orientated knowledge management data base.

For the planning of roadway development in German multiple seam mining in the next step the stress situation must be taken into consideration. The primary stress caused by overburden strata is diversified by longwall activities. Mining in several seams causes a very complex stress distribution in the area of mining activities and roadheading.

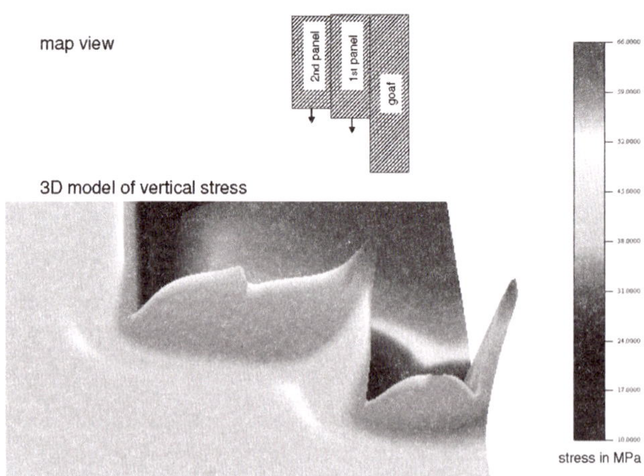

Figure 6. Vertical stress calculation in the area of longwall

Beside an expensive measurement of stress field, that requires sufficient bordering conditions and a high level of experience, the stress field can also be calculated by numerical simulation. The GEDRU software is used for this simulation. Figure 6 gives an example for the vertical stress along two longwall panels to be mined side by side following each other in a distance of 20 m. In this "single entry" – panel layout the center roadway between both panels is used twice.

The geomechanical classification takes into account stress as well as geotechnics. The classification allows a division of a prospected roadway into one or more sections depending on stress and strata conditions.

The first step of assessment for support in this planning phase results in excluding too expensive or not sufficient cross sections and support systems. The experience of German coal mining allows a rough classification of support system and a definition of bordering conditions for roadheading technique.

With the background of just-in-time delivery of roadheading systems and maintenance optimized to geological and operational conditions the planning engineer is used to solve this tasks 15 to 18 month before the roadways drivage starts.

The challenge of multiple seam mining for dimensioning of roadway support is the accurate prediction of time-depending stress field and subsidence erasing from longwalls in up to 20 seams mined in several decades.

4 STRATA AND ROADWAY DEFORMATION

The high level of stress induces fracturing of strata in the area around the roadways. Very often the fracturing starts in front of the heading and includes clastic or post fracturing movement of roof, floor and sides. The observation of strata and roadway deformation in German coal mines is documented for a long period. By the use of physical models in addition to the monitoring results on side a lot of influencing parameters are quantified. Especially tectonical faults and geotechnical elements are important because a high degree of separation and the orientation of separation planes must be taken into consideration for the prediction of roadway deformation (Figure 7).

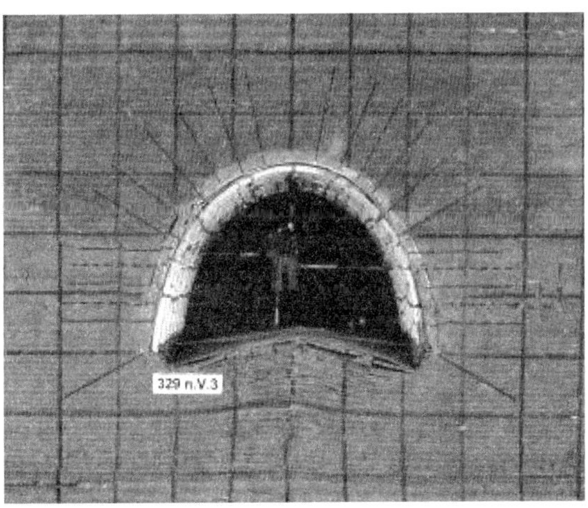

Figure 7. Example for strata deformation in a roadway simulated in a physical model

Beside numerical modelling of roadway deformation a body of rules and formula had been developed by analyzing of monitoring results. The base of this statistical analysis is founded in the 1970-ies.

Not only the roadway deformation in front of the longwall can be estimated but also the deformation after the passage of the longwall face. By installing a roadside package it is possible to control the roadway deformation and use it for haulage, transport and ventilation. This gives the opportunity for a single entry system and for minimizing the operating expense of roadway drivage. The planning procedure includes a prediction of roadway deformation for the whole lifetime of a roadway. Based on this result the support layout as well as the optimized panel layout can be specified.

Model trials have been carried out over a period of more than 30 years for numerous geological requirements and roadway geometries used by RAG. The results of these trials are incorporated into the planning by means of knowledge management.

As a worldwide service-provider for planning, RAG Mining Solutions offers this experience for use in international projects.

Many years of experience are therefore used for the benefit of operators who as yet have no such qualified experience of their own. They help to save costs and improve productivity during the introduction of innovative support technologies.

Special planning steps and measures are taken for the prevention of rock bursts.

Sudden roadway deformation by rockburst had been a problem in those mines, where high stress and sandstone where met in combination. The high strength of the sandstone is able to accumulate the energy caused by high stress and causes sudden reaction in case of overloading. To handle the rockburst problem a strictly regulation was defined based on analysis of several documented rockburst incidents. This regulation includes a testing and un-stressing program following step by step appending to test results. In most cases up to now a drilling program is the basic measurement for this. But also some other measures e.g. water-injection in front of the longwall front were undertaken successfully to avoid a rockburst.

5 SUPPORT ELEMENTS AND SUPPORT SYSTEMS – CLASSIFICATION AND QUALITY MANAGEMENT

More than 100 years of mining activities in Germany did breed a lot of different support elements and experiences. Up to now the materials steel, wood, concrete and plastics are in use.
The requirements of multiple seam mining in a great depth often demand combined support systems because simple and cheap support is not sufficient even for the phase of drivage under conditions of high stress or intensive separation. For minimizing the wide range of elements the support in German coal mines is standardized. Today e.g. one standard bolt and only a few additional types for special applications as flexible bolting and injections are common, where in the past a lot of different rockbolts where used.

The technical characteristics of all support elements are proofed according to German standards system DIN. This includes testing of material properties as well as tests on installed support. Individual test rigs also for standing support as well as for installed rockbolts were developed for optimization. The support systems available are classified according to rock-mechanical preconditions. A simple support system best fitted to high performance development is the rockbolting system for rectangular roadways. A lot of research and development work in Germany was done to enlarge the range of application in friable or squeezing strata conditions. The actual state-of-the-art is a successful application in a depth of approximately 3.600 ft. Ambitious efforts of application in a depth of up to 4.250 ft did proof a high expense for maintenance and reinforcement and advantages of other, high capacity, support systems. Highest capacity is achieved by combined support with rockbolts and additional yielding steel arches, backfilled with concrete (Figure 8).

Beside the technical description of the support system the processing of support work including transport and installation of elements are important. Especially the installation of rockbolts or the engineering and transport of concrete require special educated operators and some experience in handling quite complex systems for mixing, pumping, transport and installation.

So the German coal mines did introduce a detailed quality management for support. It is representing elements of education and responsibilities as well as examination programs for roadheading and face end T-junction. Of course the quality tests on materials and support elements by the manufactory and by RAG are elements of quality management.

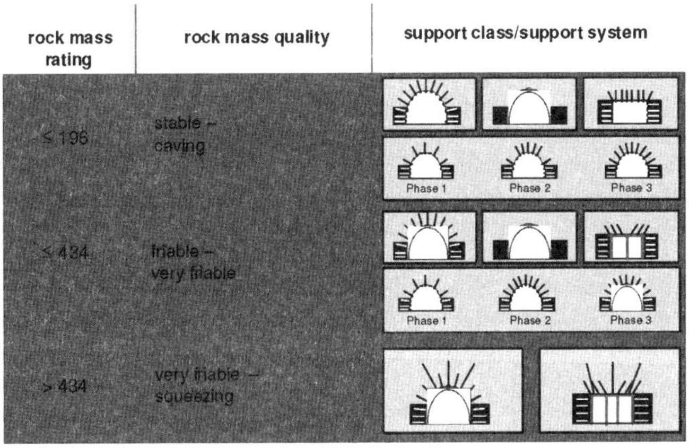

Figure 8. Classification of support systems

We offer the implementation of our expertise for support design in international projects. The range of services include development of e.g. rockbolt design, glueing cartridges and recommendations for the installation.

6 FACE END T-JUNCTION

In the area of main entrance the complete machinery for transport and haulage meets the point of highest stress concentration. This therefore is the point of interests, for both the scientist and the operator. While mining under sufficient rock-mechanical conditions never has to think about roadway deformation and necessities for machinery it is of main interest to take care of under condition of high stress and multiple seam mining. While the situation of retreat mining (Figure 9) can be handled by shield support within the roadway some additional measures are required when the roadway must kept open (e.g. for ventilation) behind the face passage. In this case a roadside package is installed at the face entry for minimizing the roadway deformation.

Figure 9. Face entry for retreat mining with shield support

Figure 9 shows an example of a fully-mechanised T-junction of a bolted rectangular roadway, which is used for retreat working.

For the full exploitation of the deposit, the roadways must be kept open after the 1st working for the 2nd use of the roadway. No residual support pillars therefore remain between individual panels, and the conditions for working the underlying parts of the deposit are optimised. This form of face layout is based on the many years of experience of German coal mining, and requires comprehensive know-how with regard to the overall understanding of interactions of rock pressure, support technology, working technology and ventilation.

The planning procedure already described in Figure 4 today provides the optimum solution for every case in German coal mining, and is also available on the international level for every mine operator, since RAG Mining Solutions offers its planning expertise throughout the world. The technical support solution is briefly described below.

This results in the technical support task of making a connection between the advancement of the hydraulic face support and the differently-designed support system in the roadway, and controlling the strata at the T-junction (Figure 10).

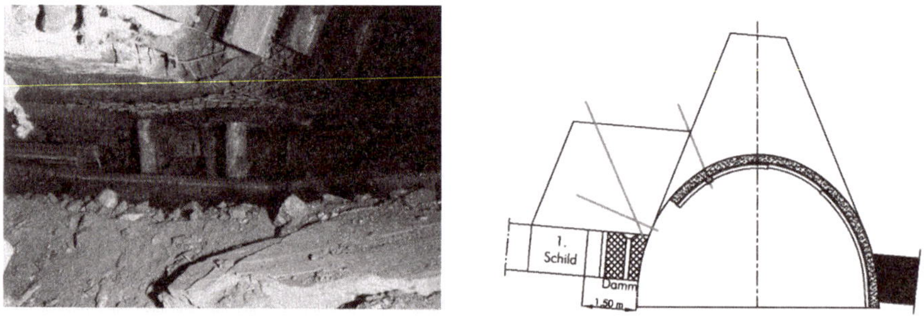

Figure 10. Arched roadway with fringe pinning and wooden props/support T-junction

The left part of Figure 10 shows a T-junction with anchor rockbolts to secure the released arch support and fringe pinning to secure the surrounding strata directly above the face entrance. The right part shows the cushioning bolts and the fringe pinning schematically by the example of an arch support. The rockbolting enables the stabilisation of the strata areas illustrated. These are dimensioned on the basis of a stability calculation. The basic requirements for safe workplace design are therefore created at the planning stage, facilitating the approval by the mining authorities with logical arguments.

RAG Mining Solutions offers the experience for designing the roadways safely and efficiently after the 1st use for the 2nd use.

In addition to the actual planning, the approaches required for this purpose of:

- Heading technology,
- Organisation,
- Support planning and
- Construction materials technology for the roadside pack

are offered by RAG Mining Solutions to all operators of coal mines throughout the world. The planning solutions are adapted to the specific regulations and general technical conditions in the course of the project work with the customer.

7 SERVICE FACILITIES OF RAG MINING SOLUTIONS FOR ROADWAY SUPPORT

RAG Mining Solutions offers various consulting and engineering services, as well as specialised trainings, for the implementation of the planning steps for roadway support described in detail above. These services are based on the experience gained in German coal mining. Planning cycles and trainings have been proven in many cases, and a component of the day-to-day work within the RAG Group. From this experience, RAG Mining Solutions has developed a comprehensive modular and flexible range of services for the handling and execution of international projects in the field of roadway support.

These services include planning, training and qualification measures, supervision of operations on site and operating consultation with regard to "lean processing", i.e. the long-term performance optimisation, and therefore cost reduction due to improved processes, within the customer's company.

7.1 Roadway support planning

The planning tool offered by RAG Mining Solutions is based on the system of standard technical support planning of RAG. The planning task is first individually formulated in consultation with the customer, and afterwards implemented in the planning steps illustrated in Figure 11 and described above.

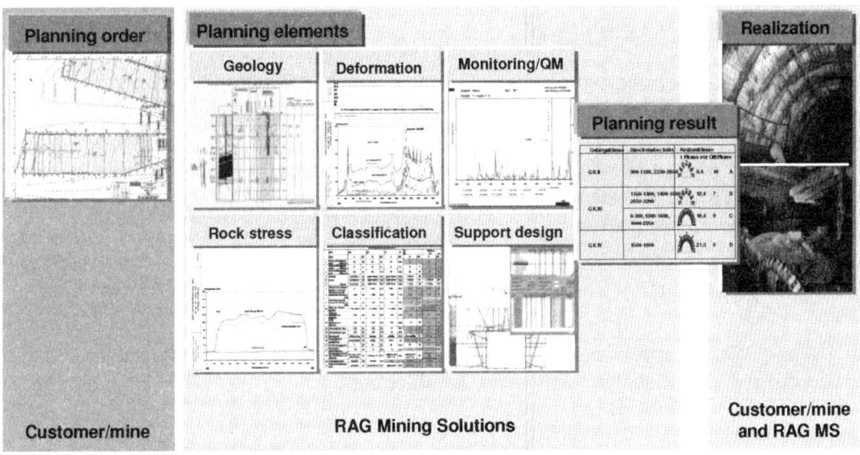

Figure 11. Procedure and subdivision of roadway support planning

One particular focal point of the planning result is that for the benefit of the customer, and taking into account geotechnical parameters, safe support variants are planned, which comply with a mathematical stability verification. If required, the planning can also provide monitoring concepts for the metrological monitoring of the roadway support.

For the gate roads themselves, different usage variants can be covered. These range from one-side retreat working through to double usage of a roadway by two operations, if permitted by the general conditions and planning specifications.

Operators who as yet have no dedicated experience in this matter can therefore also profit from the many years of experience of RAG Mining Solutions in the planning of roadway support. They can thus be helped to save costs and improve productivity by the introduction of innovative support technologies.

7.2 Training, seminars and workshops

The range of services of RAG Mining Solutions for training, seminars and workshops is directed at different target groups. There are distinctions between offers for specialists such as planners and support experts, management and heading crews. The training components offered for different target groups are shown in Figure 12.

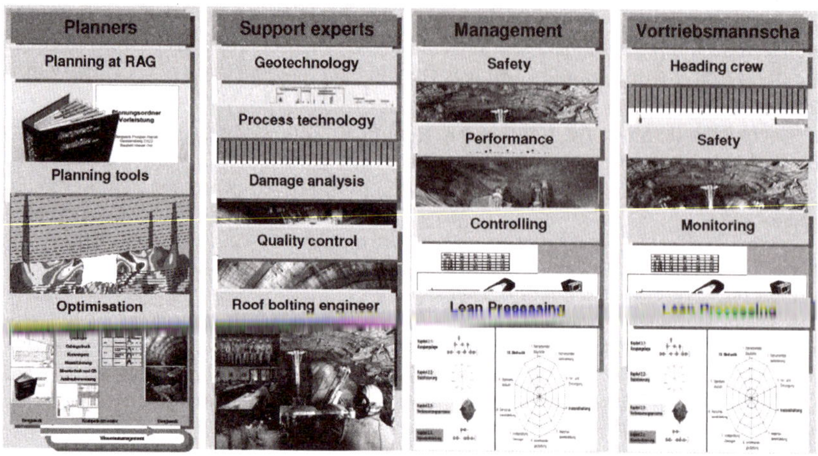

Figure 12. Modular training and qualification

For example, the block for management training contains components on operating optimisation of safety, performance optimisation, optimum controlling instruments and lean processing.

The component of lean processing appears both in the management block and the heading crew training block, since this element depends on the involvement of all personnel and also contributes to increasing motivation.

7.3 *Operating supervision*

Operating supervision by specialists of RAG Mining Solutions on site with the customer is provided in accordance with Figure 13 with different objectives.

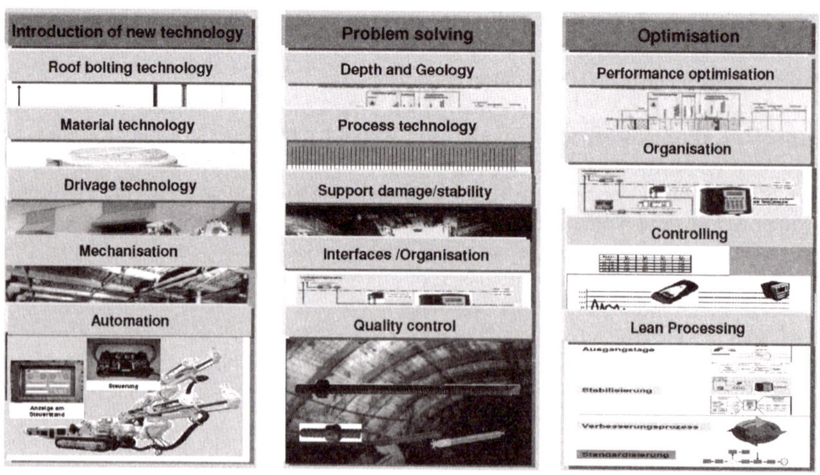

Figure 13. Scope of operating consultation and supervision

Access to the experience of German coal mining during the introduction of new technology (e.g. roof bolting technology or material technology) therefore usually leads to cost savings and performance optimisation for the customer in the relevant operations.

Individual support can also be provided in particularly difficult operating situations: this may include unknown effects of the depth-related rock pressure in new deposits, in cases of damage to the support or, if required, the improvement of the quality of support equipment and processes.

The exchange of experience through operating supervision during such projects leads to optimisations particularly in the areas of performance improvement, organisation, controlling and lean processing.

7.4 *Efficiency improvement by lean processing*

Lean processing should be understood as an aid to improving success "from the inside out". RAG Mining Solutions offers the individual process steps of analysis, stabilisation, optimisation and standardisation of processes (Figure 14). The benefit for the customer lies in further efficiency improvement and resulting cost reduction within the company.

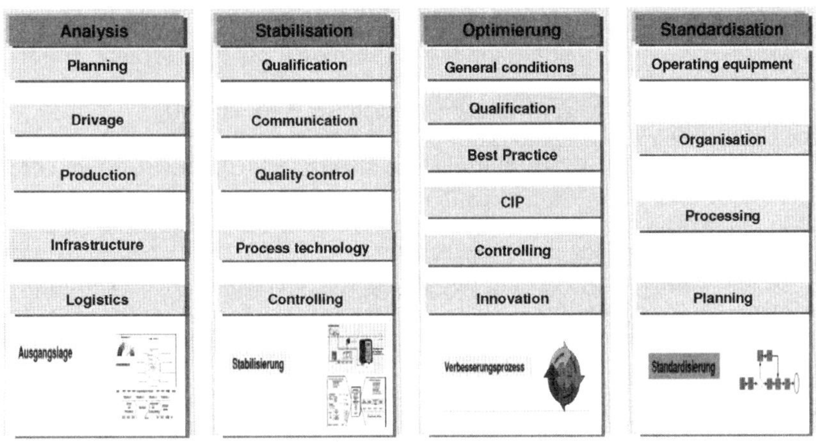

Figure 14. Planning components of lean processing

The experience of the RAG Group in all technical mining fields – from preliminary performance through to the following logistics – enables the structured implementation of the planning services of the specialist engineers of RAG Mining Solutions in the operating practice of the customer, as well as consultation from the field of mining for the field of mining.

8 CONCLUSIONS

Planning work, roadway development and excavation are the main intervals of the lifecycle for a roadway in longwall coal mining. The standard system of planning, which is offered by the German RAG Mining Solutions GmbH follows this phases and is focused on the geological and geotechnical preconditions, the stress distribution influenced by multiple seam working and a wide range of support elements and systems available. Continuous enhancements of monitoring and equipment within the past 25 years give the opportunity of development and use of the roadways with minimum technical risk.

RAG Mining Solutions offers various consulting and engineering services, as well as specialised trainings, for the implementation of the planning steps for roadway. These services are based on the experience gained in German coal mining. Planning cycles and trainings have been proven in many cases, and a component of the day-to-day work within the RAG Group. From this experience, RAG Mining Solutions has developed a comprehensive modular and flexible range of services for the handling and execution of international projects in the field of roadway support. These services include planning, training and qualification measures, supervision of operations on site and operating consultation with regard to "lean processing", i.e. the long-term performance optimisation, and therefore cost reduction due to improved processes, within the customer's company.

REFERENCES

Amadei B., Stephansson O. 1997. Rock Stress and its Measurement. London: Chapman & Hall.

Bräuner G. 1981. Gebirgsdruck und Gebirgsschläge. Essen: Verlag Glückauf.

Consol Energy Inc., Pittsburgh, Pennsylvania, USA

EGKS. 1993. Abschlussbericht zum Forschungsvorhaben 7220-AD/140. „Bessere Gebirgsbeherrschung durch Abbau und Ausbauplanung II". Essen: DMT Gesellschaft für Forschung und Prüfung mbH.

Everling G. 1960. Der Spannungszustand im unverritzten Gebirge. Glückauf. No. 19, pp. 1199-1202.

Everling G., Meyer A.-G. 1972. Ein Gebirgsdruckrechenmodell als Planungshilfe. Glückauf Forschungshefte. No. 2, pp. 81-88.

Jacobi O. 1960. Der Druck auf Flöz und Versatz. Glückauf. No. 7, pp. 409-418.

Junker M. 2009. Strata control in in-seam roadways. Essen: Verlag Glückauf.

Kammer W. 1981. Der erforderliche Stahlaufwand für Strecken mit Gleitbogenausbau. Glückauf. No. 15, pp. 933-934.

Klein J., Krahe J. 1993.Die Ermittlung der Ankertragkraft bei zusammengesetzter Beanspruchung. Glückauf Forschungshefte. No. 2, pp. 62-65.

Müller W. 1993. Mathematisches Modell zur Beurteilung von Streckenverformungen. Glückauf Forschungshefte. No. 3, pp. 101-106.

Müller W. 1992. The stress state in the Ruhr coalfield. Vorgetragen auf der Gebirgsdrucktagung in Aachen. Tagungsband. pp. 1707-1711.

MWMT. 1995. Abschlussbericht zum Forschungs- und Entwicklungsvorhaben NW 65.11-326-2. „Der Einfluss des Horizontalen Gebirgsdruckes auf die Grubenbaue". Essen: DMT-Gesellschaft für Forschung und Prüfung mbH.

MWMT. 2002. Abschlussbericht zum Forschungs- und Entwicklungsvorhaben NW 65.11-809-2; „Abbauzuschnitt für Hochleistungsbetriebe". Essen: Deutsche Montan Technologie GmbH.

N.N. 2001. Türstockrichtlinien der Bezirksregierung Arnsberg, AZ 18.24.1-2000-2, Stand 2001.

Nienhaus K., Lautsch Th. 2000. Strebbau und Örterbau – Wettbewerb und Ergänzung. Glückauf. No. 6, pp. 315-322.

Xiao Sh. 1993. Gebirgsmechanische Untersuchungen zu Bruchmechanismen und zur Gebirgsdruckverteilung an Abbaurändern im Steinkohlengebirge. Dissertation Clausthal.

International Mining Forum 2011, Dyczko et al. (eds) © 2011 Taylor & Francis Group, London, UK. ISBN 978-0-415-68938-0

Selected Methods of Forecasting Gate Roads Deformation

S. Prusek
Central Mining Institute GIG, Katowice, Poland

ABSTRACT: This paper presents selected methods of forecasting gate roads deformation in countries where underground mining of hard coal takes place. Two of those methods, developed at the Central Mining Institute, have been presented in more detail. The methods are based on the results of the research on horizontal and vertical convergence in gate roads maintained behind the longwall face with roof caving. First method is based on empirical relationships determined by statistical methods. The second one relies on the modified Hoek's Brown failure criterion and allows to predict the deformation of gate roads by means of numerical calculations with Phase software.

KEYWORDS: underground mining, gate road, convergence, prediction

1 INTRODUCTION

The increased value of stress, caused by the impact of abutment load, leads to intensive rock mass movements in gate roads. Roof sag, floor upheaval or horizontal convergence should be taken into account during the process of design and selection of the support. Thereby, stability of the gate roads and their functionality in the overall extraction process should be guaranteed by appropriate support. On the one hand, due to specific geological-mining conditions, a support of proper capacity should be chosen. On the other hand, during the design of roadway support, it is very important to take the roadway convergence into account in the estimated period of the working maintenance. It may turn out that due to low capacity and unsatisfying deformation ability, the support will not be able to reduce the deformation of the particular working as required by miners. The intense convergence would force to perform expensive, even dangerous works (floor dinting, support reconstruction etc.) to ensure satisfying cross-sections of workings.

The paper describes the selected methods of forecasting gate roads deformation in countries where underground mining of hard coal takes place. Two of those methods developed at the Central Mining Institute, have been presented in more detail. First method is based on empirical relationships determined by statistical methods while the second one relies on the modified Hoek's Brown failure criterion and allows to predict the deformation of gate road by means of numerical calculations with Phase software package.

2 SELECTED METHODS OF FORECASTING GATE ROADS DEFORMATION

The research on phenomena in gate roads, particularly rock mass movements caused by the impact of abutment load, have been performed by scientists from different countries where underground mining of hard coal takes place. As a result, many methods have been developed which allow to predict the deformation in roadways and optimize the particular supports exposed to overloading. In the last century, empirical methods were predominating. Nowadays, the most common methods to predict the deformation in roadways are based on numerical models.

The achievements within empirical methods shall be attributed to the scientists from France, Germany and Poland (Schwartz 1960, Götze, Kammer 1976, Jacobi 1967, 1976; Noltze 1981, Kammer 1977, 1980, Kulassek 2004, Biliński 1968, Biliński, Kostyk 1992, Chudek et al. 1987, Duży 2001). The long-term in situ research led to the development of many empirical equations which now allow to perform calculations of convergence in roadways. As an example, there is an equation which allows to calculate the value of basic horizontal convergence of roadways in case of advance mining (Junker et al., 2006):

$$K_o = -78 + 0{,}066\,T + 4{,}3\,SV \cdot M \ + 24{,}3 \cdot \sqrt{-4 \cdot \ln\left(\frac{\beta_{\mathrm{Dmin}} + 34}{168}\right)} \tag{1}$$

where: K_o – convergence of a roadway expressed in initial height percentage, T – depth of mining, m; M – seam thickness, m; SV – quantity, depending on the kind of side package system inby the face, β_{Dmin} – minimum compression strength of rocks in the floor and roof of the roadway, MPa.

The equation (1) allows to calculate the basic horizontal convergence of gate road which is developed in front of the advance face. This gate road is supported with steel arches manually backfilled with stone.

The recent development of new technologies allowed to substitute or supplement the empirical methods of forecasting the gate roads deformation with specific software enabling numerical modeling to be used for the phenomena in rock mass. Numerical calculations are usually based on finite difference (MRS), finite element (MES), boundary element (MEB) and distinct element (MEO) methods. The most common software used for the purposes of numerical calculations are: FLAC, UDEC, 3DEC, PHASE, PFC.

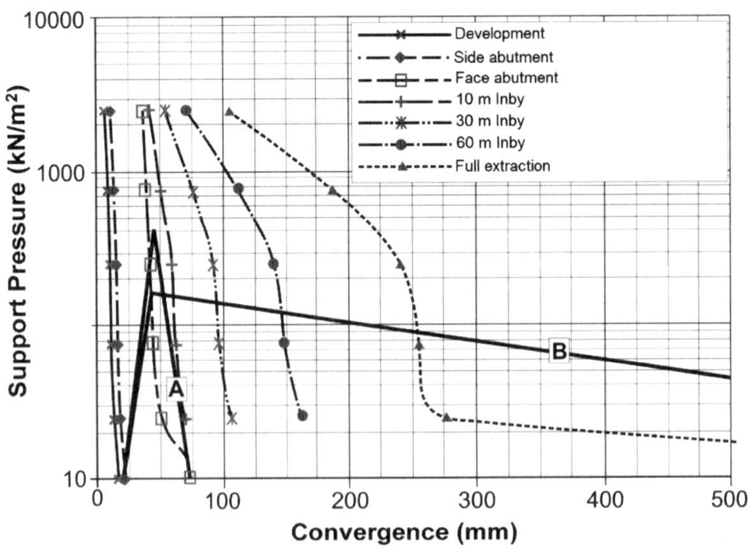

Figure 1. Ground Reaction Curves (GRCs) for different phases of the tailgate obtained from numerical calculations by the use of FLAC software. There are also depicted brittle support (A) and softening support (B) characteristics (Esterhuizen, Barczak 2006)

The numerical methods for calculations of gate roads deformation apply in many countries. In the United States of America the development of gate road supports is largely based on Ground Reaction Curve (GRC). The Curve which is indicated by numerical calculations, shows the relationship of loading and convergence that may appear in a given gate road (Mucho et al., 1999; Mark, Barczak 2000; Esterhuizen, Barczak 2006). Figure 1. shows an example of numerical calculation results within rock mass Ground Reaction Curve (GRC) indicating the tailgate driving phase and its various locations in the mining panel (Esterhuizen, Barczak 2006) carried out with FLAC software.

As the result, brittle support 'A' will reach the maximum value of its capacity at longwall face approaching with the convergence of 50 mm. At this point the support will be destructed. Alternatively, the softening support 'B' will be working properly in the whole area inby the longwall face with the convergence of 260 mm. This example reflects the methods of support selection with the use of numerical modeling for Ground Reaction Curves indication. The supports of optimized capacity and deformation characteristics are matched with particular curves by means of STOP software package (Support Technology Optimization Program) – (Barczak 2000).

Other examples of numerical modeling in the calculations of roadways deformation are the following publications of British, German, Spanish and Polish scientists: (Bowler, Robinson, Altounyan 2009, Hucke et al., 2006, Ruppel, Scior 2008, Torano et al., 2002, Majcherczyk, Małkowski 2003). Figure 2 presents the calculation results of deformation of rectangular and arch-shaped roadways supported with rock bolting.

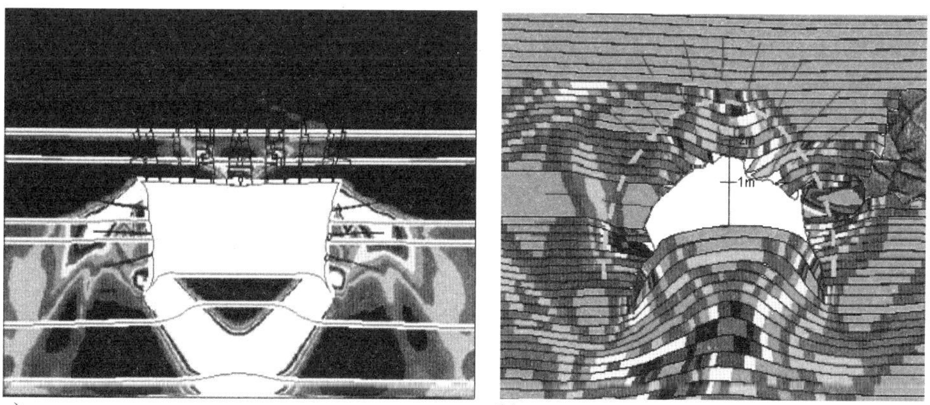

a) b)

Figure 2. Numerical modeling of roadway deformation: a – rectangular roadway supported with rock bolting (Bowler, Robinson, Altounyan 2009); b – arch-shaped roadway supported with rock bolting (Ruppel, Scior 2008).

3 FORCASTING GATE ROADS DEFORMATION METHODS DEVELOPED AT CENTRAL MINING INSTITUTE (CMI)

At this point, two new-founded methods developed at CMI will bee described. Those methods are based on the results of underground measurements which were performed in 23 roadways of the mines located in the conditions of Upper Silesian Coal Basin. The gate roads were maintained inby the longwall faces operated with roof caving (Prusek 2008b). The first method was described as an empirical-statistical model of roadway deformation. The model is based on polynomial function extended by additional parameters, as the result of profoundly tested research material. The function approximates precisely the obtained courses of roadway deformation (Prusek 2010). The second method is based on numerical modeling with the use of Finite Element Method and Hoek-Brown failure criterion (Prusek 2008a).

3.1 Empirical-statistical model of roadway deformation

During exploration of the research material in the form of convergence course of roadways, the possibility of their reflection by means of various mathematical functions was considered. It was assumed that the dependent variables are the values of roadways deformation in the form of floor upheaval, roof sag and horizontal convergence. The distance to the longwall face (outby or inby the face) – (d) is assumed as independent variable. It was decided that the gate road deformation model will include three segments: two polynomial functions connected by a linear function which may be determined so that the continuity of the first derivative of adjacent segments is maintained (Figure 3 – dashed line).

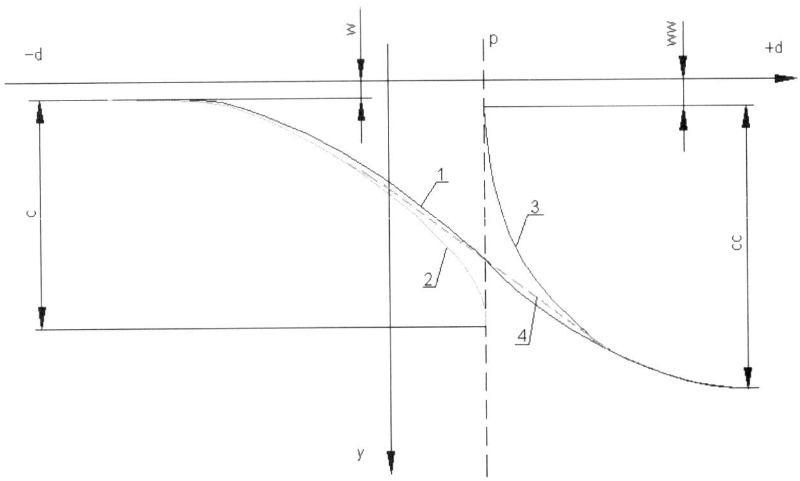

Figure 3. Model of gate road deformation along with the prime parameters, based on polynomial functions: 1 – example of real deformation course, 2, 3 – polynomial function curves, 4 – line linking interval – Prusek 2010

Using STATISTICA 8.0 software package, the values of parameters (c, cc, p, w, ww) of the assumed model have been determined separately for floor upheaval, roof sag and horizontal convergence which occurred in 23 roadways tested. The publication (Prusek 2010) presents the results of those calculations in detail.

Following the above-mentioned, the calculated parameters of the deformation model have been acknowledged as dependent variables which are influenced by geological-mining conditions occurring in the roadways. Basing on the detailed analysis of convergence underground measures results as well as study of literature and experience it was decided that the geological-mining factors influencing the roadways deformation will consist of (Prusek 2010):

- uniaxial compressive strength of roof rocks Rc roof, MPa,
- uniaxial compressive strength of floor rocks Rc floor, MPa,
- uniaxial compressive strength of coal Rc coal, MPa,
- roof number Lst, calculated up to a height of 100 m,
- floor number Lsp, calculated up to the depth of 30 m,
- product of partial coefficients mc, modifying the value of rock mass pressure,
- depth of measuring base location G, m,
- longwall height h, m,
- longwall average daily advance v, m/day,
- slope of seam, degree,
- longwall length D, m,
- average load-bearing capacity of supports ahead of the longwall face, kN/m,

– average load-bearing capacity of supports behind the longwall face, kN/m.

Each value of the above-mentioned factors for the given roadway has been presented in the publication (Prusek, 2008b).

The influence of particular factors on the deformation model parameter values has been determined with the use of regression method. At this point the method of multiple regression of many variables has been adopted using STATISTICA 8.0 software package. The developments of equations between all deformation model parameters in each factor are described in the publication (Prusek 2008b and 2010). The equations were fundamental for creating the algorithm konwergencja_gig_v1. It permits to make fast predictions of roadway deformation within the distance from longwall face, with the all geological-mining factors taken into account.

3.2 Prediction of roadway deformations by means of numerical modeling

For development of the method which allows to predict roadway deformation numerical calculations have been made, aiming at precise mapping of rock mass displacements in the particular roadway measured underground. For mapping with the help of numerical calculations of the measurements, the software based on Finite Element Method, Phase 2, was used, assuming that the rock mass is an elastically-plastic and isotropic medium. The rock mass effort state has been calculated according to Hoek-Brown failure criterion (Hoek 2006):

$$\sigma'_1 = \sigma'_3 + \sigma_{ci}\left(m_b \frac{\sigma'_3}{\sigma_{ci}} + s\right)^a \qquad (2)$$

where: σ'_1 *i* σ'_3 – effective maximum and minimum stress at destruction, [MPa]; m_b – value of Hoek-Brown parameter for the rock mass; s and a – parameters determined on the basis of rock mass properties; σ_{ci} – uniaxial compressive strength of rock sample, [MPa].

It was assumed that for underground measurements results of roadway deformation mapping, the influence of the operated longwall had an impact on strength-strain characteristics of rock mass, in the surroundings of a particular roadway. Naturally, this occurs as rock mass displacements. Therefore, for the purposes of obtaining roadways deformation measured in underground conditions, in the numerical models the parameters of rock strata have been reduced. Change in the values of parameter mb and s which in the assumed numerical strength criterion were fundamental for the devaluation of strength-strain characteristics of rocks (equation 2). The executed changes in the values of parameters mb and s influenced the course of stress-deformational rock characteristics in models, changing it from elastically ideally plastic to elastically-plastic with residual strength (Prusek 2008a and b).

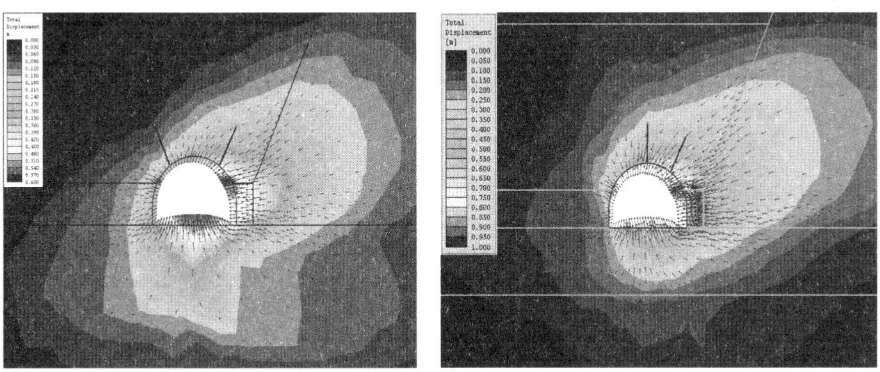

a) b)

Figure 4. Examples of calculation results of roadways deformation inby the face: a) – roadways in 'Piekary' mine, b) – roadway in 'Murcki' mine

In each case, the numerical calculations have been performed at three different distances outby and inby the longwall face. The models comprised the road located in the two-sided coal solid surroundings (outby longwall front) and the one-sided caving goaf surroundings (inby longwall face). Figure 4 presents examples of calculation results of two roadways deformation inby the face in 'Piekary' and 'Murcki' mines. Additionally, in the excavations roof bolting has been applied whereas along the gob line protective packs made of mineral materials have been executed.

Basing on the results of numerical calculations of 138 roadways models and statistical analysis, performed by means of non-linear segment regression method, it was possible to indicate the values of parameters mb and s for various rocks, depending on the distance of longwall face (Prusek 2008a, b):

– for coal:

$$m_{bz} = 0,6278 \cdot e^{-0,0024 \cdot d} \tag{3}$$

$$s_z = 0,000494 \cdot e^{-0,00393 \cdot d} \tag{4}$$

– for shale:

$$m_{bz} = 0,3197 \cdot e^{-0,0148 \cdot d} \tag{5}$$

$$s_z = 0,00041 \cdot e^{-0,01592 \cdot d} \tag{6}$$

– for sandy shale:

$$m_{bz} = 0,5093 \cdot e^{-0,0127 \cdot d} \tag{7}$$

$$s_z = 0,000762 \cdot e^{-0,01881 \cdot d} \tag{8}$$

– for sandstone:

$$m_{bz} = 0,7568 \cdot e^{-0,0127 \cdot d} \tag{9}$$

$$s_z = 0,001647 \cdot e^{-0,01697 \cdot d} \tag{10}$$

where: d – distance from the longwall face (d value is negative/positive outby/inby the longwall respectively).

In the presented equations the parameters mb and s received a 'z' index which means that they owned post-destruction features.

The equations (3–10) allow to calculate post-destruction parameters of coal seams and their surrounding rocks in the conditions of Upper Silesian Coal Basin. At this point, it is possible to do numerical calculations, regarding deformations which can occur in roadways at a particular distance from longwall face.

4 SUMMARY

The intensity of roadway deformation determines the possibility of regular coal seam exploitation using longwall system as well as safety of the crew. A lot of experience shows that significant convergence of roadways, where the transport of material to the longwall, haulage of output and which serve as ventilation track for longwall panels, may cause serious down-times in underground mining, methane and fire hazards increase or negatively affect the climatic conditions.

In order to assess the stability of roadways during the mining period, it is necessary to be familiar with the value of convergence which may occur in the particular geological-mining conditions. In countries where hard coal mining takes place by use of underground methods, various models of convergence course prediction in roadways have been developed, respecting the impact of the abutment load. In the last century many empirical methods developed and, as a result, they were applied for instance in French, German or Polish mining industry. The methods were mainly based on underground measurement results of roadway convergence. The recent development of new computer technologies allowed to calculate the roadway convergence by use of numerical modeling.

This paper presents various methods of gate roads deformation predictions developed in Poland and abroad. The empirical-statistical and numerical methods, which were developed at Central Mining Institute in recent years, have been described in detail.

REFERENCES

Barczak T.M. 2000. Optimizing secondary roof support with the Niosh support technology optimization program (STOP). Morgantown, Proceedings of the 19th International Conference on Ground Control in Mining, p. 74-83.

Biliński A. 1968. Przejawy ciśnienia górotworu w polach eksploatacji ścianowej w pokładach węgla. Zeszyty Naukowe Politechniki Śląskiej nr 221. (In Polish)

Biliński A., Kostyk T. 1992. Wpływ podporności obudowy na ruchy górotworu w chodnikach przyścianowych przy eksploatacji na zawał. Prace Naukowe GIG nr 773. (In Polish)

Bowler J., Robinson G., Altounyan P. 2009. Rib support Innovation at Daw Mill Colliery. Morgantown, Proceedings of 28th International Conference on Ground Control in Mining, p. 238-247.

Chudek M., Pach A., Żukian B., Skudlik G., Garncarz R. 1987. Prognoza zaciskania chodników przyścianowych. Zeszyty Naukowe Politechniki Śląskiej, seria: Górnictwo nr 157, p. 135-149. (In Polish)

Duży S. 2001. Zaciskanie chodników przyścianowych w procesie eksploatacji górniczej w świetle pomiarów w kopalniach węgla kamiennego. Gliwice-Ustroń, V Jubileuszowa Szkoła Geomechaniki, Część I-polska, p. 157-165. (In Polish)

Esterhuizen G.S., Barczak T.M. 2006. Development of ground response curves for longwall tailgate support design. Colorado, Proceedings of the 41st U.S. Rock Mechanics Symposium, p. 1-10.

Götze W., Kammer W. 1976. Die Auswirkungen von Streckenführung und Ausbautechnik auf die Querschnittsverminderung von Abbaustrecken. Glückauf 112 No. 4, p. 846-853.

Hoek E. 2006. Practical Rock Engineering. Rocscience Inc, http://www.rocscience.com.

Hucke A., Studeny A., Ruppel U., Witthaus H. 2006. Advanced Prediction Methods for Roadway Behaviour by Combining Numerical Simulation, Physical Modelling and In-Situ Monitoring. 25th International Conference on Ground Control in Mining, p. 213-220.

Jacobi O. 1967. Die gebirgsmechanisch günstige Führung von Abbaustrecken. Glückauf No. 26, p. 1302-1309.

Jacobi O. 1976. Praxis der Gebirgsbeherrschung. Essen, Verlag Glückauf.

Junker M. i inni .2006. Gebirgsbeherrschung von Flözstrecken. Essen, Verlag Glückauf GmbH.

Kammer W. 1977. Die Ausbauplanung durch Vorausberechnung der Endkonvergenz in Abbaustrecken. Glückauf 113 No. 8, p. 746-748.

Kammer W. 1980. Konvergenzberechnung und Ausbauplanung für Abbaustrecken mit einem EDV-Programm. Glückauf 116 No 5, p. 200-205.

Kulassek M. 2004. Erweiterung der Konvergenzberechnung für Bogenstrecken. Glückauf No. 5, p. 218-223.

Majcherczyk T., Małkowski P. 2003. Wpływ frontu ściany na wielkość strefy spękań wokół wyrobiska przyścianowego. Wiadomości Górnicze nr 1, p. 20-29. (In Polish)

Mark C., Barczak T.M. 2000. Fundamentals of Coal Mine Roof Support. Proceedings, New Technology for Coal Mine Roof Support, NIOSH Open Industry Briefing, NIOSH IC 9453, p. 23-42.

Mucho T.P., Barczak T.M., Dolinar D.R., Bower J., Byrja J.J. 1999. Design methodology for standing secondary roof support in longwall tailgates. Morgantown, Proceedings 18th International Conference on Ground Control in Mining, p. 136-148.

Noltze C. 1981. Gebirgsdruck und Druckwirkungen in Flözstrecken der Ruhrreviers. Praca doktorska.

Phase^2. 2001. User's guide, Rocscience.

Prusek S. 2008a. Modification of parameters in the Hoek-Brown failure criterion for gate road deformation rediction by means of numerical modeling. Glückauf No. 9, p. 529-534.

Prusek S. 2008b. Metody prognozowania deformacji chodników przyścianowych w strefach wpływu eksploatacji z zawałem stropu. Prace Naukowe GIG nr 874. Katowice. (In Polish)

Prusek S. 2010. Empirical-statistical model of gate roads deformation. Archiwum Górnictwa nr 2, p 295-312.

Ruppel U., Scior C. 2008. Planung von Abbaustrecken mittels numerischer Berechnungen in der Ukraine. RWTH Aachen International Mining Symposia-Rockbolting in Mining & Injection Technology and Roadway Support Systems, p. 671-684.

Schwartz B. 1960. Vorausberechnung der Bewegungen in Abbaustrecken. Internationaler Kongress für Gebirgsdruckforschung Paris, p. 383-398.

Toraño J., Rodríguez Díez R., Rivas Cid J. M., Casal Barciella M. M. 2002. FEM modeling of roadways driven in a fractured rock mass under a longwall influence. Computers and Geotechnics No. 29, p. 411-431.

International Mining Forum 2011, Dyczko et al. (eds) © 2011 Taylor & Francis Group, London, UK. ISBN 978-0-415-68938-0

Evolution of the Methods of Maintaining Longwall Galleries in the LW Bogdanka S.A. Mine

J. Chmielewski, B. Kozek
Lubelski Węgiel Bogdanka S.A., Poland

ABSTRACT: The article presents the current ways of managing and maintaining longwall galleries in the mine of Lubelski Węgiel Bogdanka. It discusses experiences related to the "advance" mining system, which was used in 1982-1990, along with maintaining longwall galleries behind the wall. It describes the presently used "retreat" mining system, and the way of exploiting longwall galleries in that system. It presents types of the gallery support system used and its modifications which took place over years of mining activity, as well as ways to reinforce the support system for the purpose of maintaining the wall-gallery crossing. It discusses new conditions concerning longwall galleries for plough walls.

KEYWORDS: deep mining, hard coal, roadway workings, gallery support system, longwall galleries, Lubelskie Zagłębie Węglowe (LZW) [English: Lublin Region Coal Basin]

1 INTRODUCTION

Hard coal mining in the Lublin Region Coal Basin started on 30 November 1982, when the first exploration longwall in bed 382 was opened. It was 100-meterlong, prepared directly by the shaft pillar and explored in the "advance" system. To date, during the almost 30 year-long period of wall mining, there have been two distinguishable periods of using different longwall mining systems and longwall gallery maintenance systems:

1. Period I – initial – from 1982 to 1990, when mining was conducted in the "advance" system, where both longwall galleries were developed along the mining progress of the face and then maintained behind the longwall,
2. Period II – since 1992 until now, when the "retreat" system with liquidation of galleries behind the wall has been used.

The years 1991–1992, when the system was changed, were an intermediate period, during which elements of both systems were employed.

The "advance" system brought immense difficulties related to the maintenance of longwall galleries, which coincided with general difficulties in maintaining all roadway workings due to their excessive convergence. It was a period of intense searching for new types of support systems suitable for the conditions of the LZW rock mass. Those difficulties, combined with the lack of mining ability in vertical transportation (a serious delay in the construction of skip shaft 1.3), resulted in very long duration of longwall mining periods. For example, mining of the first exploration wall 1/I in deposit 382, opened at the end of 1982, was finished in March 1988, i.e. after about 5.5-year-long mining period. Mining of wall 2/I, opened for mining in June 1986, ended in August 1990, which is after 4 years and a quarter. Similarly, wall 2N in deposit 382 in

Nadrybie, opened for mining in September 1987, finished its operation in January 1991, i.e. after 3 years and a quarter of operation.

During the intermediate period, i.e. in the years 1991–1992, modifications involved the mining system during longwall exploitation. This was the case for wall 3/I and 1/V, which initially worked in the traditional "advance" system with maintenance of both longwall galleries behind the wall, but from a certain moment (after passing the engineering cross-cut) went on to work in the "retreat" system with liquidation of galleries behind the wall front face. A similar arrangement was also employed for wall 0/IV, opened for mining in March 1991, which was geographically mined with "advance" system, but longwall galleries had been made earlier and were liquidated along the wall mining progress.

2 MAINTENANCE OF LONGWALL GALLERIES IN THE "ADVANCE" SYSTEM USED IN THE YEARS 1982–1990

The "advance" system, which was used in the years 1982-1990 in Bogdanka mine, was characterised by the following features:

- longwall galleries were made along with the wall with a minimum advancement of the heading against the wall face (up to 8 m),
- longwall gallery support frames were at first made of V29 section bar, later of V36, most often with 0.5 m spacing,
- longwall galleries were initially made by means of explosive materials [MW], and then a special mining and loading machine was designed,
- protection of longwall galleries employed the use of a backfilling belt, which protected the galleries from the side of gobs, built of several rows of V-shaped racks and of a crib made of railroad ties and filled with stones, of various width and various structure.

The fundamental component of this system was the extremely problematic issue of protecting and maintaining longwall galleries.

Therefore, from the very beginning of mining operations, numerous experiments and tests were performed, which led to the development, at the end of the 80s, of very costly and time-consuming but relatively efficient solutions for the purpose of exploiting and maintaining longwall galleries behind the face in this system.

In order to eliminate deformation and destruction of the support frame due to the side convergence of galleries within the zone of high impact of mining pressures along a section of up to 40 m behind the wall face, from the side of the body of coal (or from old gobs' side in case of the next longwall in the field), special frames were used, which had side arches connected by joints with roof-bar elements (the so-called jointed support system). This support system allowed cheeking off (usually double) and bending the side arch to the initial position. Within the distance of 40–50 m behind the wall, behind the zone of the most intense mining pressures which affected the gallery, the jointed side arch was replaced with common overlap type side arch.

In order to increase the support capacity of the protective belts, anhydrite binders were introduced at the end of the 80s. They were transported and supplied pneumatically (in the Polko system), by which stone was transferred within the entire protective belt.

The final stage of using the "advance" system saw the use of frames made of V36 section bar size 8/10, and the driving process employed the use of specially designed mining and loading machine (MUŁ), working with the single- or double-angle scraper conveyor, which supplied the yield from the driven gallery headings onto a wall conveyor.

Therefore, longwall galleries behind the wall were maintained by employing:

1) wooden cribs filled with stones and several rows of V-shaped racks, expanded between V-shaped straight pieces built on the floor and under the roof,
2) anhydrite binders used to seal cribs and fill spaces between the cribs and the support system. It particularly improved the performance of the frame support system from the cave-in side, providing it with full support and also full insulation of abandoned workings,

3) jointed side arches from the coal body side in order to limit the impact of creeping sidewall and reduce horizontal convergence and deformation of the support system in the roof and sidewall part,
4) cheeking off coal, floor dinting and bending side arch, which is connected to roof-bar by means of a joint.

Only the system of maintaining longwall galleries behind wall, based on the a/m elements, ensured relatively efficient protection of galleries in the "advance" mining system used. The final structure of the longwall gallery protective belt that was used at the end of the 80s is presented in Figure 1.

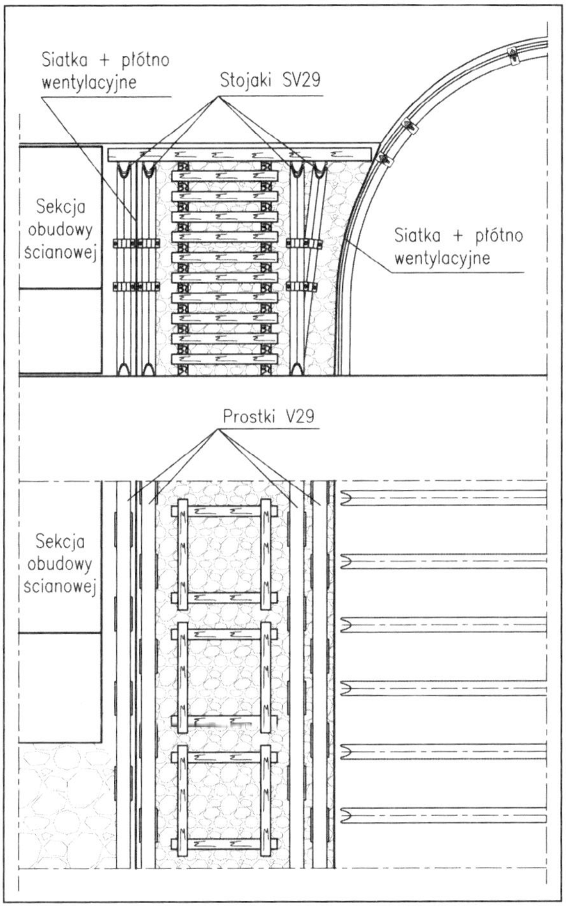

Figure 1. The longwall gallery protective belt used at the end of the 80s

Notations: Siatka + płótno wentylacyjne – *net + ventilation fabric;* Stojaki SV29 – *SV 29 props;* Sekcja obudowy ścianowej – *wall support section;* prostki V29 – *V29 straight pieces*

This complicated way of protecting the galleries was one of the key reasons why the "advance" mining system was rejected as it hindered wall mining progress.

In deposit 382, which was then mined in the "advance" system, in the years 1982–1992, five walls were mined in full length and two other walls, in the case of which the system was changed in the course of the longwall progress, were mined partially. The maximum monthly mining progress reached in this system by 1986 did not exceed 44 m, and in the mature stage, it reached its maximum value of 80 m in 1989. In 1990, which was the last year when that system was used, and when the mining process was carried at five working walls, the average monthly wall advance was about 40 m, while the maximum was 73.5 m. The last longwall to be mined along the entire length in the "advance" system was wall 1N in field Bogdanka, which finished mining in August 1992.

Despite using such complex technology of maintaining longwall galleries behind the face, maintaining those galleries, especially in the longer time, still encountered difficulties. Therefore, 1988 saw the beginning of implementation of a unique system of mining field preparation by making a network of field galleries in the gobs. Galleries in the gobswere expected to work as longwall galleries and make it possible to give up maintaining them in longer sections and in a longer period of time. The system of longwall galleries in the gobs was based on the assumption that galleries made in the caving (relaxed) zone will experience lower pressures than headings in the deposit. This assumption was confirmed by the preliminary research results as the initial intensity of their convergence was definitely lower than that of the galleries made in the deposit. In a longer period of time, when the caving was in the process of reconsolidation, the rock mass pressures gradually approached the original values and the tension relief effect was increasingly smaller. The system of galleries in the gobs was finally given up along with the change of the mining system at the beginning of the 90s. During the period of more than five years of its implementation, about 7 km of headings were made in the gobs in the Bogdanka mine, including about 5 km in field I, where this system reached the highest level of development. Apart from field I, the introduction of the system was also initiated in field V as well as in field II and VI in Nadrybie.

Development of the system of field galleries situated in abandoned workings was possible due to weak roof rocks which occur in LZW and which quickly reconsolidated in the cavings of exploited walls. This enabled "regular" driving with a progress rate which was close to that reached in undisturbed rock mass.

3 DRIVING AND MAINTENANCE OF LONGWALL GALLERIES IN THE "RETREAT" MINING SYSTEM

The first wall began its operation in the "retreat" system with liquidation of both longwall galleries behind the face front was the longwall no. 1/III in deposit 382, the mining of which started at the beginning of 1991 in Nadrybie. This was a 195 m long wall with 550 m panel length, for which one of the longwall galleries was a gallery maintained behind the front of a neighbouring wall 1/VI, which was mined in the "advance" system.

The change of the system was a turning point in the process of increasing longwall advance rates and operational efficiency in Lubelski Węgiel Bogdanka S.A. This system, which involves a single use of the longwall gallery, is still in use.

The system is characterised by the following features typical for its way of driving and maintaining longwall galleries:

- both longwall galleries are driven in advance and completed along the whole length before starting mining of the wall,
- both longwall galleries are liquidated successively behind the wall front, whereas for each next wall two new galleries are made,
- the currently used mining system does not leave longwall galleries behind the wall front and does not maintain them for the neighbouring longwalls;
- galleries are arch-shaped and have a specially designed support system made of V36 section bars, steel grade S480W, with the most often used set spacing of 0.75–0.9 m,
- an element which plays an important role in the maintenance of longwall galleries, in various stages of the maintenance of the galleries is, apart from the support lining, the bolting support which is used both as low bolting with the use of steel bar bolts (most often 2.7 m long), and high bolting with flexible bolting (string bolts, usually with the overall length of 6.0 m),

- the size of the lateral area of longwall workings used to be subject to frequent changes, whereas now the typical size for shearer longwalls is 10 and for longwall galleries for plough walls it is 11 (tailgate) and 12 (headgate),
- until 2000, the longwall gallery support was constructed as closed during the driving process. Presently, it is made as open, and it is closed only in selected cases following the undercutting of the floor.

4 EVOLUTION OF THE DESIGN OF SUPPORT FRAMES AND CROSS SECTIONS OF LONGWALL GALLERIES

Support system for longwall galleries in the "advance" mining system. During the first years of mining operations, longwall galleries were made with size 7 frame support system, made of V29 section bar, with the most often used frame spacing of 0.5 m. Those were typical 4-element frames, made as open. In the second half of the 80s, the cross section of the longwall galleries was widened by using the ŁP 8/10 frames, where roof-bars were made of size 10 frames, whereas side arches were made of size 8 frames. Those frames, the so called jointed frames, had been used until the "advance" mining system was no longer used. Figure 2 presents a view of the frames.

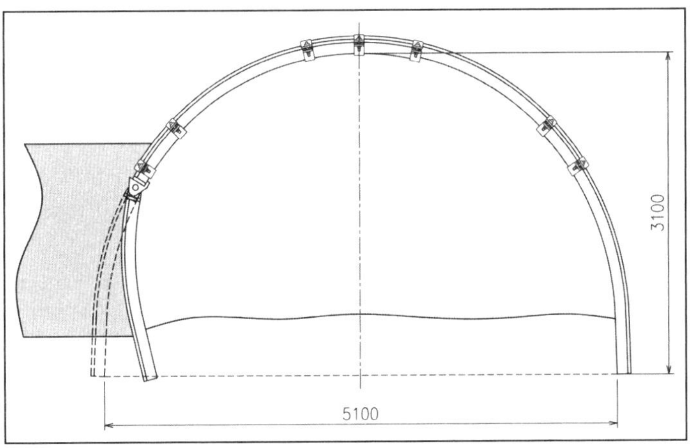

Figure 2. ŁP 8/10 frames with jointed side arch

Closed support system for longwall galleries in the 1990s. After changing the mining system and starting to make longwall galleries in advance, those galleries were made with a support system similar to that of other workings, i.e. in a special closed system made of V36 section bar, which was closed in the course of the gallery driving process in the heading zone right behind the mining machine. The most frequently used type of frames were the ŁPZS special closed frames, the same in drifts and main galleries as well as in the longwall galleries. The typical size of frames used in longwall galleries in the 90s were size 9, 5.2 m wide, the design of which is presented in Figure 3.

In that period, longwall galleries were always maintained in the closed support system. The bottom beams were removed at the distance of 60 m ahead of wall and the gallery was held in the open support only in the wall mining pressure zone.

Open support system with widened frames at the floor. Until 2001, both in standalone support systems and when spread sets with bolting were used, the closed special frames of the ŁPZS type were used as standard. The support capacity of these size 9 frames, with 1.0 m spacing, amounts to about 0.2 MPa. Those frames worked quite efficiently in the support systems, under favourable and mediocre geological and mining conditions, and with set spacing of 0.75

m. There were rare cases when foot pieces were destroyed, particularly in more difficult geomechanical conditions. Along with the implementation of mixed type support systems, the support system frames were moved apart to the spacing of 1.0–1.2 m, and even 1.5 m. Using the bolting support system in the roof partially provided an efficient reinforcement of the rock mass and compensated the reduced support capacity of the frame support system. Regarding the floor and the side walls, no proportional reinforcement of the rock mass was obtained, which resulted in frequent breakages of the foot pieces, even under mediocre geological and mining conditions.

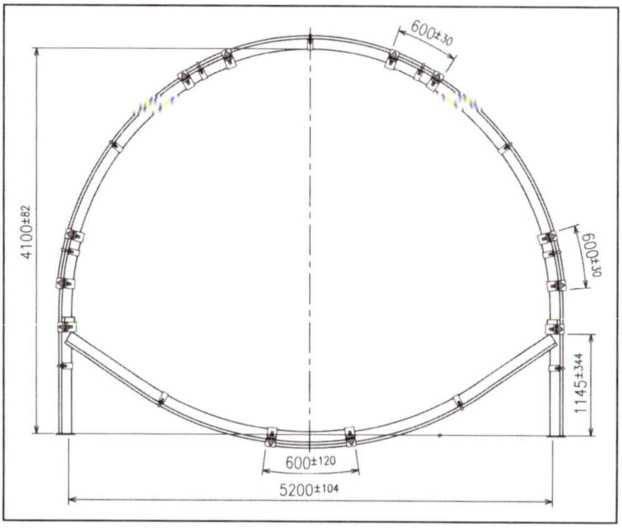

Fig. 3. ŁPZS support system frame, size 9

Attempts to maintain the floor by means of bolting, undertaken from 1996, have not brought positive results; they proved, however, that the upheaval of the floor was not as serious as it had been considered to be, and that there were no special difficulties related to the maintenance of the gallery. The most serious nuisance was not the upheaval of the floor, but the horizontal convergence of galleries, because the frame support whose side wall arches are perpendicular to the floor was excessively narrowed at the floor even at a small horizontal convergence. The open frames were, as a matter of fact, highly susceptible to the horizontal convergence.

Those circumstances formed the basis for the development of a new support system, i.e. 5-element frames of the ŁPS/C type, which offered a certain "design" width reserve for the horizontal convergence, which made them suitable to be built-in the heading of the driven gallery as open (Figure 4). Type C frames were used in the Bogdanka mine since July 2001. They quickly began to be the main type of the support system used in longwall galleries in the mine.

The ŁPS/C type support system frames are built up in the heading, principally as open. However, the technology employed to maintain workings in this support system comprises two options:

1. Option I – frame are left as open until the end of the gallery's existence. It concerns the working whose existence is shorter (mainly longwall galleries) under mediocre and favourable geological and mining conditions. In that case, only floor dinting is carried out to maintain the gallery.
2. Option II – floor is closed with certain delay and after the undercutting is completed. This option is used in workings whose existence is longer or in longwall galleries under more difficult geological and mining conditions when there is intense horizontal convergence, for example, when after one month after the gallery had been made, it lost all of its spare width (the side wall arches are already perpendicular to the floor).

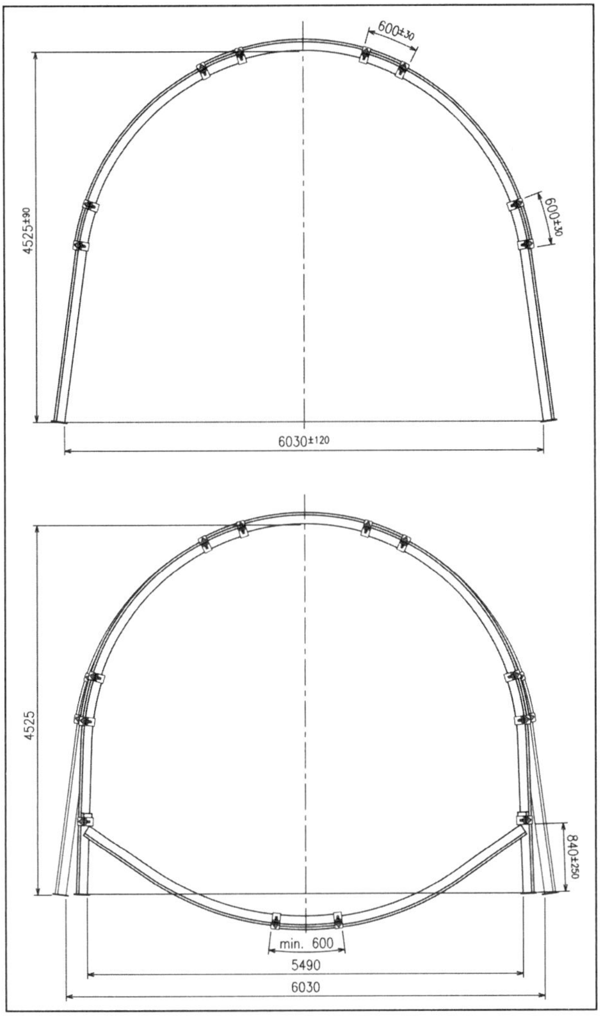

Figure 4. The ŁPZS support system frame, size 10

In the current scope of using the ŁPS/C support system, it is estimated that in 30–50 % of workings, the floor was closed with a delay, whereas in the remaining cases, frames were left open.

The provided comparison of the convergence of galleries in the ŁPS/C support system and other workings in similar conditions, in the ŁPZS support closed in heading shows that the type C support system ensures better or at least comparable level of their static stability. Maintaining the floor and dinting it does not cause any major problems. Frames were deformed more rarely. The ŁPS/C support system with 5-element frames was used as basic in the Bogdanka mine until 2009.

5 NEW ŁPSC FRAMES FOR LONGWALL GALLERIES FOR SHEARER AND PLOUGH WALLS IN BOGDANKA MINE

In the years 2008–2009, in Bogdanka mine, works were undertaken to develop a new system of frames which would be better adjusted for the purpose of use in longwall galleries. At that time,

100 % of those were longwall galleries for shearer walls. While those works were under way, a problem occurred with regard to the support system for longwall galleries for plough longwalls, which was related to the preparatory works undertaken to introduce the plough technique in the mine. In the case of plough walls, new requirements had to be met:

a) increase of the size of galleries and dimensions of frames,
b) spreading the frames in order to reduce the volume of works at the crossing with the wall (those works required stopping the plough).

Meeting those two requirements caused the need to introduce the bolting support as a fixed component of the gallery support system. Those works resulted in 6-element arch-shaped frames with transverse cross section, designed to lead and maintain the crossing with the wall. Those frames were designed to be built in the heading of the driven gallery as open, with a possibility to close the floor later after undercutting (as in the case of the ŁPS/C frames). The use of the arch-shaped frames was preceded by an unsuccessful attempt to introduce elliptical frames, the shape of which was close that used in Germany in accordance with the DIN standard.

Attempt to introduce elliptical frames of the ŁPSE type. In order to determine an optimal shape of the cross section of the newly designed frames of the gallery support system, changes of the current cross section shape which occurred in course of the support system convergence process were analysed. Following the assumption that as a result of the convergence of gallery workings underground the support system acquires a shape which is natural for it, resulting from the proportion of vertical and horizontal pressures as well as their values and distribution around the circumference of the working, the acquired shape can tell a lot about the shape of the cross section of frames built up in the heading, which should be proper in terms of geomechanics. The analysis comprised dozens of cases illustrating the state of the support system of the converged workings, mainly made in longwall galleries and in reconstructions, mainly on levels 920 and 960 m. In most cases, the converged support frames acquired an elliptical shape, which provided a basis for recognizing the elliptical shape to be optimal (Figure 5).

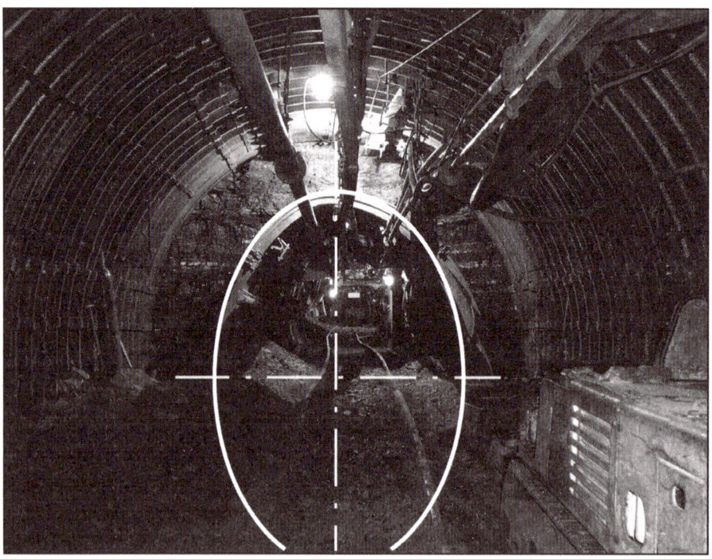

Figure 5. The elliptical shape of the converged working visible in the place of reconstruction

On that basis, in 2008, ŁPSE type frames were developed, a test lot in the amount of 500 tons was contracted and built up underground in three sections, with the major longest one in the longwall gallery of wall 3/IV in deposit 385/2. Comparison of the state of the support system in the experimental sections against the ŁPS/C type support system which was in use (while taking into consideration many additional circumstances) clearly showed that the state of the gallery with the ŁPSE support is definitely worse. In that case, in July 2008, it was decided to give up further tests of the ŁPSE support system and to consider the cross section shape as unsuitable for longwall galleries in the conditions of the Bogdanka mine.

Analysis of the causes of that situation leads to a conclusion that the assumptions and the analysis of performance of the support systems which had been developed earlier and which had led to designing the elliptical support system and considering it to be the optimal for our conditions were correct and right, however, only in case when there are prevailing roof pressures on the support system frames. On the contrary, in the conditions of the Bogdanka mine, it is horizontal pressures that most often occur in longwall galleries, while there are no roof pressures at all or they are insignificant. Furthermore, introducing the use of open frames has eliminated the vertical floor pressures, generated by the floor upheaval process and transferred to the roof by floor arches.

In the circumstances of high sidewall pressures, the elliptical frames which are converged from the side quickly lose their width, not only by the gallery floor, but also in the roof part. Frames featuring the shape used so far are to a greater extent converged only in the floor part.

As a result of those tests, the arch shape of frames has been brought back into the works to develop new support system frames.

ŁPSC type frames in the 6-element version. The newly developed type of frames has the following features:

- The number and length of the frame elements have been adjusted in order to obtain a favourable position of the roof-bar arch against the roof of the deposit being undercut in the heading of the driven gallery so as to provide a convenient wall inlet after removing side wall arches, without the need to cut the roof-bar.
- The frames used short side wall bottom elements (floor elements), whose length was designed to suit the floor upheaval values so as to enable eliminating those elements when the formerly removed side wall elements are built up again (after passing of the longwall), creating non-symmetrical frames without unnecessarily undercutting floor within the section behind the wall inlet.
- Overlaps of frame elements have a fixed value, except the connection between lower side wall elements and upper side wall elements of frames. The overlap can be changed individually within the range of 500-700 mm, depending on the actual size of the break made by the mining machine while driving the gallery. It enables setting the frame support feet on stable, solid floor, without having to provide a bed of loose rocks or sub-plates to correct the break height.
- Frames have been designed separately for longwall galleries for shearer walls of smaller frame size, distinguished with mark K, and for longwall galleries for plough walls of bigger frame size, distinguished with mark S.
- Variants K and S differed from each other in the relation of frame height to frame width. In the S frames, height no longer increased in proportion to frame width.

The frame type offered six sizes – three for longwall galleries for shearer walls (variant K – size 9, 10 and 11), and three for longwall galleries for plough walls (variant S – size 11, 12 and 13).

The new 6-element support of the ŁPSC type was used for the first time in longwall galleries of wall 1/VI in deposit 385/2, driven for the purpose of the first plough wall by the division GRP-4.

ŁPSC type frames in the 7-element version. The new ŁPSC frames for longwall galleries, introduced to be used in the Bogdanka mine, were designed for deposits which were more than 1.6 m thick. In the case of plough walls and mining deposits of small thickness (from 1.2 m to 1.6 m), which is the case with longwalls in field VIII in deposit 385/2 in Stefanów, ensuring a proper width of galleries and a proper position of the frames against the deposit at the gallery and wall crossing required further modification of the frame design. In thin deposits, it is reasonable to drive a longwall gallery with little floor dinting, specified to be about 0.8 m. In that situation, building up the roof part of working situated above the deposit requires three- instead of two-frame roof-bar elements which have been used so far, in order to simultaneously maintain all the favourable solutions used in the 6-element ŁPSC frames. This gave birth to the concept of a 7-element frame. In addition, the moving (changeable) overlap on the connection of side wall arch with the straight floor element was expanded in the 7-element frame. This overlap can be matched individually within the range of 500–900 mm, depending on the actual thickness of the deposit and the size of break made by the mining machine while driving the gallery.

The ŁPSC type of 6-element frame, which was designed earlier, was additionally expanded by three sizes (11, 12 and 13) of the 7-element door in variant S.

This frame was introduced for use in a longwall gallery of wall 7/VII in deposit 385/2 (size 12) in November 2009, and slightly later in a top gallery of wall 7/VII/385 (size 11). In December 2010, longwall galleries began to be driven in that support system for walls 1/VIII/385 and 2/VIII/385 in Stefanów.

For plough walls, the concerned frames in longwall galleries are most often built with spacing of 0.9 m and additional bolting of the rock mass with the use of usually 7-8 bolts in every inter-frame space.

The example of the ŁPSC Bogdanka frame, size 12/S, in 6-element and 7-element version, is presented in Figure 6.

This support system is also presented in the work [10] on longwall galleries for plough walls.

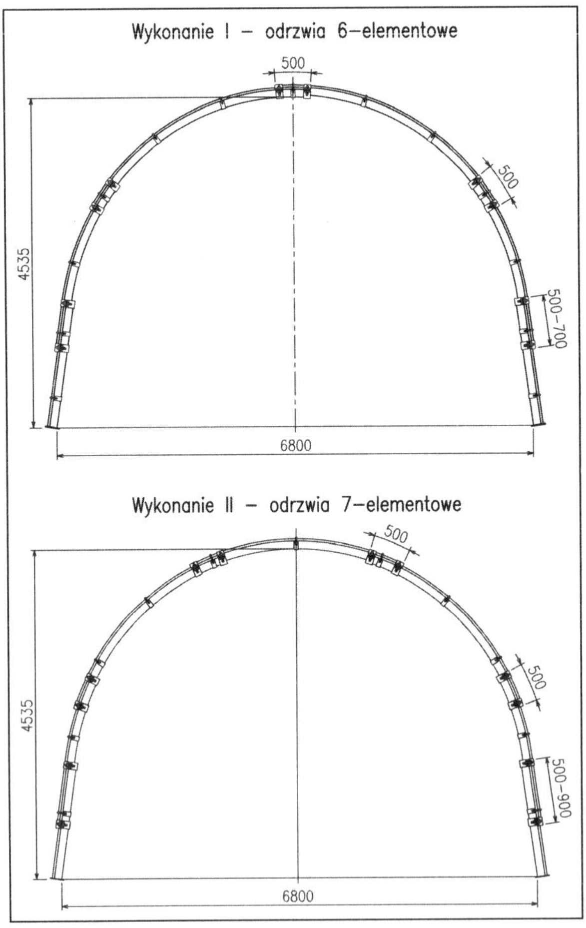

Figure 6. ŁPSC Bogdanka support system frame, size 12

Notations: Wykonanie I – odrzwia 6-elementowe – *Variant I – 6-element frame;* Wykonanie II – odrzwia 7-elementowe – *Variant II – 7-element frame*

6 EVOLUTION OF THE METHODS OF REINFORCING LONGWALL GALLERY SUPPORT SYSTEMS FOR THE PURPOSE OF MAINTAINING CROSSING WITH THE WALL

The standard way to maintain the crossing of a longwall gallery with a longwall in the 80s and at the beginning of the 90s was the support system (horseheads and props of various types and lengths). The method of unsupported leading of crossing with the longwall by means of bolting system in Bogdanka began to be used in 1993, when in the longwall gallery of wall 3/I in field I of deposit 382 in Bogdanka, an experimental section was made with *bolting the gallery support roof-bar*. The tests were successful and since that time, the method of unsupported bolt-based maintenance of wall-gallery crossing was gradually but continuously implemented. At first, this method was implemented only in the galleries which were made in the coal body and using two pairs of bolts per each frame as a mandatory standard. Later, the range of application was expanded also to comprise galleries in the gobs. In favourable geological and mining conditions, only one pair of bolts was used for bolting. In fact, already since the middle of the 90s, bolting roof-bars of the frame system for the purpose of unsupported maintenance of the crossing with the wall became a standard solution, employed in all longwall galleries.

In the over 15-year-long period for which the method of unsupported maintenance of galleries at the crossing with wall has been used in Bogdanka mine, it is possible to distinguish two variants.

a) employing the use of low bolting with steel bolts through yoke by means of:
   - one pair of bolts in case there are favourable geological and mining conditions,
   - two pairs of bolts in case there are difficult geological and mining conditions,

b) employing a horsehead type bolt embracing two adjacent frames and high bolting by means of:
   - one row of horseheads in the gallery axis, with bolting in every second inter-frame field, in case there are favourable geological conditions,
   - two rows of horseheads, which are symmetrical against the gallery axis, with alternating bolting in every inter-frame field, in case of difficult geological and mining conditions.

In Bogdanka mine, it is a standard procedure to bolt roof-bars with one pair of steel bolts and, presently, one row of horseheads with string bolt built up every second frame. Versions with two pairs of bolts or two rows of horseheads with strong bolts are used only in more difficult conditions.

The currently used method of unsupported maintenance of galleries at crossing with the longwall with the use of single and double bolt horsehead and high bolting is presented in Figure 7.

Over the last several years, Bogdanka mine has changed the way of bolting the support system frame for the purpose of maintaining the wall-gallery crossing. Before, for more than 10 years, the method used as standard involved bolting by means of a yoke and a pair of rod bolts, usually 2.7 m long (low bolting). Presently, high bolting with the use of string bolts is used as standard. Implementation of high bolting for the purpose of the maintenance of gallery and longwall crossing enabled a noticeable reduction of the roof convergence within the crossing area (the gallery roof is maintained about 20 cm higher than in case when low bolting is used).

In cases where there are difficult or very difficult geological and mining conditions, it is necessary to apply additional methods of reinforcing the gallery's support within the area of crossings with the wall, namely:

- reinforcing the roof with additional steel bolts or reinforcing the roof and the floor with workable bolts from the side of the wall and steel bolts from the opposite side,
- einforcing the roof with additional long string bolts,
- einforcing the support system by building up horseheads made of V section bar or rails fixed to frames with hook bolts, or by building up bolt horseheads,
- supporting the frames with the Valent or SV racks, or wooden racks,
- building up elements of a wooden or steel polygonal support system,
- reinforcing the rock mass by using adhesive binding agents in the crossing area, within the distance not shorter than 20 m ahead of the wall face,
- applying other solutions when required.

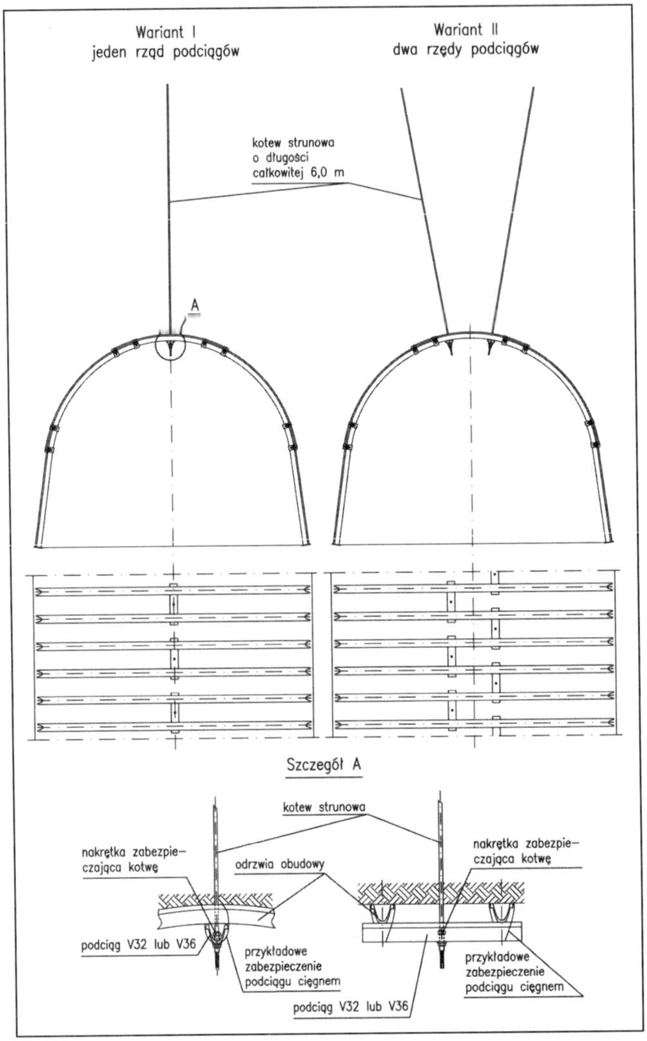

Figure 7. Typical reinforcement of the support system with the use of high bolting

Notations: Wariant I – jeden rząd podciągów – *Variant I – one row of horseheads;* Wariant II – dwa rzędy podciągów – *Variant II – two rows of horseheads;* kotew strunowa o długości całkowitej 6,0 m – *string bolt of the total length of 6,0 m;* Szczegół A – *Detail A;* kotew strunowa – *string bolt;* nakrętka zabezpieczające kotwę – *bolt securing nut;* odrzwia obudowy – *support system frame;* podciąg V32 lub V36 – *V32 or V36 horsehead;* przykładowe zabezpieczenie podciągu cięgnem – *exemplary protection of horsehead with cord*

7 TEST MAINTENANCE OF LONGWALL GALLERY OF WALL 1/VI BEHIND LONGWALL FACE

In 2010, after almost a 20-year-long break, an attempt was undertaken to maintain a longwall gallery behind the longwall face in the longwall no. 1/VI in seam 385/2 in Nadrybie, for the purpose of using it as tailgate for the next wall (no. 2/VI). This trial comprised the longwall gallery along the entire panel length, i.e. along the distance of 1.750 m. The results obtained from that trial are the subject of a separate article [6] and will not be discussed here.

8 SUMMARY AND FINAL CONCLUSIONS

1. In the almost 30-year-long period of hard coal longwall mining in Bogdanka mine, it is possible to distinguish two major periods in which longwall galleries were made and maintained in extremely dissimilar ways.
   a) initial period I, which lasted from starting the exploration wall in November 1982 until the beginning of the 90s of the previous century, during which mining was performed by means of the "advance" system, with longwall galleries driven along with the wall mining progress and maintained behind the wall,
   b) period II is when the "retreat" system was used, with longwall galleries completed before starting the wall and liquidated along the wall mining progress, which continues to be used.
2. The change from the "advance" system to the "retreat system" provided the capability of reaching high wall progress rates, high concentration of mining and efficiency of work in the conditions of Lubelskie Zagłębie Węglowe.
3. The presently used system of driving and maintaining longwall galleries involves making both galleries in advance, entirely completed before starting of the longwall mining process, and then liquidating both longwall galleries for each longwall successively along its progress. Two new galleries are made for every other wall.
4. In general, the currently used mining system does not leave the longwall galleries behind the wall front and does not maintain them for next walls, due to high operational pressures, intense convergence of the gallery's support system and floor upheaval. The exception was the trial performed in 2010, which aimed at maintaining behind wall the longwall gallery of wall 1/VI in deposit 385 to be used as top gallery for wall 2/VI/385 in this field.
5. In the conditions of Bogdanka mine, longwall galleries are arch-shaped and have a frame system specially made of V36 section bar (also partly of V32), steel grade S480W, with the most often used frame spacing of 0.75–0.9 m.
6. An element which plays an important role in the maintenance of longwall galleries, in various stages of the maintenance of galleries, is the bolting support system, which is used both as low bolting with the use of steel bar bolts (most often 2.7 m long), and high bolting with string bolts (usually 6.0 m long).
7. The cross section of longwall workings has been changed several times. At first, in 1983, it was size 7, 4.2 m wide, later 8/10, then 9, and the present standard sizes are size 10/K for shearer walls and 11/S and 12/S in the case of galleries for plough walls.
8. The attempt to introduce the elliptical shape of frames in longwall galleries brought negative results. In conditions of Bogdanka mine, under prevailing side wall pressures, longwall galleries underwent increasingly intense convergence as it has been the case with arch frames so far.
9. Until 2000, the longwall gallery support system was always closed with frame floor elements directly behind the mining machine during the driving process. Presently, in the heading, it is built as open and kept open, whereas it is closed only in selected cases, with delay and after dinting.
10. For more than 15 years, a method unsupported maintenance of wall-gallery crossing has been used, first with roof-bars bolted with pairs of 2.7 m long steel bolts through yokes, and in the recent years, by means of bolt horsehead made of V32/36 section bar with the use of string bolts whose total length is 6.0 m.
11. The new conditions with regard to driving and maintaining longwall galleries result from the plough technique, which is being introduced in Bogdanka mine, and from the extended panel length in Stefanów field (over 5 km).

REFERENCES

Chmielewski J., Kozek B. 2009. Poprawa postępów drążenia wyrobisk przygotowawczych w świetle doświadczeń kopalni LW „Bogdanka" S.A. [*Improvement of the preparatory working tunnelling progress in the light of experiences of the mine of LW Bogdanka S.A.*] Materiały Szkoła Eksploatacji Podziemnej, Szczyrk. (In Polish)

Chmielewski J., Kozek B., Masiakiewicz M. 2006. Drążenie chodników w LW „Bogdanka" S.A. [*Driving roadways in LW Bogdanka S.A.*] Materiały Szkoła Eksploatacji Podziemnej, Szczyrk. (In Polish)

Chmielewski J., Kozek B., Stopa Z., Masiakiewicz M. 2006: Doświadczenia Lubelskiego Węgla „Bogdanka" S. A. w zakresie drążenia wyrobisk korytarzowych [*Experiences of Lubelski Węgiel Bogdanka S. A. in driving roadways*] Przegląd Górniczy No. 10. (In Polish)

Kozek B. Ruchel A. 2011. Wykonywanie i utrzymywanie chodników przyścianowych w kopalni Lubelski Węgiel „Bogdanka" S.A. – dotychczasowe doświadczenia [*Driving and maintaining longwall galleries in the mine of Lubelski Węgiel Bogdanka S.A. – Experience to date*], School of Underground Mining 2011, conference materials. (In Polish)

Kwiatkowski K. 2011. Wykonywanie i utrzymywanie chodników przyścianowych dla ścian strugowych w kopalni Lubelski Węgiel „Bogdanka" S.A. [*The making and maintenance of longwall galleries for plough walls in the mine of Lubelski Węgiel Bogdanka S.A.*] IMF Materials at Bogdanka. (In Polish)

Stopa Z., Masiakiewicz M., Kozek B. 2010. Doświadczenia Lubelskiego Węgla "Bogdanka" S. A. w zakresie drążenia obudowy kotwiowej. Perspektywy stosowania obudowy kotwiowej w polskich kopalniach węgla kamiennego [*Experiences of Lubelski Węgiel Bogdanka S. A. in using bolt support system. Prospects of using bolt support system in Polish hard coal mines*] Prace naukowe GIG Górnictwo i Środowisko. Quarterly edition No. 2/1/2010. (In Polish)

Stopa Z. 2011. Technika strugowa w LW Bogdanka S.A. – stan i perspektywy rozwoju. [*The plow technique in LW Bogdanka S.A. – present state and prospects of development*] IMF materials at Bogdanka. (In Polish)

International Mining Forum 2011, Dyczko et al. (eds) © 2011 Taylor & Francis Group, London, UK. ISBN 978-0-415-68938-0

Maintaining Longwall Headings in the Mines of JSW S.A.

A. Tor, R. Dobrzyński, K. Koliński
Jastrzębska Spółka Węglowa S.A., Jastrzębie-Zdrój, Poland

ABSTRACT: The article presents the problem of stability of longwall headings behind the exploitation face and the reason for their maintaining. Furthermore, exemplary methods of maintaining headings behind longwall exploitation face in the mines of JSW S.A. were shown depending on geological and mining conditions and the role of the heading.

KEYWORDS: hard coal mining, JSW S.A. mines, longwall headings, stability, maintaining

1 INTRODUCTION

Coal exploitation using longwall system with fall of the roof in deep mines causes significant pressures put on the mining excavations. This state is reinforced if difficult mining and geological conditions occur and frequently there is a need to drive two new longwall headings for every new longwall.

Therefore maintaining longwall headings behind the exploitation face is one of the most important problems in underground mining practice.

Every mining excavation changes the original state of equilibrium due to the removal of rock mass. This causes the increased tendency of rock movement towards the excavation and consequently increased pressure put on the support. The increased load on the support is intensified by mining, which causes destruction of rocks and additional deforming pressures. The increased vertical pressures are transferred to the side-walls of the excavation causing convergence and elevation of the floor rocks. Elevation of the floor towards the center of the headings is a serious difficulty in maintaining of these excavations and ensuring their functionality for transport and ventilation and leads to numerous complications and hindrances in the mine operation.

Together with the increase of depth these adverse phenomena intensify and are accompanied by serious damage to the heading support, resulting in fragile cracking, plastic bending of the arches of the heading support and destruction of clamps between the elements of the support. This leads to a decrease in cross-section area of the excavation and in extreme cases even to its total tightening.

In longwall headings located in the exploitation fields the total pressure on the support is a sum of the pressure resulting from depth and the exploitation pressure. The support of these excavations is a subject of significant pressures which usually result in its deformation.

To restore the original functions of the excavations, reconstruction works such as remodeling of the headings and dinting of the floor are necessary. These works disturb the normal underground production processes, influence the total efficiency, increase production costs, apply additional load of gangue on horizontal and vertical transport.

Negative experiences with maintaining the headings in one-sided surrounding of the gobs forced numerous mines to resign from these practices, especially if the deposits lay deep.

2 MAINTAINING LONGWALL HEADINGS IN ONE-SIDED SURROUNDING OF THE GOBS

2.1 The purpose of maintaining longwall headings behind the longwall exploitation face

Maintaining longwall headings behind the longwall exploitation face results from the following reasons:

1. Safety reasons:
   - The necessity to ventilate the longwalls using "Y" system due to the existing methane hazard,
   - Pure exploitation of particular part of the deposit without leaving natural protective belts in which adverse concentrations of pressures occur and without the possibility of emergence of endogenous fires which is advantageous in the case of multi-seam exploitation of the deposit with occurrence of natural hazards like crumping, methane, rock and methane breakouts.
2. Organizational and economic reasons – decrease in the number of driven headings. Maintaining the heading for the next longwall allows to decrease the numer of driven headings which allows to decrease the costs related with their functioning and to optimize the conducted heading driving in the aspect of possessed technical and workforce potential.

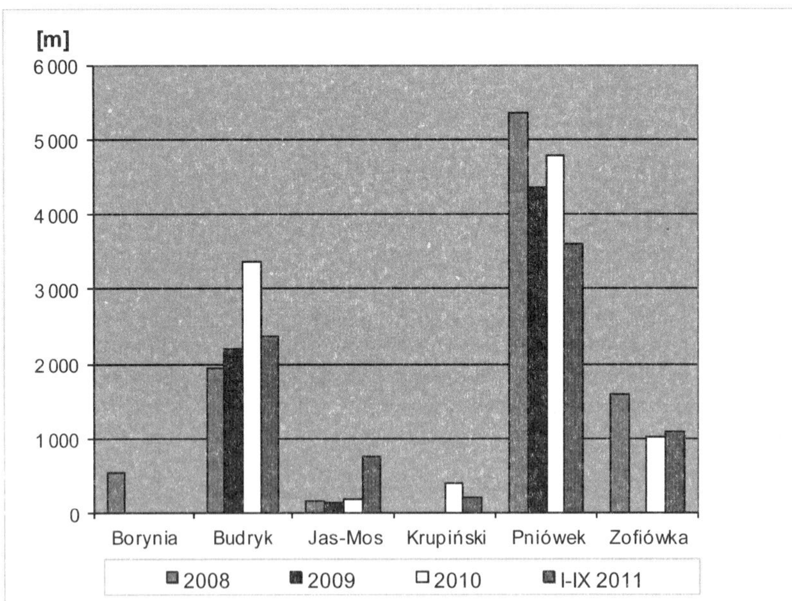

Figure 1. The length [in meters] of longwall headings maintained in JSW S.A. mines between 2008 and 2011

The actions undertaken towards the maintaining of longwall headings behind the exploitation face in the mines owned by JSW S.A. should result in:

- Decrease in longwall heading convergence,
- Decrease in the deformations of heading support,
- Efficient isolation of gobs,
- Decrease in methane and fire hazards,
- Maintaining the dimensions of the headings required by the mining law regulations and by the function of the excavation,
- Increase in work safety.

During the period of 2008–2010 and in three quarters of 2011 over 34 thousand meters of longwall headings were maintained using different methods. The chart (Figure 1) shows the length of the maintained headings in particular mines.

Different numbers result from the specific geological and mining conditions existing in particular mines of JSW S.A., especially concerning tectonics and inclination of seams, the necessity to ventilate the longwalls using "Y" system and the capability to design the development of headings to take the re-use of the excavations into account.

2.2 Methods and manners of protecting longwall headings behind the longwall exploitation face in the mines of JSW S.A.

Every mine operates using its own unique experience of the staff, takes the existing mining and geological conditions into account together with the size and intensity of natural hazards. Exchange of experiences between mines and new technologies introduced at JSW S.A. allow to partially standardize the methods and techniques of exploitation. This situation is also applied to maintenance of headings behind the longwall exploitation face. Individual solutions developed in particular mines were modified and adapted to the conditions and situations existing in other plants. This allows to use similar methods, similar materials, machines and devices as well as to draw current conclusions and estimate the effectiveness of the applied solutions.

The method of maintaining the heading behind the exploitation face depends on:

1. The function of the heading:
   - Maintaining the heading strictly for ventilation or for ventilation and transport,
   - The possibility of use the heading for the next longwall,
2. Existing mining and geological conditions and related methods of reinforcing and maintaining the stability of heading support.

2.2.1 Maintaining the excavation using frictional props on foot pieces

Until 2008 the longwall headings in "Budryk" Coal Mine were maintained behind the longwall exploitation face through the construction of so-called gates which contained two frictional props built under reinforcing horseheads or directly under the arches of the heading support. The props were built under every set of the support and placed on a steel foot piece. These reinforcements (gates) were constructed up to 20 m behind the fall of roof area which was associated with the necessity of dinting the floor behind the longwall face. Furthermore, to protect the heading a supporting and isolating belt was constructed along the heading from the gobs' side, about 0.5 m from the heading support. For this purpose following elements were constructed, depending on the current needs:

- Wooden horsehead supported with wooden props with minimal thickness of 0.18 m and 0.5 m spacing,
- Single wooden horsehead supported with SV29 props with 0.75 m spacing,
- Single horsehead from V29 shape supported with SV29 props with 0.75 m spacing.

In a separated area between the side-wall arches of the support and the built-in props profiled Big-Bag sacks made of filling linen were put. The fixed and attached sacks were systematically filled with mineral binding agent.

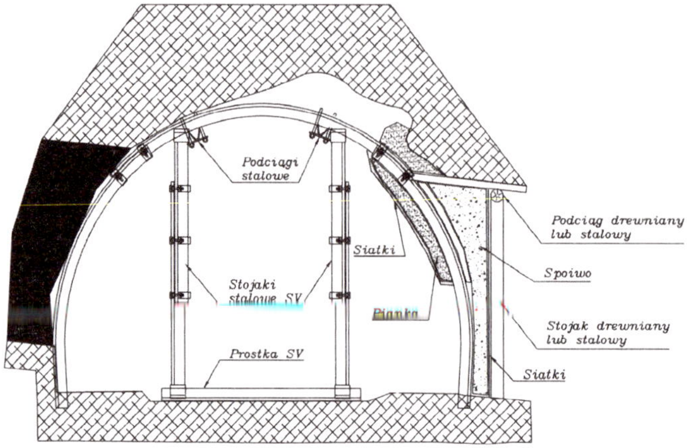

Figure 2. Maintaining the excavation behind the longwall face through the construction of so-called "gates"

Notations: Podciągi stalowe – *Steel horseheads;* Stojaki stalowe SV – *SV steel props;* Prostka SV – *SV straight element;* Siatki – *Nets (lagging);* Pianka – *Foam;* Podciąg drewniany lub stalowy – *Wooden or steel horsehead;* Stojak drewniany lub stalowy – *Wooden or steel prop;* Spoiwo – *Binding agent*

2.2.2 Maintaining the heading using filling belts

The use of filling belts meant to decrease the bending of roof rock layers and therefore cause even distribution of load put on the support of the mining excavations. Between the last section of the support and the heading two wooden horseheads supported with wooden props are built. Special containers (sacks) are attached to the props and filled with binding agent which (after it hardens) becomes a support for the bending layers. Additionally, a steel horsehead is built in the maintained heading. It is attached to every set of the support and then supported with wooden or steel props. An example of such construction is shown in Figure 3.

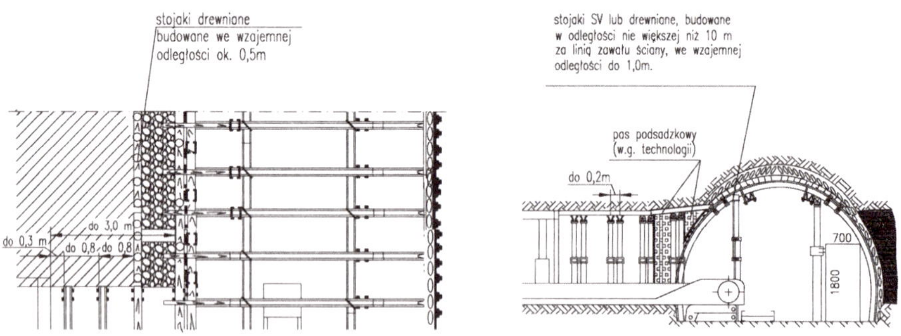

Figure 3. Maintaining the heading behind the longwall face through the use of protective supporting belt from the binding agent with the use of containers (sacks)

Notations: Stojaki drewniane budowane we wzajemnej odległości ok. 0,5 m – *Wooden props built with 0.5 m spacing;* Stojaki SV lub drewniane, budowane w odległości nie większej niż 10 m za linią zawału ściany, we wzajemnej odległości do 1,0 m – *SV or wooden props built not further than 10 m behind the fall of roof area, with spacing up to 1.0 m;* Pas podsadzkowy (w.g. technologii) – *Filling belt (according to the technology);* Do 0,2 m – *Up to 0.2 m*

2.2.3 Maintaining the excavation with the use of wooden cribs

The use of cribs, similarly to the use of filling belts, should decrease the bending of roof layers and in result cause more even distribution of pressure on the heading support. There are many known constructions of cribs. The four-point crib is a typical construction. Each layer consists of two bars and four points of contact between those bars. In this solution whole vertical pressure of the roof is transferred from one layer to the others through the support points, namely the area of contact between the bars where they cross. Whereas between the contact points the bars of the crib do not have any supporting function. This kind of construction is unstable, the bars can move relatively to one another and cause the loss of stability of the crib. Point pressure put on the nodes of the crib may also lead to a premature destruction of wood and cause significant tightening of the crib.

New solution, presently commonly used in JSW S.A. mines is to use wooden elements with notches which allow permanent overlapping of the combined elements and guarantee sticking of the contact areas along the whole length of the bars (Figure 4). This allows to transfer the vertical load through the whole length of every bar in the crib, not only in the points where the bars cross. This solution makes the crib more stable and durable.

Figure 4. Supporting crib (pile) with square basis made from wooden bars connected using quoin system

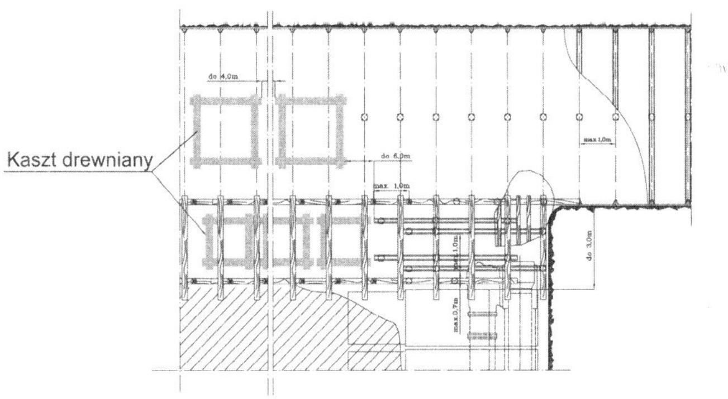

Figure 5. Maintaining the excavation behind the longwall exploitation face with the use of wooden cribs

Notations: Kaszt drewniany – *Wooden crib;* Do – *Up to*

Wooden cribs made of repeatable elements with notches for their connecting are usually square-shaped at the basis.

The cross-section of such crib may have a shape of a triangle, a rectangle or a rhombus.

An exemplary method of maintaining the heading behind the longwall exploitation face is shown in Figure 5. Cribs are built next to each other progressively behind the longwall face advance next to the last set of the mechanized support, between the double wooden horseheads built on wooden props. Additionally, in the axis of the maintained heading another crib may be easily constructed. Depending on the present needs, the cribs may be filled with a binding agent which additionally reinforces its support capacity.

The excavation behind the longwall exploitation face is isolated from the gobs through the use of ventilation linen or foil and insulation with chemical foams.

Alternatively, due to existing mining and geological conditions and the function of the excavation, the suport behind the face instead of cribs built in the excavation may be reinforced with one of two steel horseheads conncected to every set of the support (Figure 6).

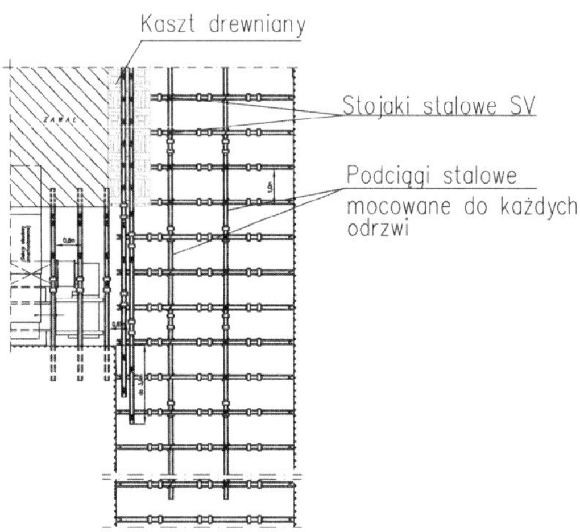

Figure 6. Maintaining the excavation behind the longwall exploitation face with the use of wooden cribs and steel horseheads

Notations: Kaszt drewniany – *Wooden crib;* Stojaki stalowe SV – *SV steel props;* Podciągi stalowe mocowane do każdych odrzwi – *Steel horseheads connected to every set of the support;* Zawał – *Fall of roof*

2.2.4 Maintaining the excavation with the use of wooden cribs and bolting

Exploitation conducted at still greater depths requires more advanced methods of maintaining the excavations behind the longwall exploitation face. Difficult mining and geological conditions cause that the aforementioned methods are not effective enough to maintain the excavation's dimensions required by law regulations and its function. Due to this fact bolting of roof rocks was introduced. Bolting is realized mainly by connecting steel horseheads with the roof using string bolts with length adapted to current conditions (bolts were from 4 to 8 meters long). This kind of bolting allows to connect the roof rocks and simultaneously stabilizes the heading support. This kind of solution is associated with building cribs in the area of longwall-heading crossing as described in chapter 2.2.3. An exemplary solution is shown in Figure 7. Depending on the needs, one or two steel horseheads are built and bolted. To avoid stratification of the roof rocks, the bolting is conducted behind the exploitation face, before the excavation goes into the area of exploitation pressure. This advance has to be at least 60 m. However in still more cases bolting of the heading is performed before the exploitation of the longwall. The target is to perform the bolting directly behind the face advance.

The use of string bolting allows to transfer the rock mass pressure deeper towards the side-wall of the excavation which results in lower convergence of roof and floor of the heading behind the longwall face. The adaptation of the heading for re-use therefore requires less work and resources.

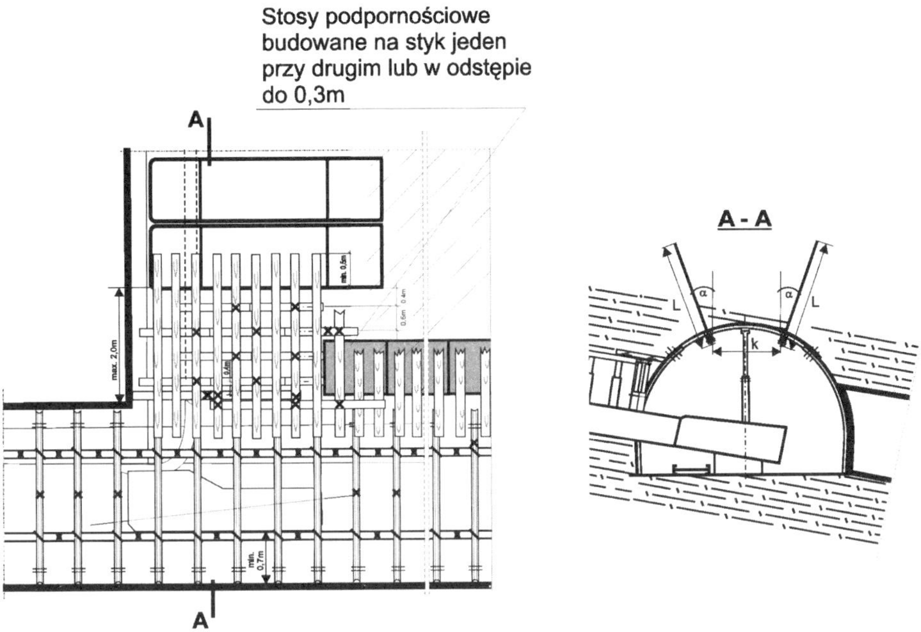

Figure 7. Maintaining the excavation behind the longwall face with the use of wooden cribs and horseheads bolted with string bolts

Notations: Stosy podpornościowe budowane na styk jeden przy drugim w odstępie 0,3 m – *Supporting cribs built next to one another with 0.3 m spacing*

Figure 8. The examples of maintaining excavations behind the face in JSW S.A. mines

Figure 9. The examples of maintaining excavations behind the face in JSW S.A. mines

3 SUMMARY

JSW S.A. mines undertake actions towards maintaining excavations behind the longwall exploitation face to re-use them during the exploitation of the next longwall as well as for ventilation reasons. Existing difficult geological and mining conditions and significant depth of exploitation result in the need to search for still better solutions in this subject.

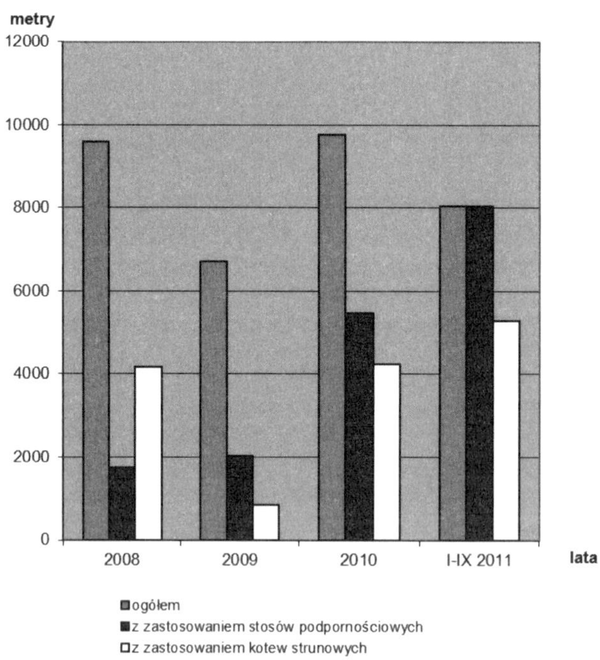

Figure 10. The length of headings maintained behind the longwall face in the years 2008–2011

Notations: Metry – *Meters;* Ogółem – *Total;* Z zastosowaniem stosów odpornościowych – *With the use of supporting piles (cribs);* Z zastosowaniem kotew strunowych – *With the use of string bolts*

At the moment cribs (supportive piles) are commonly used to maintain the headings. Cribs are made from prefabricated elements with notches what makes the work of the crib more stable and significantly increases its supporting capacity through the transfer of pressure along the whole length of the elements. Short construction time is another advantage together with the possibility to easily fill the crib with binding agents to additionally increase its supporting capacity.

To increase the capability of maintaining the headings behind the exploitation face, decrease the convergence and elevation of the floor, bolting of the roof was introduced. Bolting is realized mainly through bolting of the steel horseheads with string bolts up to 8 m long.

The length of headings maintained behind the longwall face with the use of cribs (supporting piles) and the length of headings maintained behind the longwall face with the use of string bolts is shown in the chart below.

The analysis of methods of maintaining headings behind longwall face used in JSW S.A. mines in the period of last two years allows to draw following conclusions:

1. The supporting crib made of wooden bars connected using the quoin system creates a rigid construction with high supporting capacity and the repeatability of the wooden elements allows to construction to be fast and easy
2. Additional capacity increase of a wooden crib made from bars connected using the quoin system may be achieved through filling the internal space of the crib with mineral-cement binding agent.
3. The use of supporting belts along the side-walls of the longwall headings from the gobs' side together with bolting is justified both technically and economically
4. The effectiveness of bolting together with the construction of supporting belt depends on numerous factors: organization of works, bolting spacing, advance of bolting in relation to the exploitation face, the method of constructing the supporting belt, namely the carefulness of works, well-chosen material for its filling, precision of filling the separated area, the distance between the belt and the longwall exploitation face. Width and durability parameters of the binding agent are important factors affecting the behavior of the belt.
5. Effective maintenance of the longwall headings behind the exploitation face allows to limit the length of necessary preparatory headings.
6. To maximally use the effect of bolting it is necessary to conduct the bolting simultaneously with the driving of the excavation, near the face.

REFERENCES

JSW S.A. own materials.

Straś J.R. 2007. Ochrona chodników przyścianowych w jednostronnym otoczeniu zrobami. IV Warsztaty – Popioły lotne i spoiwa mineralne w technologiach górniczych, Wisła. (In Polish)

Tor A., Dobrzyński R., Straś R., 2009. Doświadczenia w zakresie nowoczesnych technologii drążenia wyrobisk korytarzowych w kopalniach JSW S.A. – Szkoła Geomechaniki. (In Polish)

International Mining Forum 2011, Dyczko et al. (eds) © 2011 Taylor & Francis Group, London, UK. ISBN 978-0-415-68938-0

Modern Methods of Designing Supports of Excavations for Plow Longwalls

S. Prusek, M. Rotkegel
Główny Instytut Górnictwa, Katowice, Poland

ABSTRACT: Effective extraction of hard coal with the use of plow technique results in the fact that many issues within mine technologies need to be solved. One of the most important is performing a proper design of the support of excavations and its reinforcement. Generally, it means that the support may be adjusted to secure the T-junction so that the extraction process is effective and safe. In this paper the most important phases of designing the support of dog headings, by use of the method invented in the GIG, have been described. There are also examples of the supports of plow longwall gateroads, lately driven in mines "Zofiówka" and "Bogdanka".

KEYWORDS: mining, plow technique, frame support, designing

1 INTRODUCTION

A properly designed support of working should ensure safety for the crew and guarantee that an appropriate size of cross-section is maintained, regarding ventilation aspects or the necessity of transporting workers and materials and keeping the required movement space. One of the essential parameters which is taken into account while designing the support of excavations, is the rock mass loading. Its value may oscillate largely which is triggered off by characteristics of the mining-geological conditions in which a given excavation is located. In case of the Polish mines of hard coal a big impact on the value of loading which affects the supports of working have the following properties: extraction depth, the influence of residual pillars, mining edges or tremors of rock mass. It is very important to be familiar with the real parameters of rock mass in the surroundings of a working while performing calculations. For determination of these parameters there are underground researches performed which employ advanced measuring devices, allowing to indicate strength parameters of rocks or the degree of rock mass fracturing around the excavation. Loading on the support, in the given conditions of extraction, may be calculated by use of analytical-empirical or numerical methods. Given the demands of users with respect to the calculated loading value, the process of designing the working support can be started. The design process also includes post researches of the support and extended numerical analyses which aim at the optimization of the construction (Prusek et al. 2011b).

All of the above-mentioned issues are parts of the complex method of designing the support of working, which has been developed at GIG. This method is used for designing supports in hard coal mines in Poland. It was also used to design the support of plow longwall gateroads driven in mines "Zofiówka" and "Bogdanka". The main components of this method as well as the applied tools have been presented in Figure 1.

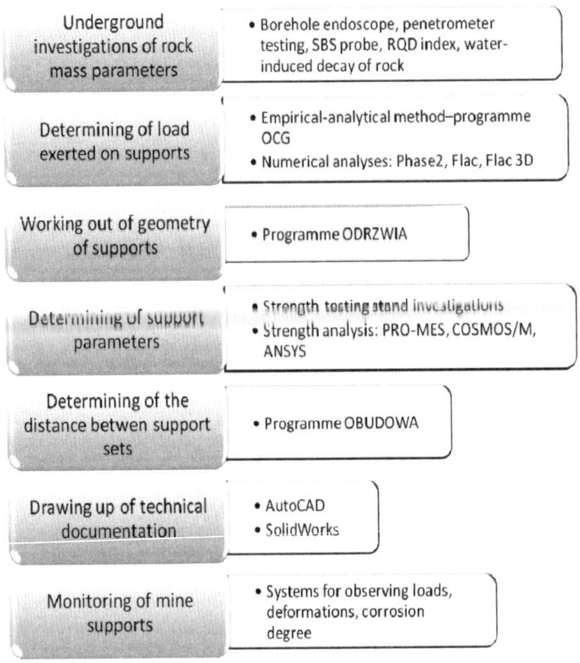

Figure 1. Main components of the complex method of designing gateroads supports developed at GIG (Prusek et al. 2011b)

2 MAIN PHASES OF THE PROCESS OF GATEROAD SUPPORT DESIGN FOR PLOW TECHNIQUE

The design of the support for plow longwalls gateroads is generally similar to the algorithms employed in case of other working supports. The process is slightly extended within the design of frame geometry, determination of the support parameters, including reinforcements and frame spacing. The phases will be described in the paper in detail.

2.1 *Research on rock mass parameters around dog headings*

The first important phase of the design method is to calculate the values of loading, which is supposed to affect the progressively designed support during the whole utilization period. Appropriate loadings allow to design an optimal support for the particular extraction conditions. For this phase, it is necessary to perform investigations on rock mass properties. The results are the input data for the calculations of the support loading both by use of analytical-empirical and numerical methods. The investigations are performed by use of a borehole penetrometer, endoscope and SBS probe. Additionally, rock soaking and RQD index are indicated.

Penetrometric investigation make the basic method of determination of rock strength parameters around the excavation. The researches are performed in boreholes made from the roadway which transect rock strata under examination. During the researches the critical penetration pressure p_m is determined which is then converted to the compressive R_c and tensile R_r strength of rocks. A resistance profile of rocks surrounding the excavation is the result of the investigations. Alternatively, for identification of the rock mass fracturing intensity around a working there are often investigations performed by use of a borehole endoscope. The investigations allow to register the view of the borehole walls. The picture is then analysed and, as a result, the rock mass fissures and fractures lay-out is obtained in the length of the tested borehole. Thereby, the reach of fracture zone can be assumed. These results are particularly useful during the design of roof bolting, especially by determining the length of anchor. Moreover, they can be used for the revision of numerical calculations of the fracture zone reach. Figure 2 presents a borehole endoscope designed at GIG as well as an example of fissure locations in the length of the borehole and a picture of one of the fissures.

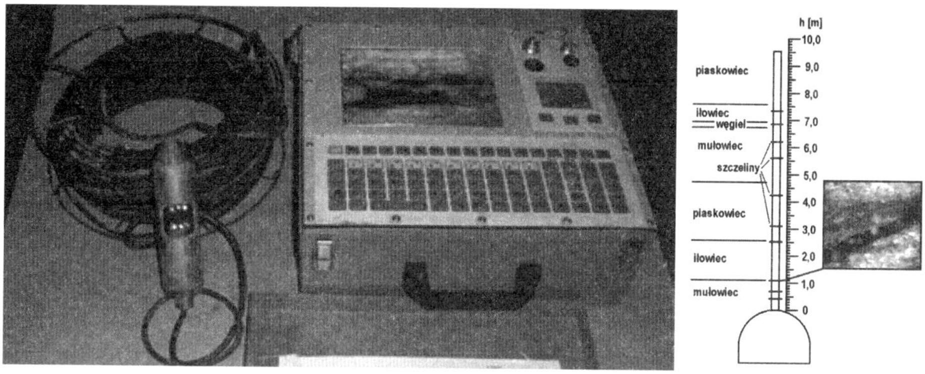

Figure 2. Borehole endoscope and an example of the investigations result of fracture zone reach around the excavation

Notations: piaskowiec – *sandstone;* iłowiec – *siltstone;* węgiel – *coal;* szczeliny – *fissures;* mułowiec – *mudstone*

An alternative for the endoscopic method is the SBS probe which allows to perform investigations in the boreholes of between 28 and 32 mm in diameter. It allows to obtain a picture of the borehole wall development along with pointed fractures and fissures, their orientation and opening.

In the researches, included in GIG methodology, on rock mass properties, it is very important to indicate the value of RQD index and water-induced decay of rock. The parameters are considered in the calculations of the support loading by rock mass performed according to the methodology.

2.2 Calculations of support loading by rock mass

Values of the expected support loading by rock mass are usually calculated by means of analytical-empirical methods. In more complicated cases numerical methods apply.

Most often, for calculations of the expected support loading by rock mass, the original program, designed at GIG, OCG is used. It is based on 'Simplified principles of support selection...' (Rułka et al. 2001). Necessary data that should be loaded in are the following: size of the support, depth of excavation localization, physical-mechanical parameters of the surrounding rocks (Rc, RQD, r), residual pillars, mining edges and tectonic dislocation. As the result of the calculations the support loading value is obtained. Figure 3 presents the interface of OCG program.

The calculations of support loading by rock mass are also performed by use of numerical models such as: Finite Difference Method (FDM) and Finite Element Method (FEM), with the help of FLAC and Phase2 software. In these programs, it is possible to determine, among others, the reach of rock mass fracture zone, its plasticization, displacements and the stresses. The rock mass analysis in FLAC and Phase2 software is performed in a planar system and allows to model the rock mass to the form of a two-dimensional disk of unit thickness. The model allows for rock strata and excavation outlines. An example of the rock mass numerical analysis results has been presented in Figure 4.

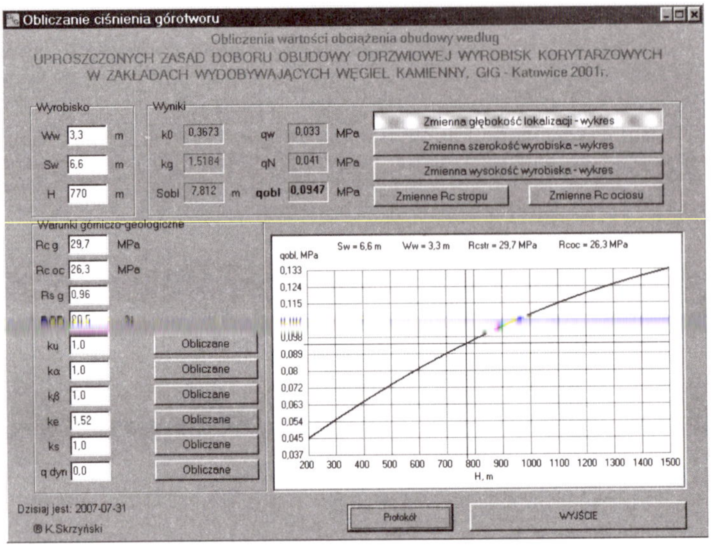

Figure 3. Interface of OCG program for calculations of the support loading

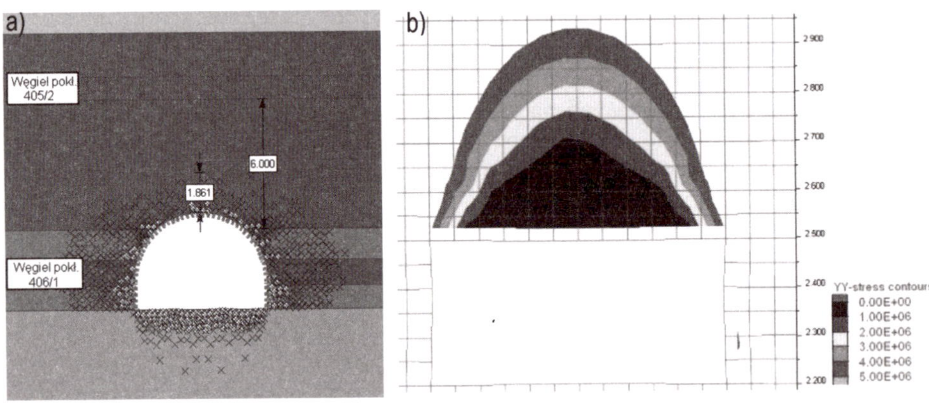

Figure 4. Numerical calculation results: a) Phase2 program – fracture zone around excavation; b) FLAC program – rock mass stresses in the roof of rectangular-shaped cross-section excavation

2.3 Design of the gateroad support

In case of using the plow technique, it is very important to design the size of excavation properly, given the overall dimensions of drives or the necessity of bringing the plow head in to the gateroad. Size and shape of the support may ensure regular operation of the plow complex during the whole extraction process. It is also important to prepare and secure the gateroads for future extraction by simplifying the construction of T-junction to the maximum and eliminating workings performed in this area during the course of longwall. It is caused, among others, due to that much less time is allowed to perform workings in the area of T-junction. The reason is the high velocity of mining which causes that the plow head which operates undisturbed is in the area of T-junction every 3÷4 minutes. In this case, in order to keep the necessary safety level and comfort of the working crews, it is necessary to reduce the workings performed in gateroads adjacent to the longwall intakes to a minimum (Prusek et al. 2011a, Rotkegel et al. 2008, Kozek et al. 2011). Therefore, it is necessary to treat the design of support set geometry of gateroads and the use of the optimal method of reinforcements of the support in the area of intersection T-junction separately. In case of designing supports of gateroads for plow longwalls, it is also important to adjust the lengths of particular arches so that their joints on the side of the longwall were located slightly above the roof of the extracted coal seam. If the above-mentioned requirements are met, it is no longer necessary to shorten the roof-bar arches or perform the time-consuming and dangerous process of securing the roof in the area of T-junction bared after disassembly a sidewall arch. A concept of this construction is presented in Figure 5.

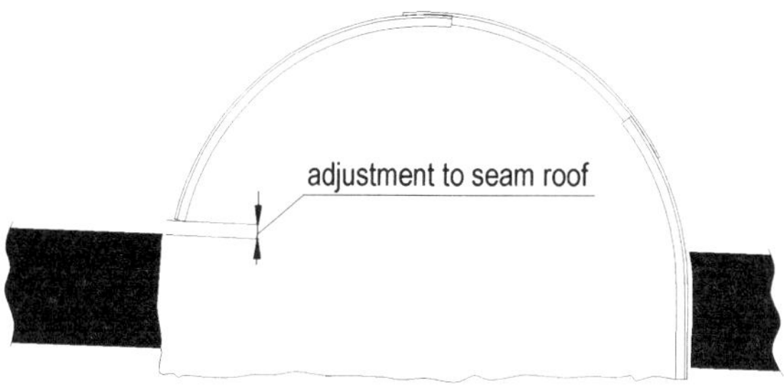

Figure 5. Concept of a gateroad support adjusted to the seam thickness

During the process of designing the frame, of which the roof-bar arch tip is above the roof of a given seam, it is necessary to take the following parameters into account: longwall height, slope of the longwall, the size of sill measured in between the floor and the bottom. It allows to state that the support is individually adjusted to the geological-mining conditions as well as indicate the extraction technology – 'made-to-measure support'.

An individual approach to the frame geometry is possible thanks to advanced computer aided designing software. An example of such software is "ODRZWIA" program (Rotkegel 2009a, Rotkegel 2004) whose interface has been presented in Figure 6.

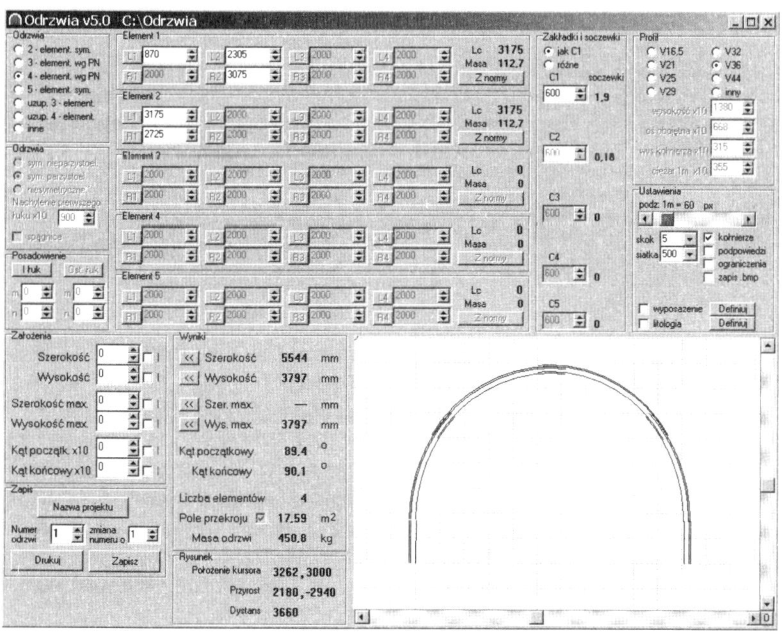

Figure 6. Interface of the "ODRZWIA" program designed at GIG

In this program, the user is supposed to determine the geometrical parameters of the particular components of the frame – arches and laps which may be connected to the sets. The results are presented in the form of a printout including among others: width and height of the designed set, inclination angle of the sidewall arch, section area of the set as well as the gross mass of the set. The program also allows to generate pictures in 'dxf' format, which can be read and edited be means of majority CAD software. This method of notation allows to prepare drawing documentation quickly as well as constitutes a geometric base for numerical calculations.

2.4 Post researches, resistance analyses and support selection

During the design of new set construction it is important to perform post researches. Thereby, it is possible to check the regularity of set operation and determine its load-bearing capacity parameters. The indexes of the set load-bearing capacity can be determined during the post researches according to the standard PN-92/G-15000/05. The sets are examined in two states: rigid and yield. The results of the first case allow to obtain deformational characteristics of the rigid frame and their maximal load-bearing capacity. Alternatively, in the second case it is possible to obtain working characteristics and assess the operation of the frame in the conditions close to reality. Moreover, during the research the deformation of the frame as well as yield in the frictional joints are being registered.

The determination of load-bearing capacity parameters of more frames (i.e. for a complete type series) usually takes place by means of numerical calculations using the Finite Element Method (FEM). In order to perform the analyses the following software is used: PRO-MES, COSMOS/M and ANSYS. In these programs, in order to construct geometrical models, it is possible to create files in "ODRZWIA" program or in other CAD software. These programs allow to perform strength calculations of a single support set – in planar or the support segment system – in non-planar system. Figure 7 presents a part of the support model, location of particular elements in this model and their sectional parameters (Figure 7b) (Prusek et al. 2006c, Prusek et al. 2006, Prusek et al. 2008b).

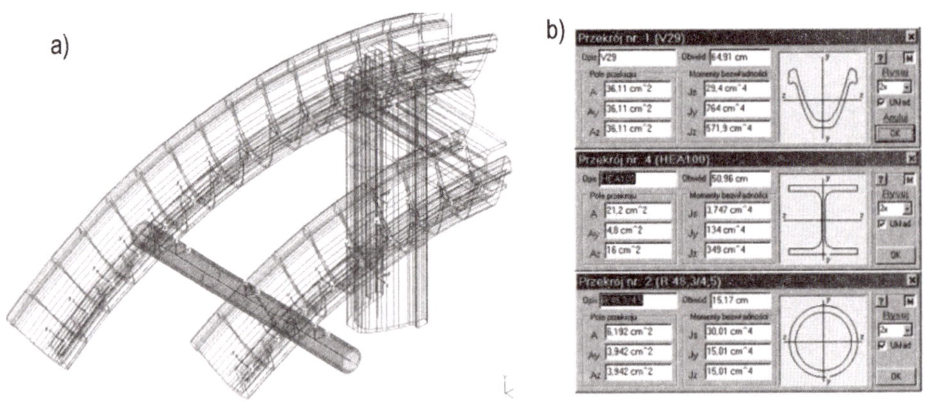

Figure 7. Fragment of a gateroad support model: a) location of the elements in the model; b) sectional parameters of particular elements

From the user point of view modeling relies on inserting the geometry of the whole system and determining parameters of its particular elements, such as: material properties, sectional parameters and, in case of non-linear analysis, material curves. Disposing a comprehensive base of results of post and underground investigations, it is possible to calibrate the tested models very precisely.

In basic cases the support selection comes down to calculation of the support set. Nevertheless, more and more often, it is necessary to consider in additional analyses the reinforcements like props, anchors and stringers, especially in case of the support analysis in the area of T-junction.

An example of the numerical analysis result is the distribution of the reduced stress indicated according to Huber – Mises – Hencky theory for the support in the area T-junction, reinforced by means of steel stringer has been presented in Figure 8.

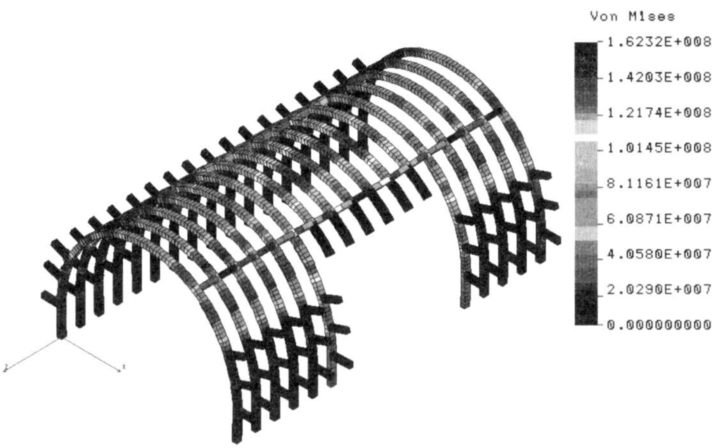

Figure 8. Example of calculation results of intake support to longwall in the program COSMOS/M – reduced stress according to Huber – Mises – Hencky theory (stresses [Pa], deformation scale x10)

2.5 Monitoring of the support operation

The methods of construction design are verified on the basis of underground investigations on loading and deformation of the support (Prusek 2006a). In the last few years in the GIG there was work performed to construct a system that allows to continuous monitoring of the gateroad support. In addition, many sensors and devices allowing to measure the loading and deformation of steel arching and roof bolting have been invented. The works of the monitoring system were performed within the "MONSUPPORT" project (Prusek 2008a) which was funded by Research Fund for Coal and Steel (Contract RFCR-CT-2005-00002).

3 EXAMPLES OF GATEROAD SUPPORT FOR PLOW LONGWALL

The set ŁPZof which is adjusted to the conditions of plow longwalls in "Zofiowka" mine as well as the ŁPSC appropriate for the conditions in "Bogdanka", are the examples of supports designed basing on the methodology described in point 2.

3.1 ŁPZof support

The set of support ŁPZof has been designed particularly to secure the gateroads of plow longwalls in area F, seam 405/1 and area B, F, seam 406/1 in KWK "Zofiówka". The construction involves both seam thickness in the mentioned areas and longwall inclination. The support may serve for securing gateroads of longwalls of between 1.0 and 1.6 m. in height and with inclination of 0° to 10°. The set has been adjusted to the parameters of longwall by applying a roof-bar arch on the side of extraction and increased laps, which allows to build the arch in using various methods in different configurations of the laps. ŁPZof-type set is asymmetrical with arc profile features. It consists of two identical sidewall arches and two roof-bar arches with different lengths. The curvatures of all these arches are identical. In the heading face the set is built in with a longer roof-bar arch on the side of the longwall. From this side, the roof-bar arch is 'mobile' and its location in the frame is determined during the phase of driving and securing the excavation. Thanks to the possibility of adjusting the laps (sidewall arch – roof-bar arch and roof-bar arch – roof-bar arch), it is possible to match the location of the lower tip of the roof-bar arch to the level of roof of the extracted seam. Size of the laps depends on thickness and inclination of the seam.

The ŁPZof-type set has been used the first time to secure the gateroads of plow longwall B-1 in seam 406/1 in mine "Zofiówka". As a support, the tailgate and bottom road contained a frame of size 12 (width 6100 and height 4225 mm) made by use of V32 shapes from steel of increased mechanical parameters. The set gauge in both excavations came up to 0.8 m (Prusek et al. 2006b, Rotkegel et al. 2009b). Figure 9 presents an outline of ŁPZof support set.

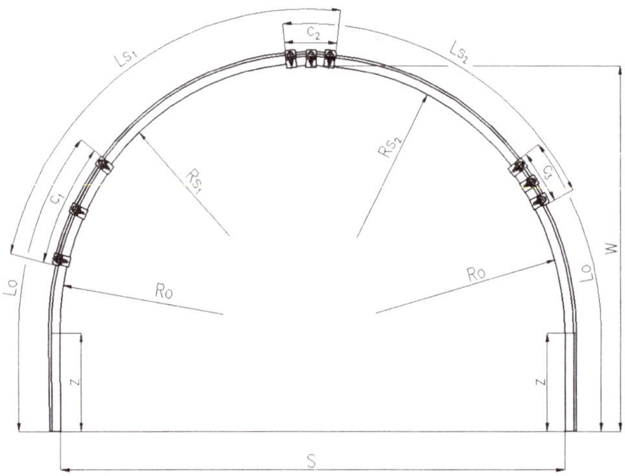

Figure 9. The set of ŁPZof support

The ŁPZof support of gateroads in longwall B-1 has been reinforced by use of an additional roof bolting. In order to reinforce the support, the roof-bar arches have been anchored by use of steel anchors of 2.5 m. in length. Each support set has been anchored by a clamp with pair of bolts. Moreover, in order to ensure stability of the support in the area of T-junction, where bending of sidewall arches on 8 m. length was planned, the roof-bar arches have been anchored by use of a stranded rockbolt of 4.5 m. in length with the help of a steel stringer from the shape V29. The stringer has been anchored between each set of the gateroad support. Additionally, on the side opposite to longwall the set was being stabilized by use of two steel stringers from V29 shape. The outline of the support of the T-junction has been presented in Figure 10.

The presented above outline of the support of T-junction longwall-maingate has been applied after performing series of calculations and analyses (Prusek et al. 2011a, Prusek et al. 2006b, Rotkegel et al. 2009b). The assumed initial anchor lengths has been compared to the calculation results of the rock mass facture zone reach around the gateroads. Reach of this zone has been calculated regarding the influence of longwall B-1 by use of Phase 6.0 software which is based on the Finite Element Method (FEM). The results have been also used to indicate the value of loading affecting the support. This loading has been taken into account in the next phase – strength analysis of a support driven by use of the program PRO-MES. As a result of the modeling researches, the distribution of reduced stresses in the model has been determined. Comparing the maximal values of stresses with the acceptable values for the particular components of the support, it was possible to verify its construction as for resistance positively.

The presented outline in figure 10. was successfully applied in practice with positive experiences.

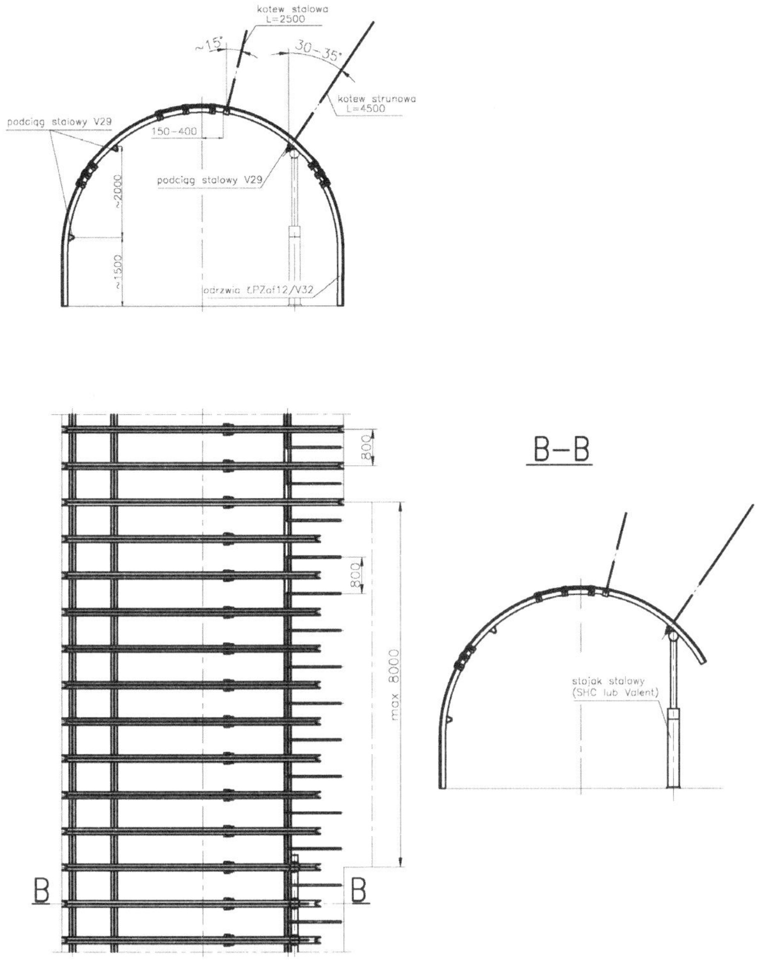

Figure 10. Outline of the support of T-junction longwall B-1 with maingate

Notations: Podciąg stalowy – *Steel horsehead;* Odrzwia – *Set;* Kotew stalowa – *Steel bolt;* Kotew strunowa – *String bolt;* Stojak stalowy – *Steel prop*

3.2 ŁPSC support

Between 2008 and 2009 in "Bogdanka" mine there was work done on a new series of types of the support for gateroads. The results provided a support set marked as ŁPSC, of which the type series have included both the current requirements for shearer longwalls and the designed plow longwalls in this mine.

The support ŁPSC sets are performed of six- or seven- elements with arc-shaped cross-section. They are also adjusted to drive and maintain the T-junction. The frame have been provided to building in the excavations as the open one with the possibility of closing the floor after a horizontal convergence and the floor taking. Similarly to ŁPZof support, the set construction allows to obtain advantageous location of roof-bar arch tip towards the roof of seam being extracted so that it was not necessary, after disassembly the sidewall arches, to shorten the roof-bar arch (Kwiatkowski & Leś 2010, Rotkegel et al. 2008). The outline of a six-element set of ŁPSC support is presented in Figure 11.

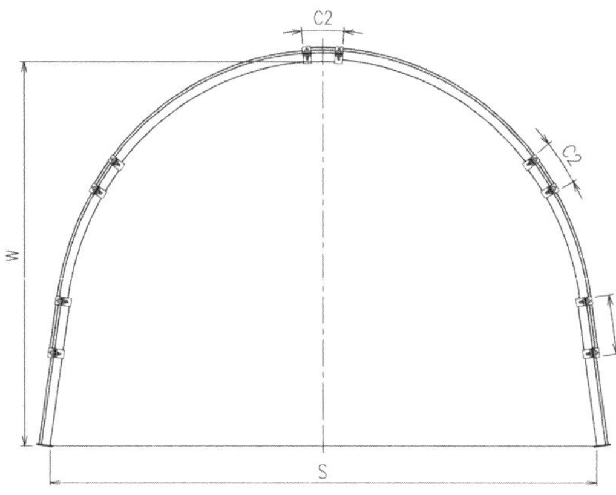

Figure 11. Six-element set of ŁPSC support

In the set of ŁPSC support there have been applied short under most side wall elements (in-floor) at the length approximately equal to the value of the predicted floor heave, increased by the size of a lap. Thereby, after longwall transition and floor heave, it is possible to cease the building of these elements by designing an asymmetrical support behind the T-junction. It eliminates the floor ripping for its rebuilding. Laps of the set elements mostly have a constant of 500 mm. The exception is the connection of sidewall elements (straight connections) with upper sidewall arches. These laps (marked as C1 in Figure 11) can be changed separately between 500 and 700 mm, depending on the real seam thickness and the size of breakout driven by a read-header during the process of drifting the gateroad.

The series of types includes 6 sizes of frames from which 3 are intended for gateroads of shearer longwalls (design version "K" – sizes 9, 10 and 11) and 3 for gateroads of plow longwalls (design version "S" – sizes 11, 12 and 13). The ŁPSC-type support has been applied for the first time in gateroads driven for the purposes of the first plow longwall no. 1/VI in seam 385/2 in "Bogdanka" mine.

In "Bogdanka" mine a roof bolting was additionally applied in gateroads. In the mainroad a ŁPSC 12/S support in the spacing of set every 0.9 m was applied. Between frames, 6 or 7 steel anchors of the length up to 2.7 m have been applied in turns. They were built in the face area. The tailgate was driven in ŁPSC 11/S support with set spacing of 0.9 m. Between the sets, 3, 4 steel anchors of the length up to 2.7 m have been applied in turns. They were also built in the face area. In both gateroads the sets were made from V36 shape of S480W steel. The sets were anchored during driving the gateroads (Rak et al. 2010).

In order to ensure the stability of the T- junction with maingate there was additionally applied a set anchoring by use of two rows of anchored stringers with stranded rockbolt 6.0 m in length which are arranged in turns every two sets as well as anchoring the roof-bar arches of the set on the longwall side with a couple of steel anchors 2.7 m in length thank to the clamp from J profile, which allows to disassembly the sidewall elements earlier given the work of machine loader in the T-junction itself (Rak et al. 2010).

Moreover, two stringers have been applied; along the right and left sidewall about 0.7 m. above the seam roof. Alternatively, in order to ensure the required load-bearing capacity and stability in the area of the T-junction with tailgate there was applied a set anchoring by use of two rows of anchored stringers with string anchors 6.0 m. in length which are arranged in turns every two sets as well as anchoring the roof-bar arches of the set on the longwall side with a couple of steel anchors 2.7 m in length thanks to the clamp from J profile. In this case one steel stringer on the side of the longwall has been applied and assembled about 0.7 m above the seam roof.

It should be mentioned that the mainroad of longwall 1/VI was maintained inby the longwall along its entire length. Thereby, a special protection pack has been designed as the venture was experimental in the conditions of "Bogdanka" mine. The tailgate was gradually removed along with wall advance, according to the current technology. Figure 12 presents an outline of the support of T-junction of longwall 1/VI/385 with the mainroad which includes the above-mentioned reinforcements.

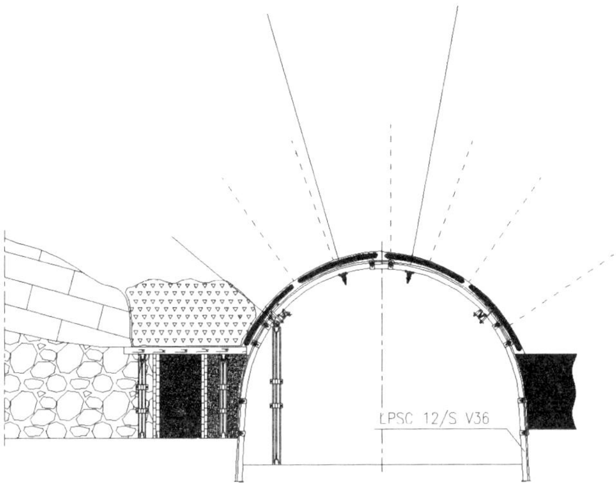

Figure 12. T-junction of longwall 1/VI/385 and the mainroad

4 SUMMARY

The presented methodology of designing supports, developed and successfully applied in the GIG, has been an important tool for the designer of gateroad supports as well as its components, mostly frames. By extending some phases, it has been successfully applied in designing gateroad support with plow technique. Because each case is treated separately, it is possible to design a support optimal for the particular geological-mining and technological conditions. This fact can be proved by looking at the type series of ŁPZof and ŁPSC support sets, which performed well during driving the plow longwalls, ensuring appropriate sizes necessary for the devices of plow complexes and functionality within T juncion. According to this methodology, it is possible to perform the design process effectively by use of advanced tools in the form of modern CAD systems and professional software for simulating phenomena which take place in the rock mass exposed to the influence of extraction pressure.

REFERENCES

Kozek B., Prusek S., Rotkegel M., Zabój K. 2011. Obudowa wyrobisk przyścianowych dla ścian strugowych – wymogi konstrukcyjne oraz doświadczenia praktyczne. Szkoła Eksploatacji Podziemnej. Kraków. (In Polish)

Kwiatkowski K., Leś A. 2010. Wdrożenie pierwszej zautomatyzowanej ściany strugowej w Lubelskim Węglu „Bogdanka" S.A. XII Konferencja Zarządu Głównego SITG „Górnictwo wczoraj i dziś". Ustroń. (In Polish)

Prusek S. 2006a. Monitoring chodnika przyścianowego oraz podporności obudowy. Przegląd Górniczy nr 2. Katowice. (In Polish)

Prusek et al. 2006b. Projekt obudowy wyrobisk chodnikowych dla możliwości efektywnego zastosowania techniki strugowej w KWK „Zofiówka". Dokumentacja pracy GIG nr 411 56205-150. Katowice (unpublished). (In Polish)

Prusek S., Rotkegel M., Skrzyński K. 2006c. Komputerowe wspomaganie projektowania obudowy wyrobisk korytarzowych i ich połączeń. Przegląd Górniczy nr 3. Katowice. (In Polish)

Prusek S. 2008a. Rozwinięty system monitoringu obudowy wyrobisk korytarzowych. Prace Naukowe GIG nr 3. Katowice. (In Polish)

Prusek S., Rotkegel M., Tor A. 2008b. Przebieg kompleksowego procesu projektowania nowej konstrukcji obudowy wyrobisk korytarzowych. Szkoła Eksploatacji Podziemnej. Kraków. (In Polish)

Prusek S., Rotkegel M., Zabój K. 2011a. Obudowa chodnikowa dla efektywnej eksploatacji pokładów węgla kamiennego techniką strugową. Wiadomości Górnicze nr 3. Katowice. (In Polish)

Prusek S., Rotkegel M., Witek M. 2011b. Comprehensive method for designing roadway support. World Mining Congress. Turkey.

Rak Z. et al. 2010. Projekt wzmocnienia istniejącej obudowy i sposobu utrzymania chodnika podścianowego ściany 1/VI w pokładzie 385/2 w Bogdance na odcinku około 10 m przed skrzyżowaniem ze ścianą, uwzględniającego uwarunkowania technologiczne ściany strugowej. Praca Fundacji Nauka i Tradycje Górnicze z siedzibą przy Wydziale Górnictwa i Geoinżynierii AGH. Kraków (unpublished). (In Polish)

Rotkegel M. 2004. Komputerowo wspomagane projektowanie odrzwi obudowy chodnikowej. Charakterystyka programu CAD i przykłady projektowe. Prace Naukowe GIG nr 862. Katowice. (In Polish)

Rotkegel M. et al. 2008. Projekt odrzwi obudowy chodnikowej dla efektywnego prowadzenia eksploatacji techniką strugową w warunkach LW „Bogdanka”. Dokumentacja pracy GIG nr 581 27718-151. Katowice (unpublished). (In Polish)

Rotkegel M. 2009a. Program komputerowy do wspomagania konstruowania odrzwi stalowej obudowy wyrobisk korytarzowych. Wiadomości Górnicze nr 12. Katowice. (In Polish)

Rotkegel M. et al. 2009b. Wykonanie projektu obudowy podporowo-kotwiowej skrzyżowania wyrobisk korytarzowych ze ścianą strugową B-1 pokład 406/1 w KWK „Zofiówka”. Dokumentacja pracy GIG nr 581 25209-151. Katowice (unpublished). (In Polish)

Rułka K. et al. 2001. Uproszczone zasady doboru obudowy odrzwiowej wyrobisk korytarzowych w zakładach wydobywających węgiel kamienny. Główny Instytut Górnictwa, Seria Instrukcje, Nr 15. Katowice. (In Polish)

International Mining Forum 2011, Dyczko et al. (eds) © 2011 Taylor & Francis Group, London, UK. ISBN 978-0-415-68938-0

The Practicalities of Optimizing Automated Longwall Plough Systems under the Conditions Prevailing at LW "Bogdanka" S.A. Mine – Special Longwall Cavity Lining and Output Self-Loading System at the Longwall-Heading Crossing

M. Masiakiewicz
LW "Bogdanka", Puchaczów, Poland

A. Szyszka
"SIGMA" S.A., Barak, Poland

ABSTRACT: The decision to mine thin coal seams was a natural outcome of the rational and balanced approach LW "Bogdanka" S.A. mine has adopted towards exploiting coal deposits. These resolutions resulted in deploying the first automated plough longwall (manufactured by Bucyrus International). The plough had previously worked in "Nadrybie" field. After observing the system in operation and gaining experience with the selected technology, a team of experts from LW "Bogdanka" S.A. coal mine, along with representatives of "SIGMA" S.A. and "HAJDUK Group" Sp. z o.o., was set up to find ways of optimizing the automated plough longwall. The result is a highly innovative solution, viz. a Special Longwall Cavity Lining and an Output Self-Loading System.

KEYWORDS: Hard coal mining, thin seam, automated plough longwall

1 PREFACE

The constantly growing demand for energy requires a new approach to all energy sources. The classification of energy sources as renewable and non-renewable indicates an increasing awareness and understanding of the opportunities and threats posed by natural resources.

With respect to the theme of the conference, attention needs to be focused on issues related to non renewable energy sources, e.g. fossil fuels, the exploitation of which leads, by definition, to their depletion. Two crucial questions therefore need to be asked: Can we afford to manage them in an unreasonable manner? Can we afford to squander them?

It is generally held that LW "Bogdanka" S.A. coal mine extracts coal in a rational, efficient and economically justifiable manner. This is the approach that led to the decision to start mining thin coal seams.

Mining a thin coal seam requires the implementation of a longwall system adapted to certain specific conditions. As mentioned above, the company selected the automated plough longwall manufactured by Bucyrus.

The efficiency and cost-effectiveness of coal plough technology depend on a number of factors, including:

- nominal equipment parameters,
- mutual compatibility of components,
- frequency of system failure,
- work organisation,
- work cycles,
- auxiliary devices required,

– technological down time, etc.

An analysis of how this system actually operated in a longwall (in „Nadrybie" field) revealed that it was feasible, and even necessary, to optimize the automated longwall coal plough system by making certain changes.

LW "Bogdanka" S.A. coal mine therefore invited "SIGMA" S.A. and "HAJDUK Group" Sp. z o.o. to draw up guidelines and implement changes to improve the operation of the longwall system. These changes were meant to extend the effective working time of the coal plough during a twenty-four-hour period, especially given the assumed production level of 20,000 Mg/day. The team focused on the following issues:

- upgrading works to make the longwall cavity secure,
- eliminating the need to use Load Haul Dumpers to load the output accumulating in the longwall gatcroads as a result of coal plough operation.

2 SPECIAL LINING OF LONGWALL CAVITY

A longwall cavity, understood as the space crated between an outermost longwall section and a heading lining, is an inconvenient and troublesome phenomenon. Longwall cavities are caused by longwall mining conditions, including:

- the top and tail gates not being parallel,
- the longwall axis not being parallel to the axes of the longwall gates,
- the system moving along the longwall in the direction of the gates, etc.

There exist certain known mechanisms of minimizing the width of the cavity by applying a hydraulically-adjustable width to the outermost sections and shifting its relative position along the axis of the longwall conveyor and/or by erecting an additional linear lining section. These actions, however, are very labour-intensive and not technically warranted given that the width of the cavity tends to increase and decrease along the longwall. Therefore, the necessity of supporting the roof with a specified force requires the introduction of additional technological operations.

During the operation of the coal-plough system in „Nadrybie" field longwall, the roof was secured with moulded profiles that worked with hydraulic props and timber components to both seal and prop up the roof. This method of securing the roof – although it is effective and meets the required load-bearing capacity – involved a great deal of expenditure in terms of labour and resources and was excessively time-consuming. That is because it required a timber structure to be constructed under the roof, and the supporting beams and hydraulic props had to be shifted manually as the longwall progressed.

The team therefore focused on preparing a conceptual design of a supporting device that would require minimum manual intervention and accelerate the rate of cavity lining work, while increasing crew safety.

The work of the team resulted in the design and constructing of the Propping-and-Supporting Device presented in Figure 1.

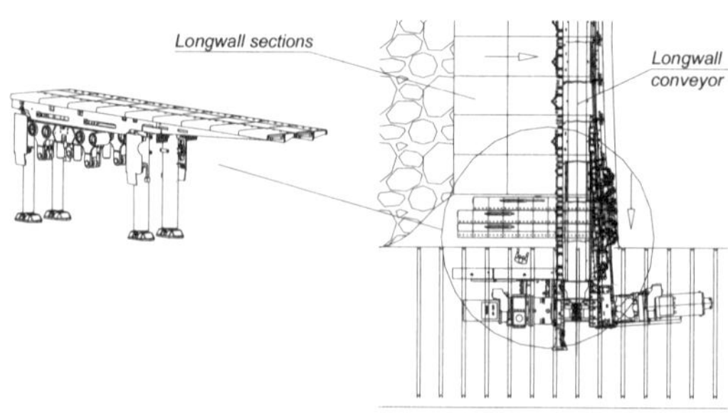

Figure 1. Diagram of Propping-and-Supporting Device to work with longwall system

The following characteristics of the device are especially worth mentioning:

- the support width can be adjusted by expanding the roof-bars,
- the construction is modular,
- a wide range of cavity heights can be secured,
- no workers are required in the support-mounting zone while the device is being shifted,
- particular roof-bars are shifted hydraulically as the longwall progresses.

As the support width can be adjusted by expanding the roof-bars (Figure 2), the crew can adjust the device – within a set range – to the changing surface of a secured roof. Prop extension work is therefore unnecessary. This functionality compensates for the variability of cavity width without significantly increasing the number of servicing activities.

The modular construction of the device (Figure 3) is an important feature as it allows the user to secure cavities of any width by a relatively easy expansion of the system. In connection with the possibility of adjusting the spacing of the roof beams as discussed above, the range of functional operation of the device is considerably increased. The allowable cavity width is the only constraint on the overall support width.

A wide range of secured cavity heights (Figure 4) can be obtained by the right choice of hydraulic props, the travel of which depends on the expected height of a particular longwall. This makes it possible to work in the cavities of the lowest longwalls, e.g. 1.0 m high, as well as – through all the intermediate values – in the cavities of the highest longwalls (3.0–4.0 m high).

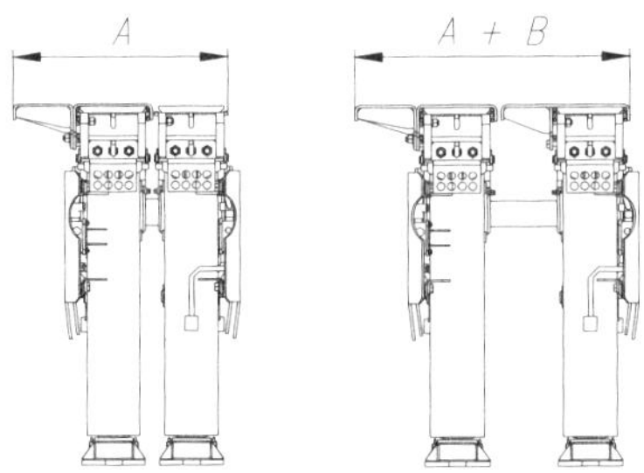

Figure 2. Functional features of the Propping-and-Supporting Device – support width adjustment

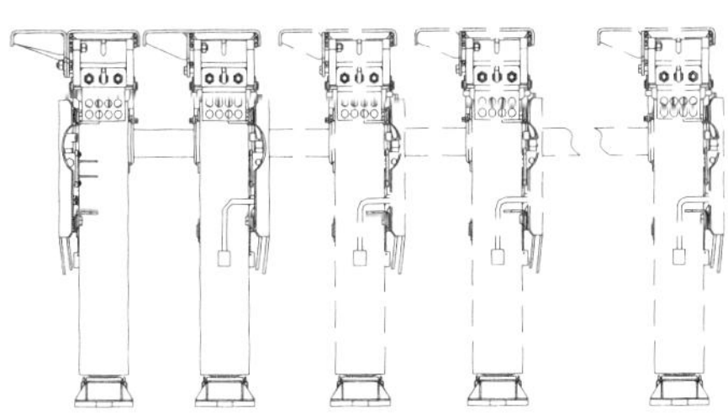

Figure 3. Functional features of the Propping-and-Supporting Device – modular structure

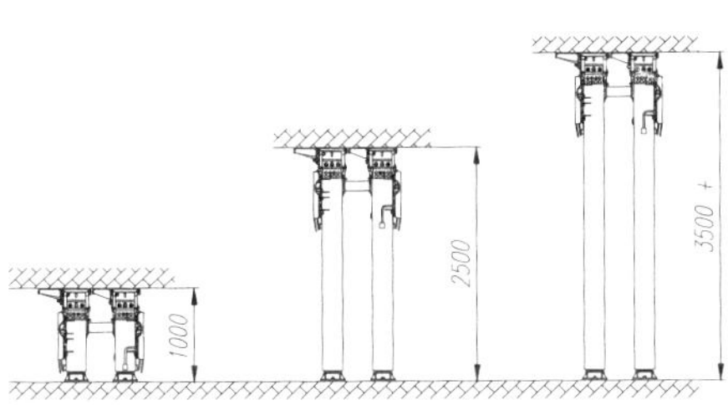

Figure 4. Functional features of the Propping-and-Supporting Device – range of secured cavity heights

Crew safety is considerably increased as a result of there being no workers needed in the support-mounting zone while the device is being shifted (Figure 5). When shifting the device in the direction of the mining, the operator is in a safe place while retaining appropriate visibility.

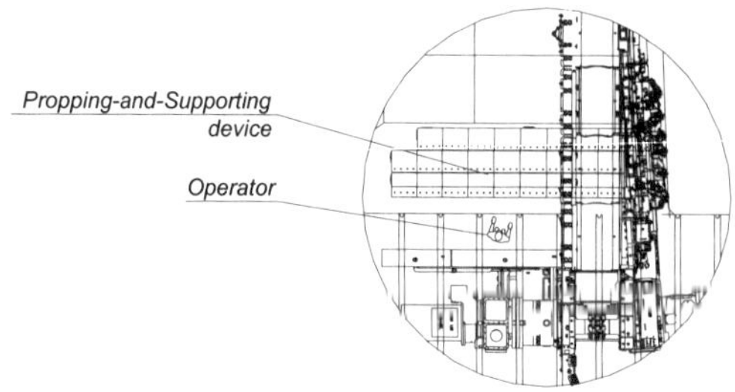

Figure 5. Functional features of the Propping-and-Supporting Device – safe work place for the operator

The hydraulic shifting of particular roof beams with the longwall meets at least three basic requirements: improved work safety (workers are no longer in a hazard zone); improved work ergonomics (physical strength is no longer required); and accelerated works performance.

The Propping-and-Supporting device shall be implemented in the longwall of LW "Bogdanka" mine in Stefanów field.

3 OUTPUT SELF-LOADING SYSTEM AT A LONGWALL-HEADING CROSSING

One of the features that distinguishes longwall mining from the coal-mining combine and coal-plough methods is that, in the first case, output loading on a longwall conveyor is realized by rotating the spiral unit (there is no need to shift a longwall coal-mining combine), while in the second case, output loading is possible only during a run of the coal plough head. As a result, a longwall coal-mining combine can "clear" headings with the cutting unit, while a coal plough pushes the output into headings when it reaches turning points.

The effect of littering longwall corners with output in the coal-ploughing system used in „Nadrybie" field was eliminated by means of two load haul dumpers that worked in the head gate and the tail gate, between the corresponding drives of a longwall conveyor and a roadside (Figure 6).

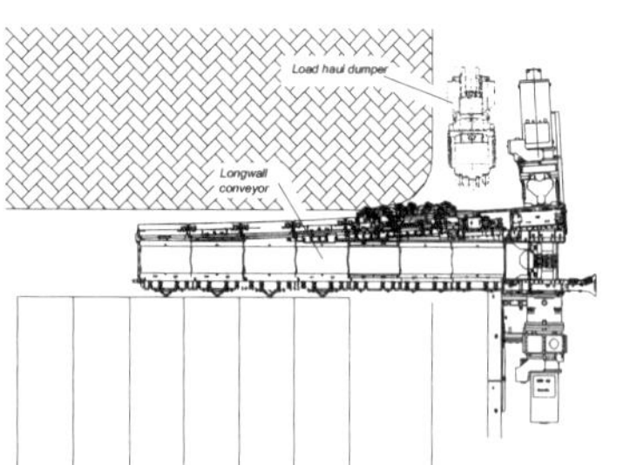

Figure 6. Current technology of clearing accumulated output from a longwall corner

The main purpose of the machines was to level (dint) the piles on the floor in the corners of the heading and in loading the accumulated output on a longwall conveyor.

The disadvantages of this solution include having to stop the operation of the coal plough once the load haul dumper starts to work, and having both machine and operator in a hazardous zone. Additionally, the remaining work – e.g. that related to dismantling the side wall arches before the longwall head – could only be done when these machines had finished their job. Therefore, it was impossible to perform certain technological operations simultaneously, which prolonged the process of carrying out the preparatory work for a plough head run. This reduced effective work time.

The task faced by the team (consisting of representatives of LW "Bogdanka" mine, "SIGMA" and "HAJDUK Group" companies) in preparing solutions aimed at rationalizing the technological activities discussed above required a meticulous analysis of coal plough operation, especially during the moments when the plough head reaches crossings.

The following observations were made during system operation:

- about 30% of the output that accumulated in the headings was transported together with the plough head, while the remainder accumulated as a result of the cracked longwall corner being torn away (Figure 7),
- variable distance between the corresponding drives of the longwall conveyor and the roadside (Figure 8) may, under certain circumstances, prevent the load haul dumpers from entering the longwall-heading crossing zone. As a result, output will not be cleared from the heading and the dismantling of the heading lining before the longwall head will be made more difficult.

Figure 7. Sources of output littering of a longwall crossing (left: a corner littered as a result of the plough head; right: littering due to scattering of the longwall corner)

In connection with the above, the following conclusions were formulated before commencing work on optimizing the system:

- a method of continuously clearing headings of output needed to be devised,
- cracked longwall corners needed to be prevented from being scattered into headings in order to minimize the amount of rocks transported into headings,
- the solution had to be able to handle the variable width of the space between the roadside and the structural elements of the longwall conveyor.

The following features of the system are especially worth mentioning:

- the coal plough separates the headings from the output transport space,
- the appropriately-profiled construction allows the output to be continuously loaded onto the longwall conveyor in the crossing zone,
- a part of a cracked corner can be supported and prevented from being scattered,
- the device can be adjusted to the variable width of the space between the roadside and the construction of the longwall system by hydraulically shifting the sealing plate,
- access to the servicing points of the longwall conveyor is ensured by a hydraulic mechanism that shifts the Self-Loading System.

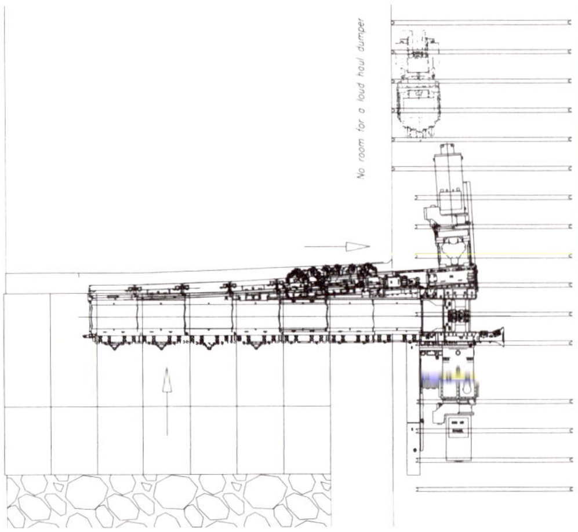

Figure 8. View of the space between the roadside and the structure of a longwall system

The result was the creation of the Output Self-Loading System, presented in Figure 9.

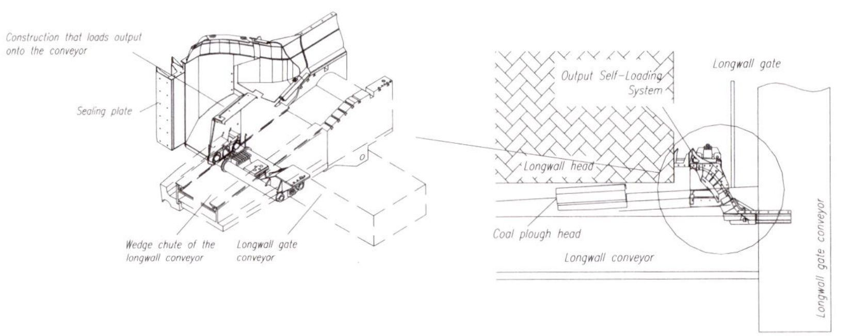

Figure 9. Output Self-Loading System

As headings are separated from the output transport zone (Fig. No. 10), it is practically impossible for the output from a longwall to accumulate in headings. As a consequence, load haul dumpers can be withdrawn from the crossing zone and used to level floors at about 50 m from the longwall head. This eliminates collisions during the dismantling of side wall arches and during coal plough downtime.

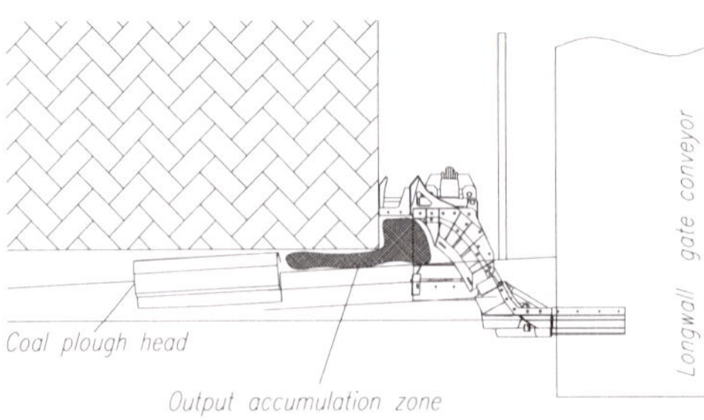

Figure 10. Functional features of the Output Self-Loading System – separation of headers from the output transport space

The appropriately profiled construction makes it possible to continuously load output onto the longwall conveyor in the crossing zone (Figure 11), when the coal plough reaches the turning point. Thus, the output does not accumulate anywhere, but it is removed in a timely manner by means of successive haulage devices.

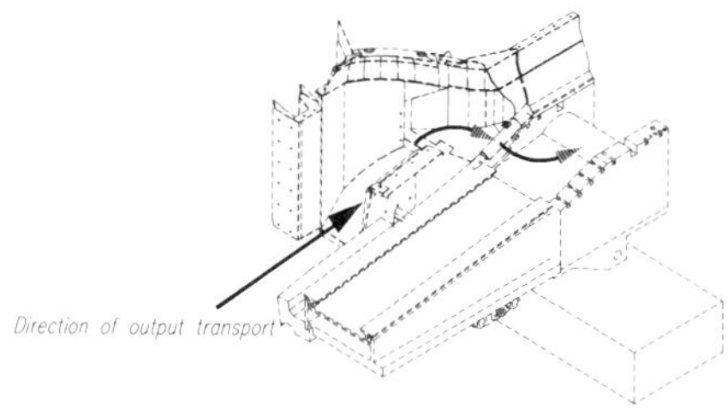

Figure 11. Functional features of the Output Self-Loading System – method of loading output on the longwall conveyor

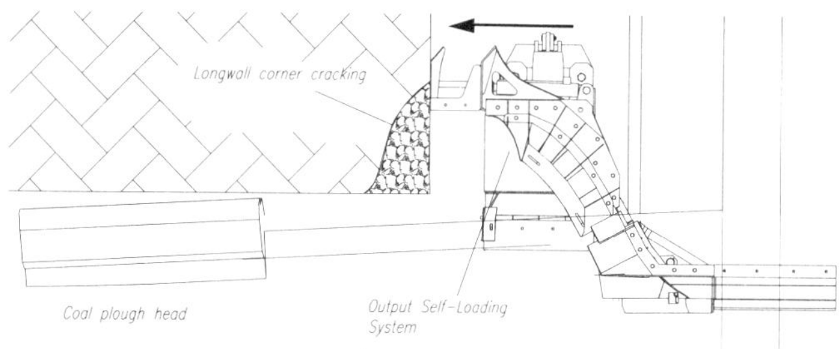

Figure 12. Functional features of the Output Self-Loading System – method of supporting cracked longwall corners

As it is possible to support a part of a cracked corner and prevent it from being scattered (Figure 12), a considerable part of the output that originally accumulated in headings is eliminated at the very source.

Adjustment (by hydraulic shifting of the sealing plate) to the variable width between the roadside and the construction of the longwall system (Figure 13) makes it possible to ensure the required tightness of the entire system. This additionally contributes to the realization of the fundamental goal of eliminating factors that slow down mining work.

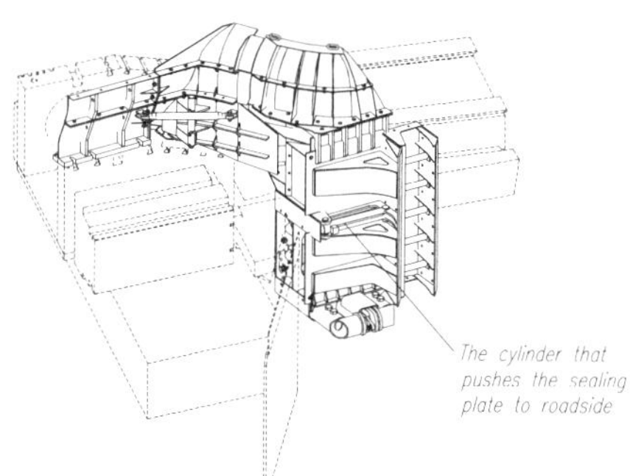

Figure 13. Functional features of the Output Self-Loading System – adjustment to the variable width of the space between the roadside and the longwall conveyor construction

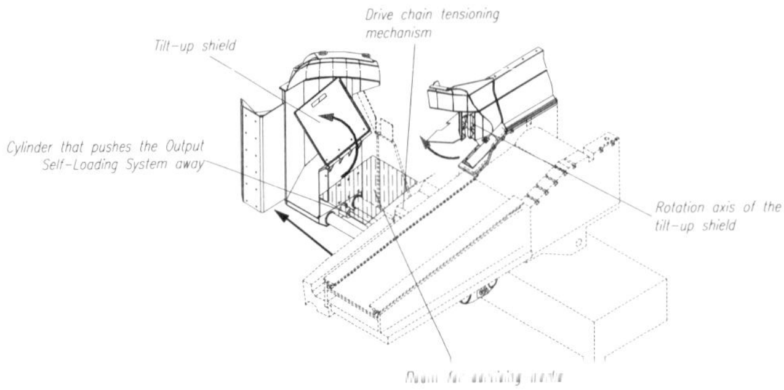

Figure 14. Functional features of the Output Self-Loading System – access to longwall conveyor components for servicing purposes

Access to the servicing points of the longwall conveyor is ensured by a hydraulic mechanism that shifts the Self-Loading System (Figure 14). This is a very important function, as the elements for adjusting and tensioning the driving chain of the coal plough head are located where the device is installed.

The constructed Output Self-Loading System shall be implemented in the coal-plough longwall at LW "Bogdanka" mine in the Stefanów field.

4 CONCLUSION

Optimization (i.e. searching for the best solutions according to a selected criterion) is indispensable, especially when certain experiences indicate that it is possible to improve a process.
The way of thinking that only attempts to improve efficiency by increasing, e.g. the power of machines, nowadays seems to be insufficient and lacking in innovation. In many cases, the existing solution should be examined first. New possibilities might then be found by modifying work organization and/or by introducing structural changes and/or new equipment. Technological progress opens up new options every day.

The extensive operational experience of the team of experts of LW "Bogdanka" mine and the design and production resources of the "SIGMA" and "HAJDUK Group" companies resulted in the creation of solutions that are bound to significantly impact the results achieved by the longwall system. The elimination of unnecessary activities, the improvement in performance, the removal of the causes rather than the effects of certain phenomena – all this guarantees the success of the new approach to the longwall work carried out using the existing coal plough system with the aid of only a few additional peripherals.

International Mining Forum 2011, Dyczko et al. (eds) © 2011 Taylor & Francis Group, London, UK. ISBN 978-0-415-68938-0

Prototype of Temporary Mechanized Mine Roadway Support of the Dog Headings

K. Krazue, Ł. Bołoz, T. Wydro
AGH University of Science and Technology, Krakow, Poland

ABSTRACT: Technology of the dog heading driving with use of roadheaders or explosive material has been described in the present study. Difficulties occurring during dog heading driving, particularly those related with the support installing, have been presented. Constructional solutions of different mechanized mine roadway supports, including their advantages and disadvantages, have been presented. Based on the executed analysis of the mine dog heading driving, including construction of mechanized roadway supports, a new concept of temporary mechanized roadway support has been developed and its technical parameters have been cited

KEYWORDS: coal mining, dog heading driving, roadheader technology, mechanized roadway support

1 INTRODUCTION

Mine dog heading driving is one of the key processes applied in hard coal mining industry. These headings, development and preparatory, should be planned on advance and consequently realized.

The dog headings can be mined in parallel cycle, step by step cycle, or step by step – parallel cycle (Klich et al. 1998). However, each of these cycles comprises mining, loading and supporting phase. In the case of mining and loading phases the works can be currently executed in parallel (roadheader) or in step by step mode (drilling, blasting, loading). Whereas, in the case of the support installing (arc yielding support) this process is realized exclusively step by step and practically manually. That is why, during the support installing, roadheader cannot be operated, because personnel responsible for mounting arcs of the roadway support is present in front of the roadheader. It is obviously related with the personnel working safety, because the hazard is related with rock sliding during operation of the roadheader with movable arm and pick, as well as with great dustiness. However, even in a case when the roadheader is not operated, different hazards occur (rock sidle), which are dangerous for the personnel working near the working face (Krauze 2010). Application of the temporary support during the roadheader operation assures safe support installing. In such case, the mining and loading processes could be realized in parallel with installing the permanent support. In such case the personnel responsible for installing the support is present near the border between temporary and permanent support. In result, surface of unprotected roof is minimized and the permanent support is installed in a distance of about 20 meters from the working face. Roadway supports do not satisfy expectation of hard coal mines and are not used because of his reason.

Application of temporary mechanized support in mine dog heading during its driving calls for development of such constructional solutions that satisfy expectation of potential users in the future. Such assumptions will be a base of the support designing phase (Gospodarczyk et al. 1999, Krauze et al. 2011).

2 CURRENT STATE OF THE TEMPORARY MECHANIZED SUPPORTS APPLICATION

As was mentioned before, in the case of mechanical driving of the mine dog headings with use of roadheaders, after execution of the web up to permissible roof exposure, the mining works should be stopped and the support should be installed. The process of the support installing, irrespective of the type, is the most time- and work-consuming process in the roadway mining works. Traditional installing of the permanent support results in shutdowns, as well as it decreases driving advance with use of high efficiency roadheaders. That is why concepts of introduction of mechanized supports serving as temporary supports moved along the working face appeared already in the 80s of the former century. Depending on the roof conditions, the permanent support is installed few meters behind moving temporary support, independently on the works done in the working face. It results both in working safety improvement in the working face and during installing the permanent support. In addition, elements of the permanent support are transported only to the end zone of the temporary support, without the necessity of usually manual transport to the working face in the roadheader vicinity. In this case it was related with great personnel experience and mounting abilities of the personnel.

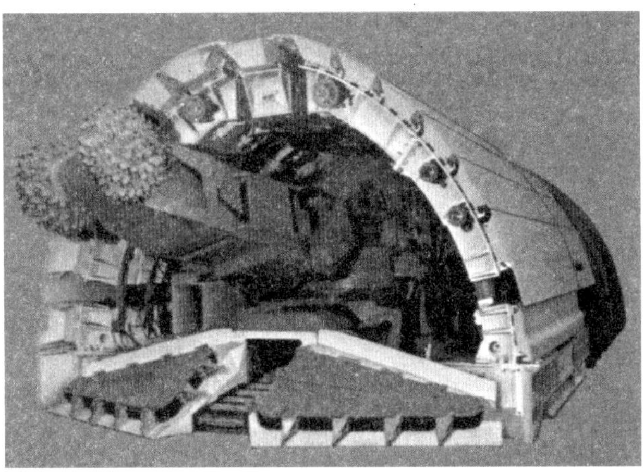

Figure1. View of the mechanized shield support by Westfalia Lünen with roadheader manufactured by the same company – version with central loading elevator (informational materials – Westfalia Lünen)

Different constructional solutions of hydraulic support are known. Support manufactured by Westfalia Lünen is shown in Fig.1. Similar solution was developed by American company AEC Incorporated (informational materials – Westfalia Lünen, informational materials – AEC Incorporated).

However, the mentioned supports possess reduced possibilities to be adapted for variable geometrical parameters of the roadway (informational materials – Westfalia Lünen, informational materials – AEC Incorporated). The other solutions comprise supports mounted on roadheader, for example support Bert 110D manufactured by Eimco or DSS-S by Voest Alpine. However, in such solutions the support plays role of the roadheader protection, along its arm length. In such case, the working face does not cooperate with the support along whole circumference but only in its roof part, what constitutes the next essential disadvantage (information materials – EIMCO Ltd).

3 ACTUAL STATE OF THE MINE DOG HEADING DRIVING

Driving cycle of dog headings comprises mining, loading and supporting phases. Actually in hard coal mines the mining is realized mechanically with use of roadheader or via drilling and blasting. But loading can be realized in parallel with use of the roadheader or, in the case of blasting-based mining, step by step with use of the loader, or occasionally manually. Obviously, in each of these cases the supporting is made step by step and almost exclusively manually (installing rod-bar arcs, clamps, sprags, shields) (Klich et al. 1998, Krauze et al. 2007). Only roadheader arm manipulator, platform or loader bucket are used for lifting the roof arcs.

Analysis of the dog heading driving process in Polish mining industry allows to conclude that presently and in the near future the headings will be driven with use of the roadheader or with use of the explosives. However, at the moment in each of the mentioned cases, installing the permanent support is accompanied with stoppage of all other works conducted in the working face. From the other side, lack of suitably efficient mechanization process results in elongation of the driving time. That is why we can state that improvement of the process is necessary, in order to reduce the operation time (Klich et al. 1998, Krauze et al. 2007).

Possibility of the improvement of installing the roadway support ŁP is seen in application of the temporary roadway support and suitable pneumatic manipulator. In such case, the installing of the support ŁP could be made in parallel, in the rear of the working face, behind the temporary support.

4 CO-OPERATION OF THE TEMPORARY SUPPORT WITH THE MACHINES OPERATED IN THE WORKING FACE

Temporary mechanized roadway support is operated in the working face of the mine dog heading. It should assure the dog heading stability until permanent support ŁP is installed.

Medium-sized roadheaders AM 75Ex are at present commonly used for mining works with use of the support ŁP 9 or ŁP 10. Overall dimensions of the roadheader co-operating with such support, is shown in Figure 2.

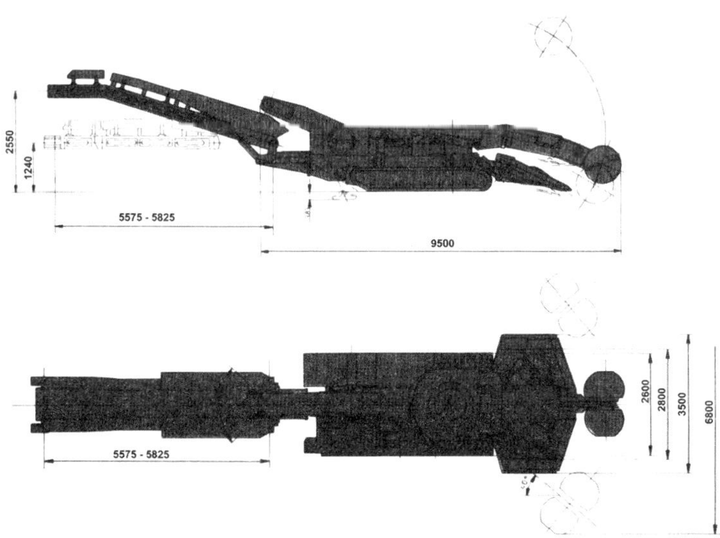

Figure 2. Overall dimensions of the roadheader AM 75 Ex, equipped with short bridge conveyor (Krauze et al. 2011)

The whole set is almost 16.0 m long, up to 3.5 m wide, and its height amounts for 2.55 m. These overall dimensions should be taken into consideration when the concept of temporary support co-operating with a roadheader of this type is designed (Krauze et al. 2011).

Irrespectively of the mentioned machines, in the working face, also de-dusting installation protecting the personnel against the dust is operated. Overall view of the roadway set, including the front and back view of the roadheader AM 75 Ex is shown in Figure 3.

Taking under consideration the above suggestions, temporary mechanized support should protect both the roadheader and bridge conveyor, up to the point where the winning is loaded onto band conveyor. Thus its length should amount for at least 16 m. Width of the free space inside the support should possess minimal width equal to the roadheader width, in his case about 3.5 m, and height should amount for 2.55 m.

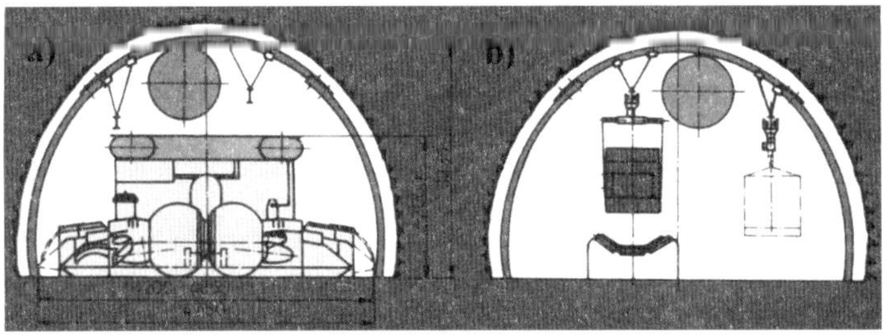

Figure 3. Roadway complex with roadheader AM 75 Ex: a) front view, b) back view (Krauze et al. 2011)

It is commonly assumed that single web of the roadheader is 0.5 to 0.8 m. Thus the support should assure possibility of supporting the web in a single step or in few consecutive steps.

In the case of application of mechanized temporary support during mining with use of explosives, both version for manual drilling of the blasting holes and version with use of drill rig should be taken under consideration.

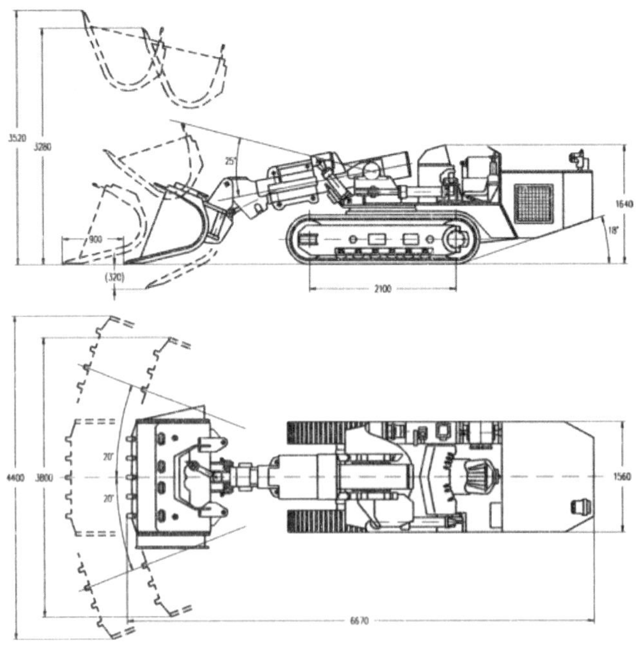

Figure 4. Overall sizes of side loader ŁBT 1200 manufactured by Bumar (Gospodarczyk et al. 1999)

In the case of the manual drilling, from sprag or pneumatic support, side loader is localized in the rear part of the roadway. When drilling and blasting is completed the loader is moved to the working face and the winning is loaded onto transporter (scraper conveyor, mining trucks). The transporter is usually located by one of the roadside walls. Width of the scraper conveyor is usually not bigger than 800 mm, whereas width of the railway with mining tracks amounts for about 1200 mm. The railway or the scraper conveyor is elongated or relocated every few webs (6 to 10 m), usually during repair shift.

Overall sizes of big, narrow face, side loaders, which are the most commonly used in Polish coal mines, are shown in Figure 4. Their width does not exceed 1.6 m, height 1.65 m and length 6,75 m respectively.

For these parameters, length of the temporary support should be assessed for about 10.0 m, and free space inside the support should possess minimal width of 3.4 m.

It is assumed that single web during blasting equals 1.5 to 2.5 m. Thus we can conclude that the support should assure possibility of supporting such web in several consecutive steps.

In the case when drill rigs are used, both loader and drill rig are operated alternatively, first the rig and next, after blasting in the working face, the loader, which is used to load the winning and also for auxiliary works during installing the support. One of the drill rigs, which is used in Polish mines, including its overall dimensions, is show in Figure 5. Its height and width are comparable with those shown in Figure 4 and amount for 1.55 m and 1.65 m respectively. However, the length of the drill rig, including its arm, amounts almost for 9.5 m. Assuming that the temporary support should protect both machines, its length should amount for at least 17 m, whereas free space inside the support should be minimally 4.2 m wide. The other functions should be similar as in the case of manual drilling.

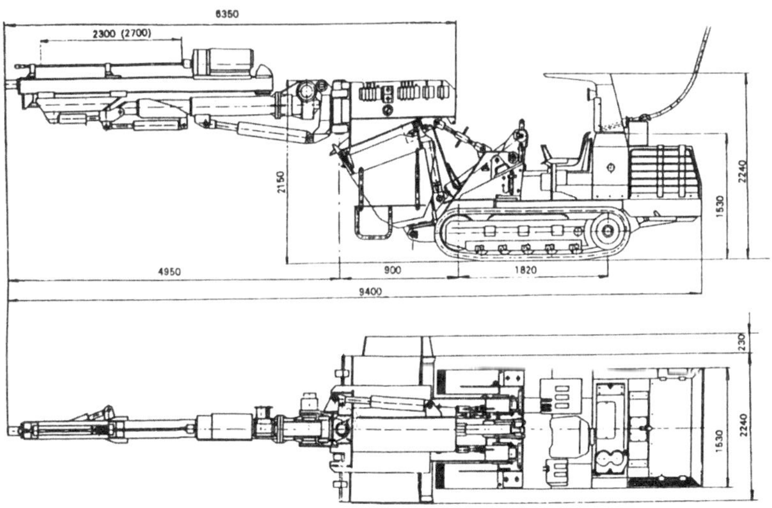

Figure 5. Overall dimensions of the single-arm drill rig manufactured by Deilmann Haniel (information materials – Deilmann Haniel GmbH)

The temporary support should assure continuous and parallel driving, loading the winning with use of one of mentioned mining techniques and installing the yielding support. Assembling and support installing should be realized right after installing temporary support with use of mechanized manipulator or installing platform. Expanding and relocation of the temporary support should in no way collide with operation of working face machines. It also shouldn't disturb process of permanent support installing. Temporary support sets should be relocated successively due to the advance of mined working face. Elements of temporary mechanized support should be supplied with emulsion from autonomous hydraulic aggregates.

5 A NEW CONCEPT OF THE TEMPORARY, MECHANIZED MINE ROADWAY SUPPORT

In the case of Polish hard coal mines, the dog headings are made in order to open and prepare the deposit for exploitation. Depending on their function, the dog headings should possess suitable cross-section, length and equipment. However, depending on their spatial orientation these headings are mined in different rocks, having different inclination and rectilinearity, what result in various natural hazards.

Based on obtained information, planned for the period 2011–2015, total length of the mine dog headings for individual coal companies in dependence on their cross-section are gathered in Table 1. The most important conclusion drawn on the basis of presented data is that we can expect growing amount of the mine dog headings driven with use of the support ŁP9 and ŁP10. Total amount of these headings is bigger than total amount of the other ones. That is why, temporary support should be considered, which will be used for headings of the cross-section ŁP9 and ŁP10.

Table 1. Roadway length for different cross-sections planned for implementation in the period 2011–2015 (Krauze et al. 2011)

No.	Coal company	Length of roadways for different types of the support ŁP [m]					
		ŁP9	ŁP10	ŁP12	ŁP13	ŁP14	ŁP15
1	Coal company 1	187,500	187,600	93,787	–	–	–
2	Coal company 2	150,000	150,000	150,000	–	–	–
3	Coal company 3	240,000	66,000	–	–	–	–
4	Coal company 4	18,805	46,200	5,476	324	680	–
5	Coal company 5	–	58,300	161,040	159,280	153,120	151,800
6	TOTAL	596,305	508,100	410,303	159,604	153,800	151,800

Based on the gathered information it was noticed that the roadway inclination is variable, particularly in a case of preparatory headings, which are usually driven along the coal seam floor. So we can assume that inclination of these headings balance within the range ±20°. However, the development headings are mined usually in case of smaller inclination – amounting for ±10°.

During driving the mine headings mentioned above, we should expect the possibility of occurrence of each of the mentioned hazards.

Taking under consideration requirements of previously mentioned hazards, parameters, functions and observations we can determine assumptions, which should be considered in development of the concept and design of the temporary mine roadway support. The support in question should satisfy the following assumptions (Krauze et al. 2011):

- relocation due to the working face advance,
- permanent support ŁP9 or ŁP10 can be installed behind the temporary support,
- roadheader or drill rig and loader, including scraper conveyor and mine cars can be operated under the temporary support,
- minimal length of the temporary support (with possibility of its elongation) should be bigger then the roadheader length, or the length of mining and loading machines (for actually used machines this length shouldn't be less than 16 m),
- free space inside the support assuring free movement and passing of the operated machines should possess minimal width of 4.2 m,
- duration of the permanent support installing (manipulator, suspended rail) should be shorter than total time of mining and loading on the whole length of the web,
- minimal rate of the dog heading advance should exceed 10 m / 24 h,
- possibility of the machines operation in dog headings with inclination ±12° and radius of curvature over 15 m should be assured,
- temporary support should be supplied with emulsion from hydraulic supplier,

- possibility of assembling and disassembling sections of the temporary support (use of manipulator or working platform for installing permanent support and suspended rail) should be assured).

Based on the mentioned assumptions and requirements, a preliminary concept of mechanized, temporary mine roadway support has been developed. The support is designed as a module construction, including possibility of individual segments adding or demounting. General view of the set consisting of six segments is shown in Figure 6.

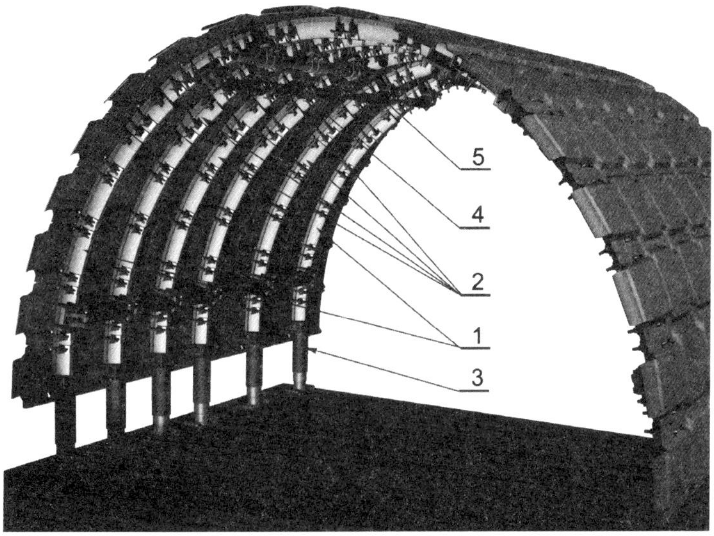

Figure 6. General view of six element support set of the mechanized, temporary support, according to the preliminary concept (Krauze et al. 2011)

The support consists of two marginal sets *1* and four identical inside sets *2*. In lower, straight-lined segment the arcs are equipped with shift cylinders having the stroke of 300 mm *3*, whereas the nominal support dimension is obtained for the shift cylinder advance of 150 mm, and another 150 mm is considered a reserve. Individual segments are joined together via four shift cylinders *4*. Because of the operational mode, the marginal shift cylinders are two-side acting, and the others one-side acting. In addition, the arcs are joined together with pipe-type guides *5* assuring their stability during drawing off phase. General view of one of such arcs is shown in Figure 7. Each support set consists of two roadside arcs *1* and one roof-bar arc *2*. The arcs are joined together with clamps *3*. It should be mentioned that the arcs are commonly available and applied in the mining industry at present. Holders of shift cylinders 4 and holders of pipe-type guides are installed on the arcs via separable joints with use of special sheet metal plates. Moreover, considerable part of the support set is covered with special shield *5*, which is designed in such manner that it protects the heading against rock waste, plays role of guiding element, as well as it allows sheet metal plates and other equipment mounting on the arcs.

In case of the solution shown in Fig. 6, the mechanized temporary support is adapted for driving the roadways of cross-sections tailored to the support ŁP 9 and ŁP 10, depending on the type of applied arcs. Just one set relative shift of the other amounts for 280 mm and it results from the assumed shift cylinders stroke. The scale, i.e. distance between individual segments during the machine operation, with advanced shift cylinder, is equal to 900 mm (Krauze et al. 2011).

After installing the support set on the length depending on the machine type, or type of mining and loading machines, the cycle of machine operation can be started. Due to the advance of the roadway working face, individual support segments are withdrawn via removal of shift cylinder located in supporting foots, beginning from the working face, and then they are relocated with use of shift cylinder toward the front. After being shifted, the segment is re-spaced and consecutive shift of the other segments is realized. Because of the relation between the web and shift cylinder stroke, control system of the cycle for 2 or 3 supporting segments is assumed. Development and making of the control system for individual supporting segments will be necessary for proper realization of the described operation.

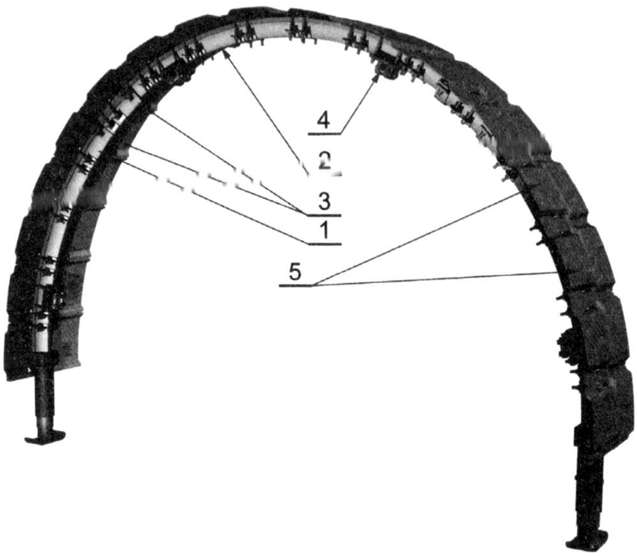

Figure 7. General view of the support set of the mechanized, temporary support (Krauze et al. 2011)

SUMMARY

Analysis of global relations in mining industry, as well as mining-geological conditions occurring in Polish hard coal mines allowed formulation of the following conclusions and notes:

1. Actually, the vast majority of the mine dog headings (about 85%) is driven in mechanical manner, with the use of roadheaders. Driving advances are on average between 8 and 10 m / 24h (coal and coal-stone headings).
2. In the case of drilling and blasting, when drill rigs and narrow working face loaders are applied, the obtained driving advances usually do not exceed 3.5 m / 24h (stone-headings).
3. Step by step character of works related with mining and loading, as well as with mounting permanent support of the dog heading considerably influences the daily driving advances.
4. At the moment there are no constructional solutions of mechanized, temporary support designed for hard coal industry, which assure the heading stability and protection of the personnel and concurrent permanent support installing.
5. Based on the analysis of parameters and operational mode of machines used in the mine dog heading driving, the assumptions of a new concept of mechanized, temporary support has been proposed. The support should be minimum 16 m long in order to assure protection of the driven heading for the arc yielding support ŁP 9 and ŁP 10. The support should be self-disassembling and supplied with hydraulic emulsion.
6. The developed preliminary concept of the mechanized, temporary support comprises module construction allowing adding or removal of individual module segments.
7. The presented concept can be easily modified (construction and amount of arcs and shift cylinders).
8. Design of a control system allowing suitable supply of the shift cylinder will be necessary, mainly for dislocation of individual segments of the support to the working face.

9. Equipping the mechanized, temporary support with manipulator or working platform allowing work mechanization during installing permanent support consisted of ŁP arcs seems to be reasonable. It should accelerate the support installing process. The manipulator or working platform moving on the rail suspended under the temporary support can be used for the support elements assembling and disassembling.

REFERENCES

Gospodarczyk P., Kotwica K., Kalukiewicz A., Klich A., Pawlik K. 1999. Maszyny i urządzenia dla inżynierii budownictwa podziemnego, Wyrobiska korytarzowe i szybowe w górnictwie. Wydawnictwo „Śląsk", Katowice. (In Polish)

Informational materials of AEC Incorporated, USA. (In Polish)

Informational materials of Deilmann Haniel GmbH, Niemcy. (In Polish)

Informational materials of EIMCO Ltd, Anglia. (In Polish)

Informational materials of Voest Alpinie Bergtechnik, Austria. (In Polish)

Informational materials of Westfalia Lünen, Niemcy. (In Polish)

Klich A., Krauze K., Gospodarczyk P., Kalukiewicz A., Kotwica K., Reś J. 1998. Niekonwencjonalne techniki urabiania skał. Wydawnictwo Naukowe „Śląsk", Katowice. (In Polish)

Krauze K. 2010. Problemy bezpieczeństwa w budowie i eksploatacji maszyn i urządzeń górnictwa podziemnego. Monografia, ISBN 83-922183-9-6. Wydawca Centrum Badań i Dozoru Górnictwa Podziemnego, Lędziny. (In Polish)

Krauze K., Bołoz Ł., Wydro T. 2011. Zmechanizowana obudowa chodnikowa dla zwiększenia efektywności drążenia i poprawy bezpieczeństwa pracy. W: Problemy bezpieczeństwa w budowie i eksploatacji maszyn i urządzeń górnictwa podziemnego, Red. Krauze K., Lędziny. (In Polish)

Krauze K., Kotwica K., Bołoz Ł. 2010. Developing tendencies in construction of locking devices for mechanical props type SV. Collective work „Selected problems of coal and cohesive rocks exploitation". Wydawca Katedra Maszyn Górniczych, Przeróbczych i Transportowych, Kraków.

Krauze K., Kotwica K., Mąsior M., Bołoz Ł., Wydro T. 2011. Identyfikacja warunków górniczo-geologicznych wyrobisk korytarzowych polskich kopalń węgla kamiennego i opracowanie założeń i wytycznych dla projektu wstępnego rozwiązania zmechanizowanej obudowy chodnikowej. Praca badawcza niepublikowana, Kraków. (In Polish)

Krauze K., Kotwica K., Wydro T. 2007. Zastosowanie tymczasowej obudowy zmechanizowanej w procesie drążenia wyrobisk korytarzowych. Konferencja Naukowa „Górnictwo zrównoważonego rozwoju", Gliwice. (In Polish)

International Mining Forum 2011, Dyczko et al. (eds) © 2011 Taylor & Francis Group, London, UK. ISBN 978-0-415-68938-0

Headgate Deformation Characteristic during Plow Exploitation of the 1/VI Longwall Face in 385/2 Seam in LW "Bogdanka" Mine

W. Korzeniowski, Ł. Herezy, K. Skrzypkowski
AGH University of Science and Technology, Krakow, Poland

ABSTRACT: The article presents the results of displacements and deformations measurements of the headgate obtained during the first plow complex exploitation in Lublin Coal "Bogdanka" mine. Together with the advance of the longwall face the headgate was maintained for use in future exploitation. For particular types of roof support and rock mass reinforcement methods (cribs, rod bolts and string bolts) the values of heading convergence, fracture zone reach and fissures have been defined, indicating the rock mass strengthening effect depending on the chosen technical solution.

KEYWORDS: plow technique, maintaining of headings, headings deformations

1 EXPLOITATION TECHNOLOGY WITH PLOW COMPLEX OF THE 1/VI FACE IN 385/2 SEAM

Exploitation field of the 1/VI longwall, referred to as research field, is located in "Nadrybie" area of LW "Bogdanka" in 385/2 seam at 910 m level. Deposition of the seam is nearly horizontal. Longitudinal inclination of the cross heading is 2°. The thickness of the seam is variable and ranges from 1.3 to 1.8 m with the average of 1.6 m. Mudstone partings of about 0.1 m thickness and 15-32 MPa compression strength exist in the seam. The strength of coal ranges between 12 and 18 MPa. Following hazards have been identified in the area:

- 1st category of methane hazard,
- 1st category of water hazard,
- 4th group of fire hazard (self-ignition),
- B class of coal dust explosion hazard.

Exploitation field of 1/VI longwall has been developed by completing the following headings (Figure 1):

- belt heading 1fN which connects with field heading 5 in 385/2 seam and with headings: headgate 1/VI, tailgate 1/VI, 2/VI heading which is an extension of the headgate 1/VI. The heading is located in the surroundings of coal mass from both sides and supported with closed support V36/9 with 0.75 m set spacing,
- headgate 1/VI which connects belt heading 1fN and with Eastern heading 4/385. The heading is located in the surroundings of coal mass from both sides and supported with open, six-element support ŁPSC V36/11S with 0.90 m set spacing,
- headgate 1/VI which connects belt heading 1fN, heading 2/VI and Eastern heading 4/385. The heading is located in the surroundings of coal mass from both sides and supported with

six-element ŁPSC V36/12S and five-element ŁPSC V36/12 support, both with 0,90 m set spacing,

- cross heading connecting the 1/VI headgate and tailgate. In the longwall reinforcement and in plow notching phase ŁPrP V32 support was used. Next, a cavity was made and secured with roof-bars made from straight V36 or V32 elements supported by SV props, laying on horseheads from rails built on KGO elements of the cross heading's support. During the exploitation the cross heading was secured by Bucyrus chocks sections.

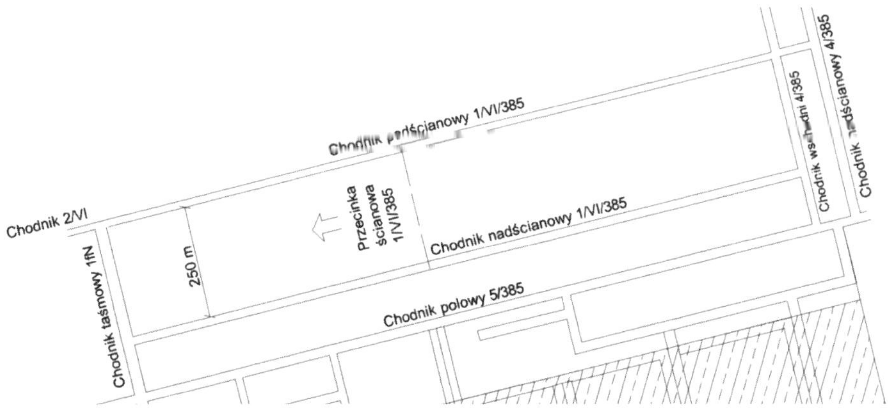

Figure 1. Heading system in the area of 1/VI longwall field

Notations: Chodnik – *Heading;* Chodnik taśmowy – *Belt heading;* Chodnik podścianowy – *Headgate;* Chodnik nadścianowy – *Tailgate;* Chodnik polowy – *Field heading;* Przecinka ścianowa – *Longwall crosscut;* Chodnik wschodni – *Eastern heading*

The designed exploitation longwall had the length of 250 m and panel length of 1750 m. In the direct vicinity of the exploitation field of the analyzed longwall exploitation hasn't been conducted. So, it may be assumed that during the exploitation of this particular longwall the displacements in the rock mass in the considered area were mainly the result of the primary pressure of the rock mass and the exploitation pressure.

The considered research longwall 1/VI was exploited with full thickness of the seam using the longitudinal system with roof collapse, from the borders of the mining area. The cutting was bidirectional with the use of Bucyrus GH 1600 plow. The getting started on 23rd March 2010. In the first phase of longwall start-up a plow notching into coal mass was performed. After full roof collapse has been reached and preliminary exploitation parameters have been set, the efficiency has been gradually increased and the parameters of the complex's work have been optimized. The start-up of the longwall and reaching target capacity of 10,000 tonnes of output per day was expected to last three months.

2 CHARACTERISTICS OF HEADINGS SUPPORT

For the safety of headgates and tailgates three types of yielding arch support were used. Six-element ŁPSC V36/11S was used in the tailgate; in the headgate six-element ŁPSC V36/12S was used in the first 650 m of the panel and in the remaining part of the heading five-element ŁPSC V36/12 was built over (Korzeniowski et. al. 2011, Rak 2011).

To secure the stability of the crossings of the longwall with the heading additional reinforcements of the support were made through bolting and building-in horseheads not further than 60 m from the face of the longwall.

In the headgate (Figure 2a) every second space between the sets of the support was reinforced with horseheads made from straight V36 or V32 elements bolted to the roof with string bolts (1) installed alternating relatively to the axis of the heading in the distance of 1,0 m from the axis. Additionally, in every second space six or seven 2.7-meter-long bolts were used.

Furthermore additional technique was used to reinforce the side walls through the injection of adhesive agents (Figure 2a) (5) to protect them during the separation of support elements for technological needs.

The protection of the entry to the longwall face from the tailgate, behind the support, was made using wooden props. Behind the exploitation front the tailgate was liquidated through the collapse of the roof rocks. Yielding arch support and the reinforcing horseheads were progressively dismantled. (100% scrap recovery).

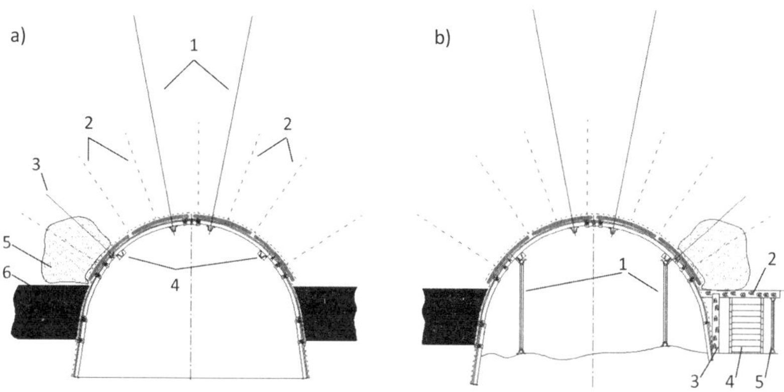

Figure 2. Reinforcement of the headgate a) in front of the longwall face, b) behind the longwall face

The headgate behind the face (Figure 2b) was maintained for future re-use during the exploitation of the neighboring longwall. For this purpose SV props (1) were situated by both side-walls under horseheads installed earlier. From the gob side, in a 2.0-meter-wide cavity preliminarily secured with a prop and a wooden roof-bar (2.3) and SV prop (5) a beech crib (4) was constructed and filled with binding mix. The space between the crib and yielding arch support was filled with the same binding agent for the better distribution of pressures at the contour of the support and for the improvement of heading ventilation conditions.

3 GEOLOGICAL AND GEOTECHNICAL CHARACTERISTIC OF THE ROCK MASS IN LONGWALL AREA 1/VI/285

Longwall research field was located at the depth of 740 m at 910 m level. During the development works no tectonic or sedimentary displacements were found.

The geomechanical parameters in the area of the longwall were assessed basing on exploratory boreholes (Figure 3) located in the following headings:

- Roof boreholes BR-11/08, BR-19/08, BR-24/08, BR-23/08, BR-25/08, BR-8/09 i BR-15/09,
- Floor boreholes BR-11/1/08, BR-19/1/08, BR-24/1/08, BR-23/1/08, BR-25/1/08, BR-8/1/09 i BR-15/1/09.

BR-11/08 i 11/1/08 boreholes were drilled at the crossing of the tailgate with belt heading 1fN,BR-19/08 i 19/1/08 boreholes were located at 300th meter (1628 m of the panel length) of the heading if counting from the crossing with belt heading 1fN, BR-24/08 i 24/1/08 boreholes were located at 602nd meter (1326th meter of the panel length), BR-23/08 i 23/1/08 boreholes at 901st meter (1027th meter of the panel length), BR-25/08 i 25/1/08 boreholes at 1200th meter (728th meter of the panel length), BR-8/09 i 08/1/09 boreholes at 1496.4th meter (432nd meter of the panel length) and finally BR-15/09 i 15/1/09 boreholes were at 1802nd meter (126th meter of the panel length).

Tailgate:
- Roof boreholes BR-10/09, BR-11/09 i BR-14/09,
- Floor boreholes BR-10/09, BR-11/09 i BR-14/09.

BR-11/08 i 11/1/08 boreholes were located at 308.1th meter of the heading if counting from the crossing with belt heading 1fN, BR-11/09 i 11/1/09 were at 599th meter, BR-14/09 and 14/1/09 were at 894.5th meter.

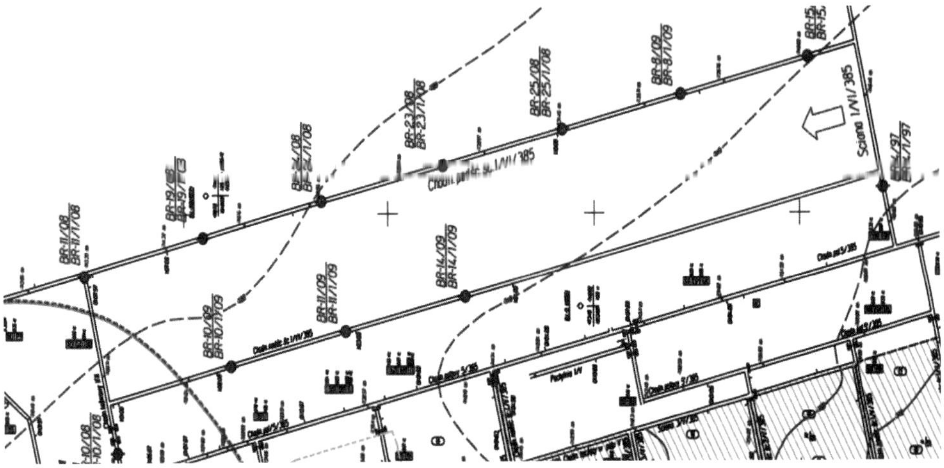

Figure 3. The location of exploratory boreholes for the identification of geomechanical parameters of the rocks in tailgate and headgate (Central Mining Insitute measurements, unpublished work)

The direct roof is a system of siltstone, mudstone and sandstone. The thickness of the mudstone located directly above the seam ranges from 0.8 to 2.0 m. Sphaerosiderite nodules were found locally. Directly above the siltstone a 0.2 to 7.0-meter layer of mudstone is deposited. The highest thickness of the mudstone is observed in the middle of the panel length. Above the mudstone lays a layer of sandstone with thickness up to 7.0 meters, locally interbedded with mudstone. The thickest layer of sandstone may be found at the beginning of the panel length. Alternating layers of siltstone, mudstone and coal are located above the sandstone together with local layers of sandstone and mudstone soil.

A layer of siltstone ranging from 0.3 to almost 1.5 meter is situated practically at full panel length in the seam's floor. Siltstone soil may be found locally under the seam. Underneath a variable system of layers may be observed, namely alternation mudstone, mudstone soil, sandstone, siltstone and locally coal. Sphaerosiderite nodules were found relatively often, especially in the mudstone.

4 RESEARCH POSTS IN THE LONGWALL AREA

For the assessment of behaviour of the headgate 1/VI research stands were established at its whole length and measurements of rock mass and support elements displacements were taken. In particular following parameters were measured:
- The change of height and width of the heading in front of and behind the longwall face (Table 1, Figure 8),
- Dislocations of the support in front of and behind the longwall face (Table 1),
- Fissure of the roof rocks behind the face in the bolted zone (Figure 7),
- Distribution of cracks in the rock mass above the crib reinforcements (Figure 7),
- Dislocations of the roof above the cribs,
- The change of thickness of the seam in the longwall area 2/VI (Table 1, Figure 9).

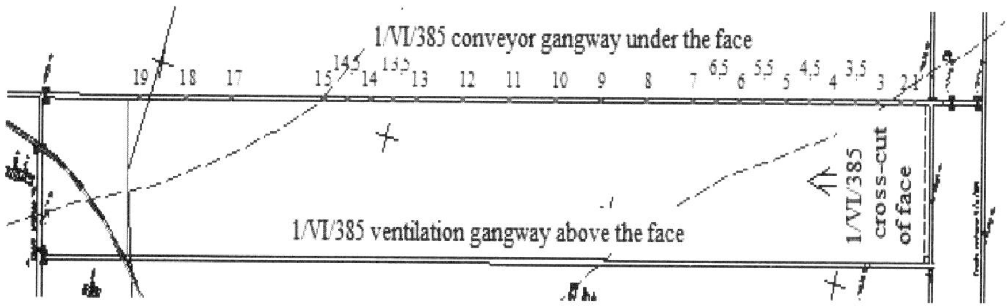

Figure 4. Localization of research stands in headgate 1/VI

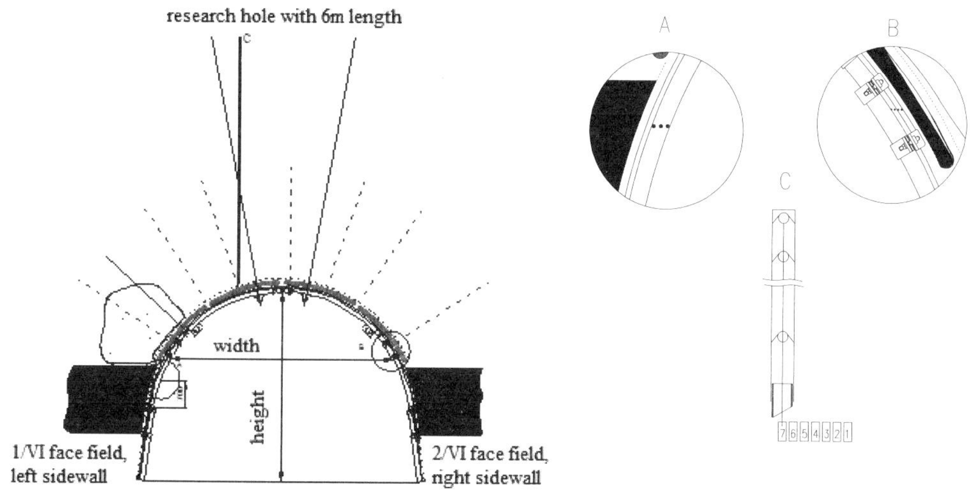

Figure 5. Research post scheme in a part of the heading with six-element support

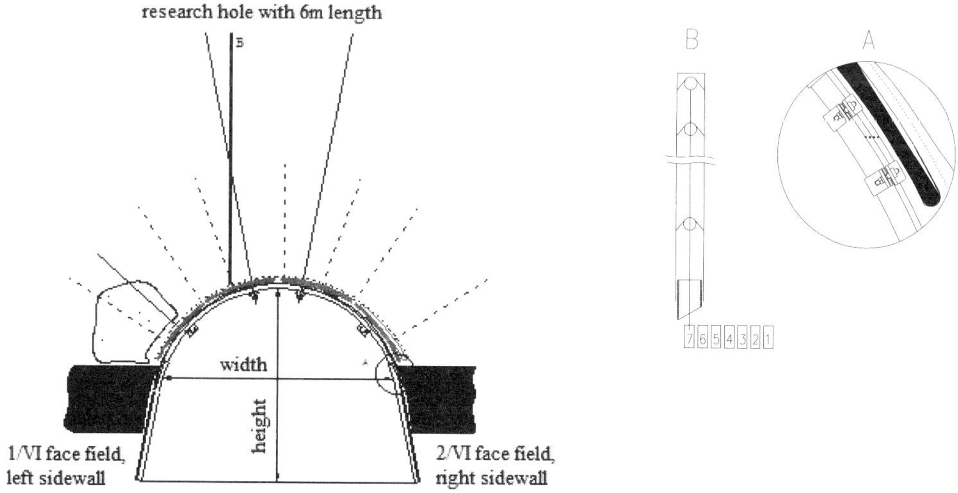

Figure 6. Research post scheme in a part of the heading with five-element support

Research stands no. 1-19 were standard stands where all aforementioned measurements were carried out. Research stands no. 3.5; 4.5–14.5 were auxiliary stands where supportive measurements were taken. The basic stands were located every 100 m starting at 100th meter of the panel length. Posts no. 1 and 2 were exceptions as they were situated at 20th and 50th meter of the panel length. Auxiliary stands were located at half the distance between standard stands. A scheme presenting the locations of the stands is shown in Figure 4. In Figure 5 and 6 schemes of research traverses are depicted.

The measurements were conducted from 25th March 2010 directly after exploitation in the longwall area 1/VI/385 was started and were continued during the whole time of exploitation and for a certain period after it was finished.

5 THE RESULTS OF MEASUREMENTS AND CONCLUSIONS

Systematic measurements conducted during the exploitation of analyzed longwall allowed to draw certain conclusions that may verify the techniques used to reinforce the roof support of the headings for the planned exploitation in other areas of the mine. Table 1 shows one exemplary spreadsheet of measurements results and Figure 8 depicts complex changes of convergence in the headings determined during the whole panel length of the longwall.

Figure 7 shows an exemplary result of cracking area measurements over the heading and location of the borderline of roof collapse. The measurement was carried out eight weeks after the exploitation ended.

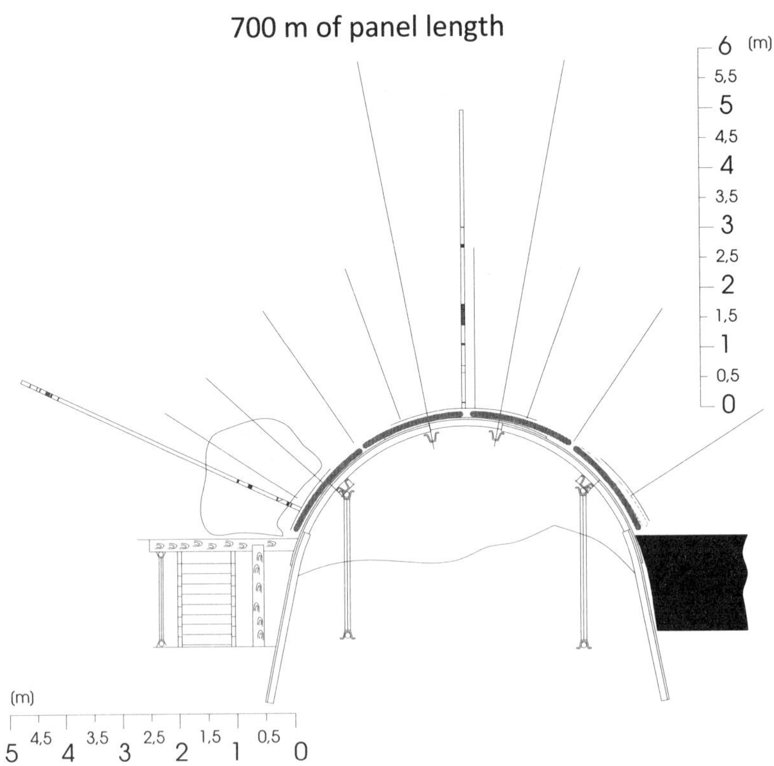

Figure 7. An exemplary localization of fractures and stratifications of rock mass around the heading behind the longwall face

Table 1. An exemplary results of measurements from the research post no. 9 located on the 700th meter of the panel length

Measurment point no. 9, 700 liner metre of panel lenght																
Distance from forehead of face [m]		290	252	222	91	61	14	-10	-93	-169	-231	-312	-364	-458	-820	-1050
Width of gangway under the face [mm]		5520	5520	5510	5510	5510		5420	5230	5230	5230	4940	4940	4940	4940	4940
Height og gangway under the face [mm]		3980	3980	3960	4320	4240		3390	2840	2600	2540	2500	2350	2320	2220	2150
Change the width of gangway [mm]		60	60	70	70	70		160	350	350	350	640	640	640	640	640
Change the height of gangway [mm]		644	644	664	304	384		1234	1784	2024	2084	2124	2274	2304	2404	2474
Slide down of arch yielding support between sidewall arch and sidewall-roof arch [mm]																
Left sidewall	Arch no. 1	0	0	0	0	-5	-80		-80	-90	-110	-110	-110	-110		-110
	Arch no. 2	0	0	0	0	-5	-10		-55	-55	-55	-55	-70	-190		-190
	Arch no. 3	0	0	0	0	-25	-80		-80	-110	-130	-160				
Right sidewall	Arch no. 1	0	0	0	0	0	-20	-220	-310	-320	-320	-320	-320	-320		
	Arch no. 2	0	0	0	0	0	-10	-190	-375	-430	-430	-430	-430	-430	-440	-440
	Arch no. 3	0	0	0	0	0	-20	-200	-410	-470	-485	-485	-500	-500	-500	-500
The date of measurement		26.05.10	2.06.10	9.06.10	21.06.10	24.06.10	30.06.10	3.07.10	12.07.10	20.07.10	26.07.10	2.08.10	9.08.10	18.08.10	26.09.10	14.12.10
Average value of slide down of arch yielding support [mm]	Left sidewall	0	0	0	0	-12	-57		-72	-85	-98	-108	-90	-150		-150
	Right sidewall	0	0	0	0	0	-17	-203	-365	-407	-412	-412	-417	-417	-470	-470
Thickness of coal seam [m]	In front of face	1,72														
	Behind the face	-														

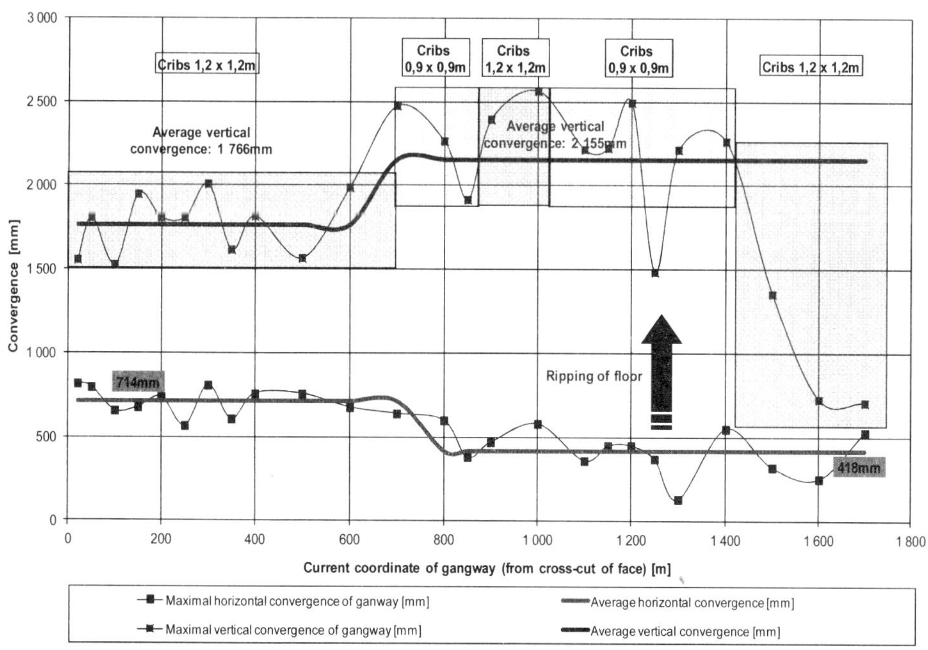

Figure 8. The convergence of 1/VI headgate

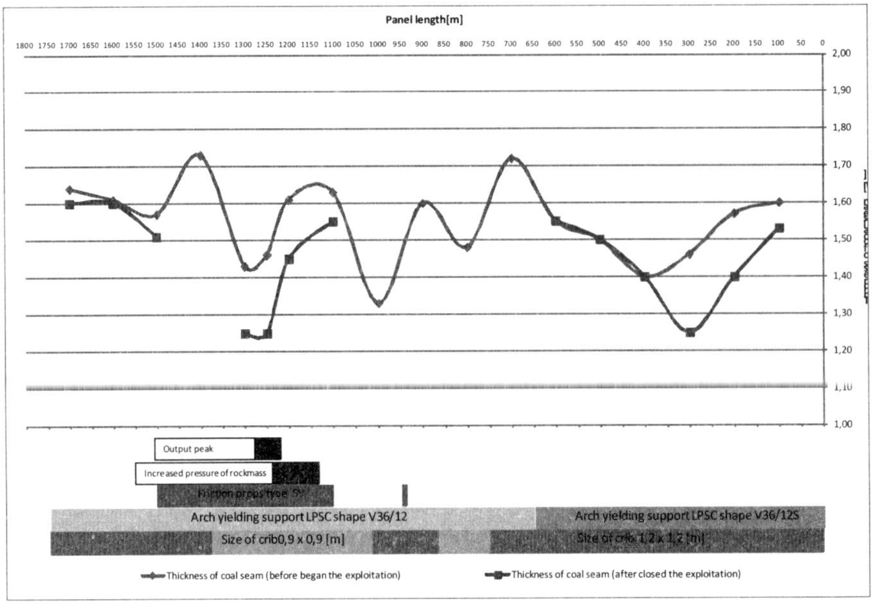

Figure 9. Change of the coal seam thickness along the panel length due to pressure of the rock mass

CONCLUSIONS

1. At the length of 700 meters, starting from the beginning of the panel length (from the longwall crossing) the heading has been protected with yielding arch support and with cribs of cross-section of 1.2 m x 1.2 m.
2. At the length ranging from 700 to 1400 meters the heading was protected with yielding arch support and with cribs of cross-section of 0.9 m x 0.9 m. Locally the dimensions were increased.
3. Horizontal convergence and vertical convergence are two most sensitive parameters allowing to assess the effectiveness of support of the headgate.
4. Convergence range variety is strictly associated with the dimensions of the cribs used for support
5. Vertical convergence in the headgate was ranging from 1500 mm do 2000 mm when cribs of 1.2 m x 1.2 m were used (Figure 8).
6. The decrease in cribs' dimensions to 0.9 m x 0.9 m resulted in higher vertical convergence: from 2000 to 2500 mm (Figure 8).
7. The average convergence resulting from the decrease of the cribs' dimensions changed from 1766 mm to 2155 mm (Figure 8).
8. The borderline of these types of cribs has also caused a change in horizontal convergence from 714 mm to 418 mm. The decrease in value is associated with the observed elevation of the floor which blocked the sidewall arches from moving towards the center of the heading (Figure 8).
9. The analysis of fissures and cracks in the sidewall boreholes has shown that the borderline of the roof collapse is located about 3-4 meters from the sidewall, at the height of 3 meters from the original heading floor location.
10. The increased pressures in the rock mass have caused a decrease in the seam thickness. This effect was observed in particular after 250 meters of the panel length were reached and also between 1150th and 1350th meter (Figure 9).

The work has been conducted as a part of statutory works no. 11.11.100.370

REFERENCES

Korzeniowski W., Herezy Ł., Skrzypkowski K. 2011. Nowoczesna technologia ścianowej eksploatacji pokładu węgla o miąższości 1,6m kompleksem strugowym. Przegląd Górniczy. Tom 66, nr 1-2. Katowice. (In Polish)

„Pomiar GiG" Przedstawiciel Głównego Instytutu Górnictwa. 2008. Badania własności geomechanicznych skał w chodniku podścianowym ściany 1/VI w pokładzie 385/2 w Lubelskim Węglu „Bogdanka" S.A. – część I. Lublin, praca niepublikowana. (In Polish)

Rak Z. 2011. Utrzymanie chodnika za ścianą w trudnych warunkach geologiczno-górniczych na przykładzie kopalni LW „Bogdanka" S.A. Część 2. Przegląd Górniczy. Tom 66, nr 1-2. Katowice. (In Polish)

International Mining Forum 2011, Dyczko et al. (eds) © 2011 Taylor & Francis Group, London, UK. ISBN 978-0-415-68938-0

Assessment of Powered Support Loadings in Plow and Shearer Longwalls in Regard to the Pressure Measurements in Props

M. Płonka, S. Rajwa
Central Mining Institute, Katowice, Poland

ABSTRACT: The results of static pressure measurements in the space under props pistons of powered support, performed during mining of seams in five longwalls with roof caving were presented. Measuring systems and methods of data collection and storage were characterized. The results of measurements were presented in the form of plots of pressure variations and the support load-bearing capacity. Sample of statistical analysis of the recorded data were presented.

KEYWORDS: powered support, longwall, pressure measurements in props

1 INTRODUCTION

In the Polish mining industry it is possible to encounter modern control and monitoring systems of longwall complexes which help to run the process of production effectively. They allow to observe and register particular parameters, for instance, the pressure in props of support section in case of powered supports, alterations in the distance of shifts in section meaning the changes of position in the excavation, exact time of measure, duration of particular technological operations etc. It is possible to watch most of the information on displays on the surface in real time. Examples of monitoring system panels have been shown in Figure 1 and 2.

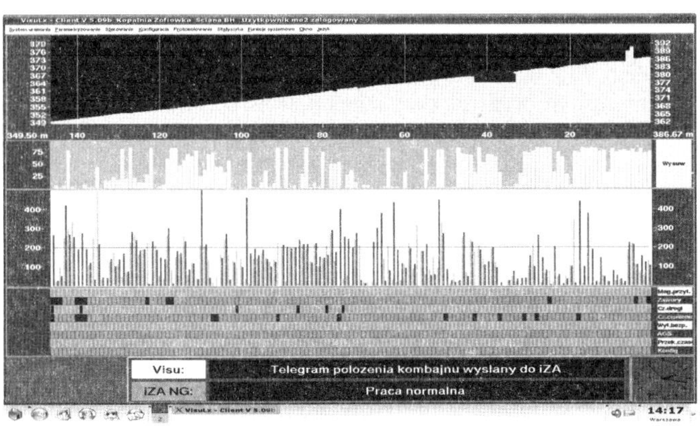

Figure 1. Example of monitoring system panel in shearer longwall

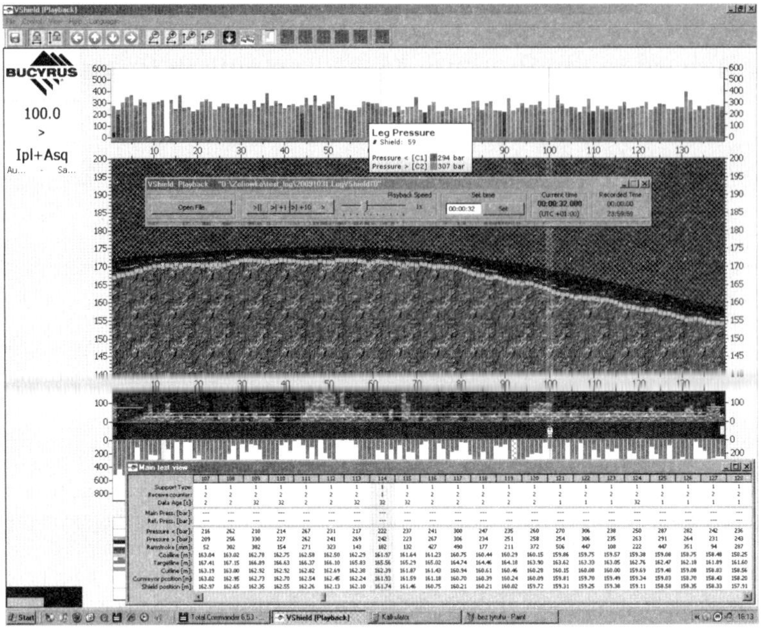

Figure 2. Example of monitoring system panel in plow longwall

As the result of co-operation between the Department of Extraction Technologies and Mining Support GIG, mines and producers of powered supports, it was possible to obtain and prepare the measurement data. Various modern control and monitoring systems helped to decide that the subjects of interest would be four longwalls with roof caving in which longwall coal combined system (marked as A, B, C and D) and longwall coal plow system (E) have been applied. The last longwall, after many years of abandonment of plow technique in Poland, is a technological novelty on account of significant changes in automatics and operating. The authors would like to thank all Companies for co-operation and the possibility of using the necessary data in this article.

2 DESCRIPTION OF THE APPLIED MONITORING SYSTEMS

2.1 *Shearer longwalls A, B and C*

The electrohydraulic control system allowed to monitor the position of the powered support sections in longwall, averaged pressure in props of each support section, current value of line feed shifter and alert warnings.

Depending on the configuration, the measurement system has gathered data on two computers: the first a measuring one placed underground and the second one for longwall visualization on the surface.

Sampling of the pressure in the space under props pistons of all powered supports took place every 1 minute.

The averaging readings were sent every 15 minutes to the computer on the surface in the form of data package. The pressure pick-ups allowed to take measurements within 0÷500 bars to the accuracy of ±1,5%. Saving the data in two-byte system allowed to receive pressure readings expressed in integral numbers. The controlled powered support of 1.4÷3.4 m in height, was equipped with props with working valves which were set to 47 MPa, to the accuracy of ±5%. Practically, the whole measurement range of the applied pressure pick-ups was used. The pressure of section feeding came up to 25 MPa. In addition to the monitoring, in order to perform supplementary analyses, some procedures which enable to read the registered data were prepared. They have been designed by use of CodeGear™ Delphi® 2007 at GIG to meet the requirements of the most popular Windows systems (XP, 2000, Vista).

The described measurement system was applied while measuring longwall A and B. Before the extraction with longwall C changes were introduced, allowing to increase the frequency of receivingand recording data. The modified system maintained the package records of all sections every 15 minutes as well as allowed to receive the information of sudden unit pressure changes in the controlled space under props pistons of the powered support.

2.2 Shearer longwall D

In longwall D a different measuring system was applied, allowing to take pressure measures with the frequency of one minute and to the accuracy of 1 MPa in each prop of 164 powered support sections. Moreover, in this measuring system the records and measures haven't been taken during the draw off cycle. The registered data comprised a date, time and the successive values of pressure in the props expressed in integral numbers. In that excavation a powered support within 1.4÷2.4 m in height was used; it was fed with the pressure of 25 MPa and equipped with props with working valves set to 42 MPa.

2.3 Plow longwall E

Control and monitoring of the automated plow complex based on independent, however, communicating measurement, data transmission and control systems. In that case it is not possible to precisely determine the frequency of measurement data registration as it is conditional on flow capacity of communication bus and other functions of the system. Nevertheless, the data being gathered (different systems) are sequentially registered at various intervals.

For the visualization of the condition of devices subordinate to those systems and their parametric representation, a software was implemented in two positions: first in the gate road, the second one on the surface, in a special room for that purpose, Figure 3.

The powered support in this longwall within 1.1÷2.2 m in height, equipped with props with working valves set on 44 MPa was observed. The initial pressure came up to 25 MPa.

Figure 3. Stand of control and visualization of the longwall coal plow system E on the surface

3 METHOD OF PREPARING MEASUREMENT DATA

Given the fact that the method of powered support selection applied at GIG (and other publications on the subject of powered support in foreign literature) generally combines the loading of the support in longwall with the variations of natural and technical conditions on its panel length, the registered measurements of pressures from the space under props pistons have been examined, in relation to the field areas of the tested longwalls (except for longwall E on account of the reasons described in 3.3), by presenting the chosen statistical graphs.

During the measurements, the coordinates of the section position have been determined basing on coal seam maps in the area of the tested longwalls, regarding the time of measurements being registered, section sequence in longwalls and their progress. After that, the highest values of the recorded pressures in props that occurred in a particular unit area were selected. Maximum pressure values, respecting the geometry of section, represent the range of the needful and developed load-bearing capacity of the support which is the reaction on the rock mass loading. Thus, it is possible to obtain maps of maximal pressure distributions.

The choice and justification of this method of data interpretation results from the analysis of a typical course of the section loading from rock mass, during its working cycle presented in Figure 4 (fragmentary data from the measurement system of one longwall). In the sections of powered support, during this course, it is possible to observe the cyclic periods of pressure increase in the props from the sprag state to the state preceding the draw off and pass.

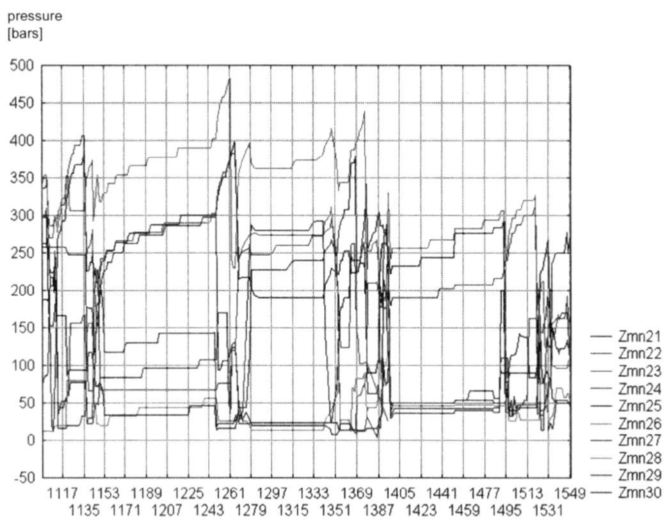

Figure 4. Example of measurement pressures from 10 props: Zmn21÷Zmn30, measurements no. 1117÷1549

3.1 Evaluation of the registered pressures in longwalls A, B and C

Longwall A, driven with 2.15÷2.45 m height and when strata deposition depth was about 660÷780 m, obtained a panel length of 480 m. Maximum registered values of pressures in props depicted on the fragments of longwall areas in the form of colourful points are presented in Figure 5. The colourless areas on the segments of the panel length result from different reasons connected with the lack of measurement signal. The presented view of measurement distribution was designed as a result of preparation of roughly 1.7 million (M) measurement readings, from which 0.18 million have been rejected on account of the routine selection of incorrect or unreliable data. The information was obtained from about 91% of the longwall area.

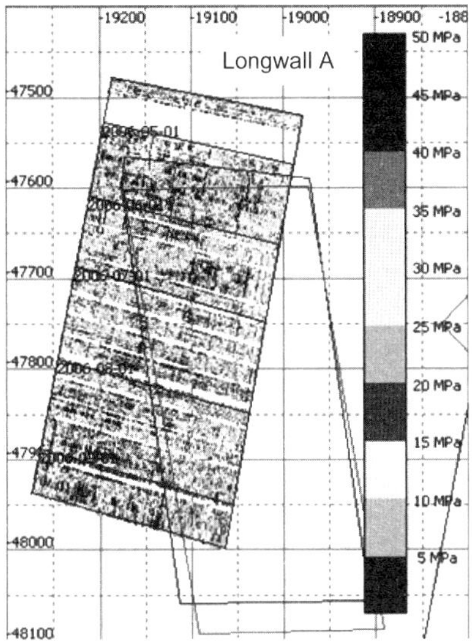

Figure 5. Maximum pressure values of the longwall A area

Longwall B, driven with 2.40÷3.20 m height and when strata deposition depth was about 820÷910 m, obtained a panel length of roughly 460 m. Maximum registered values of pressures have been determined by the examination of roughly 1.3 M measurement readings, from which 0.06 M have been rejected on account of the routine selection of incorrect or unreliable data. The information was obtained from about 91% of the longwall area.

In longwall C a modified measurement system was applied. It was driven with 2.65÷3.05 m height with strata deposition depth about 880÷960 m and obtained a panel length of 520 m. The maximum pressure values have been determined from the selection of roughly 18 M measurement readings which cover around 51% of the longwall area.

3.2 Evaluation of registered pressures in longwall D

During preparation of the measurement data, after rejecting mistakes from the process of registration, 30 M records of pressure have been obtained. It was assumed that for each section it is an average calculated from the pressure of two props. In case of the lack of pressure readings from one of these props the existing reading was taken into account.

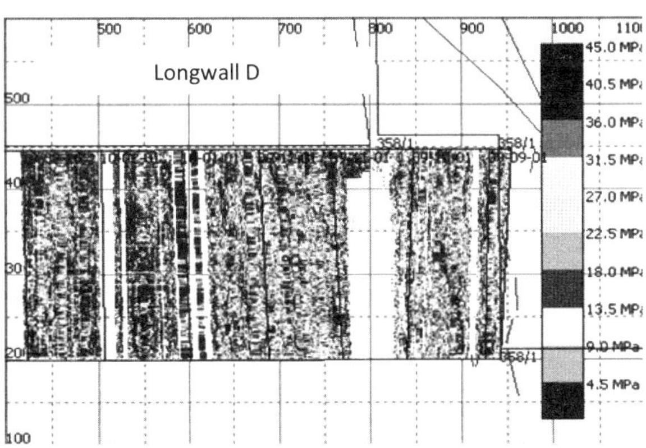

Figure 6. Maximum pressure values of the longwall D area

According to the experience, it was assumed that the measurements system error may be within 5%. Thus, the pressure values exceeding 44.1 MPa were rejected, in the amount of 8.6 thousand, which is 0.03% of the pressure records. The presented data in figure 6 cover roughly 83% of the longwall D area.

Longwall D was driven with roof caving on the panel length of roughly 980 m. Its length came up to about 250 m and its height, in the area of measurements, oscillated between 1.6 m to 1.8 m. In this case, strata deposition was at the depth of between 1000 m to 1040 m from the surface.

3.3 Evaluation of the registered pressures in longwall E

Although a large database documenting the work of caving system was gathered, some difficulties appeared with the tool serving for obtaining from database the information effectively as well as the description of record formats which may allow to perform a comprehensive analysis. Neither the data could be exported from the database to an alternative format nor copied, for instance, in the form of packages. As a result of this situation which was impossible to change during the course of research in 2010, the possibility of comparison of the course of changes in load-bearing capacity and the reaction of the supports in plow and shearer longwalls was significantly decreased.

Thus, it was not possible to draw a complete graph of pressure view in the background of the field of plow longwall E as it was in the case of the previous four shearing longwalls Nevertheless, a detailed analysis of recording system allowed to distinguish some fragments used and registered in the database. The information concerned the panel length of between 50 m and 300 m. Basing on the data two graphs have been drawn: of average pressure in props of each section in that panel length segment (Figure 7) and of average pressure relationships of all props along the panel length (Figure 8).

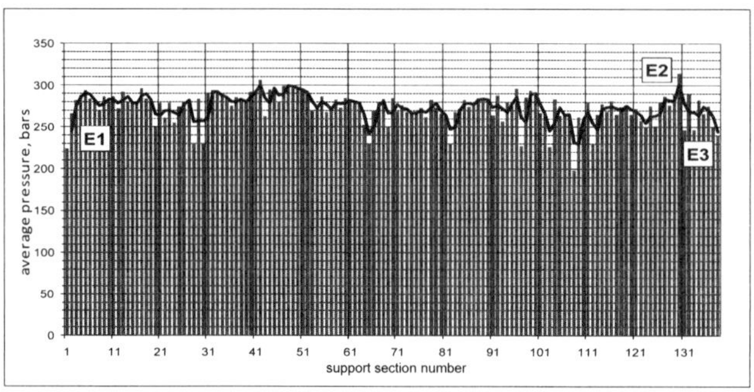

Figure 7. Diagram of the average pressure in section props along longwall E

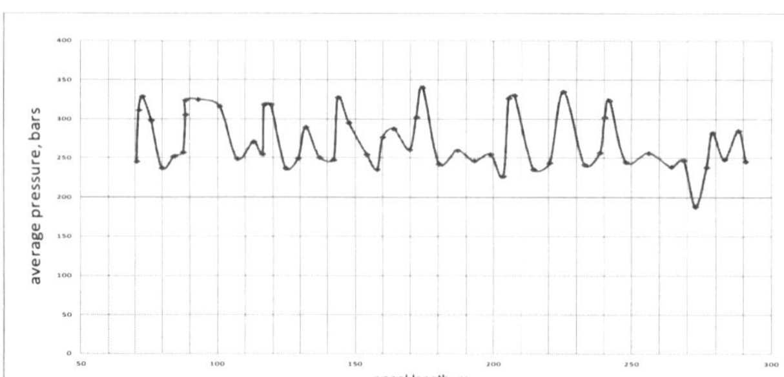

Figure 8. Course of the prop average pressure of all sections on the panel length E segment

4 DESCRIPTION OF LOAD-BEARING CAPACITY OF THE POWERED SUPPORTS

To determine the unit load-bearing capacity of a powered support with the pressure measured in the space under prop pistons, a calculation model designed at GIG was applied. Basing on the geometrical measurements of support section parts, pressure values in basic and headboard props, inclination angle of the canopy etc. it is possible to calculate the unit load-bearing capacity of the powered support. To perform the calculations, the following principles have been applied: variable working height of the section in the particular panel length segments (changing the breadth of shearing path), parallel driving of base and canopy, loading of the shield system which depends on its inclination angle and the pressure value registered while taking measurements. In the case of height range of longwalls A, B and C, the unit load-bearing capacity against the height of support section and pressure in props, obtained by the use of multiple regression is defined with the equation (1) below:

$$P_{jedn} = 0{,}0997 \cdot h + 0{,}0222 \cdot F - 0{,}257 \qquad (1)$$

where: P_{jedn} – unit load-bearing capacity [MPa]; h – working height of support section [m]; F – pressure in props [MPa].

Similarly, in the height range of longwall D the unit load-bearing capacity against the support section height and pressure in props, is defined with the equation (2) below:

$$P_{jedn} = 0{,}1422 \cdot h + 0{,}0146 \cdot F - 0{,}2841 \qquad (2)$$

In case of the plow longwall E, the unit load-bearing capacity for the support section (for working height ~ 1.2 m) was described according to the pressure in the props (3):

$$P_{jedn} = 0{,}018263 \cdot F - 0{,}05958 \qquad (3)$$

5 LOAD-BEARING CAPACITY VS SECTION LOADING

The results of maximum values of pressures in the space under prop pistons have been used to indicate the unit load-bearing capacity of the sections, developed by them during the course of the tested longwalls. The values of the developed load-bearing capacity of support sections, obtained for the longwalls as a result of the pressure measurements and calculations, may be treated as a reaction of the powered support on rock mass loading. Diagrams of the average values of unit load-bearing capacity of the particular sections during extraction of longwalls A, B, C and D have been shown in Figure 9. The red line presents the load-bearing capacity distribution along the working face approximated with polynomial function. Alternatively, the average value of the unit load-bearing capacity of all section build in a given longwall was marked with a green line. Figure 10 presents the load-bearing capacity distribution for the plow longwall E.

The highest values of the unit load-bearing capacity marked with A2, B2, C2, D2 points were located in the inner parts of the fields, most often in the areas of exploitation events influence (which have been outlined in the pictures). They were roughly 11÷25% higher from the value of the average unit load-bearing capacity developed in a particular longwall with roof caving.

On the longwall tips in the area of intersection with the excavations marked with A1, A3, B1, B3, C1, C3, D1 and D3 points, it may be spotted that the values of unit load-bearing capacity are roughly 14÷30% lower that the average values of section loading.

In case of the plow longwall the highest values of unit load-bearing capacity marked with E2' and E2" points, were located in the inner part of the field. They were roughly 13÷17% higher than the value of the average unit load-bearing capacity developed in a particular longwall.

On the tips of panel length E segment, in the area of intersection with the excavations marked with E1 and E3 points, the unit load-bearing capacity values (calculated as the average of 3 marginal sections), were roughly 7% lower than the average values of section loading in the whole longwall.

In case of the part of measurement data analysis of plow longwall E, the load-bearing capacity values are not as much diversified as the one in longwall with roof caving. However, it may result not only from the specifics of mining the plow longwalls, but also lower excavation. Moreover, the assessment refers to the smaller part of measurement readings.

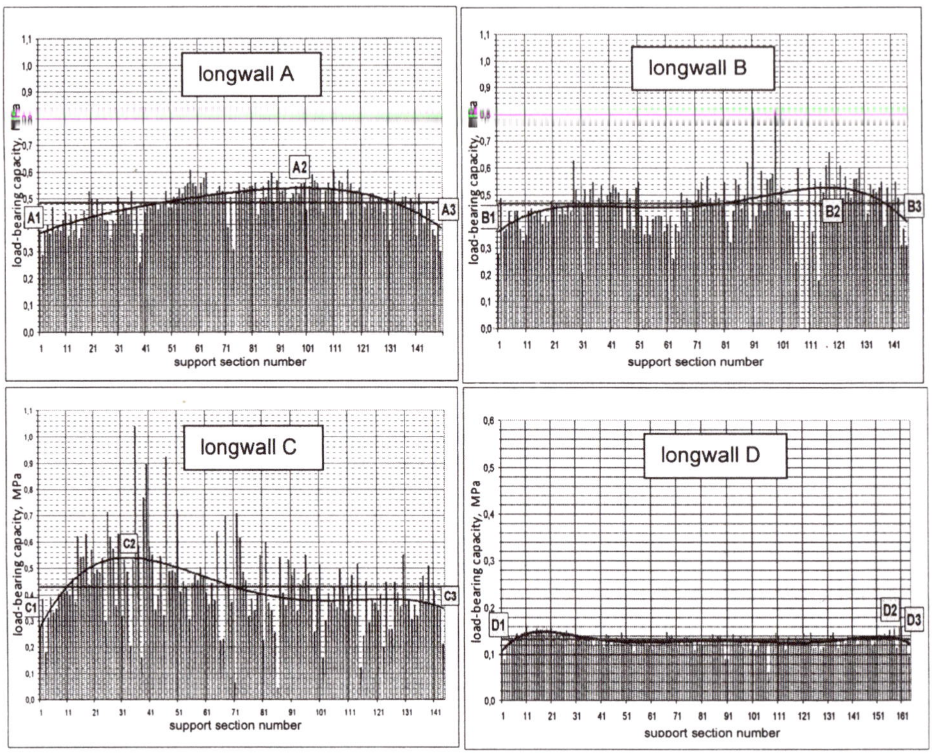

Figure 9. The average unit load-bearing capacity of section in longwall A, B, C and D

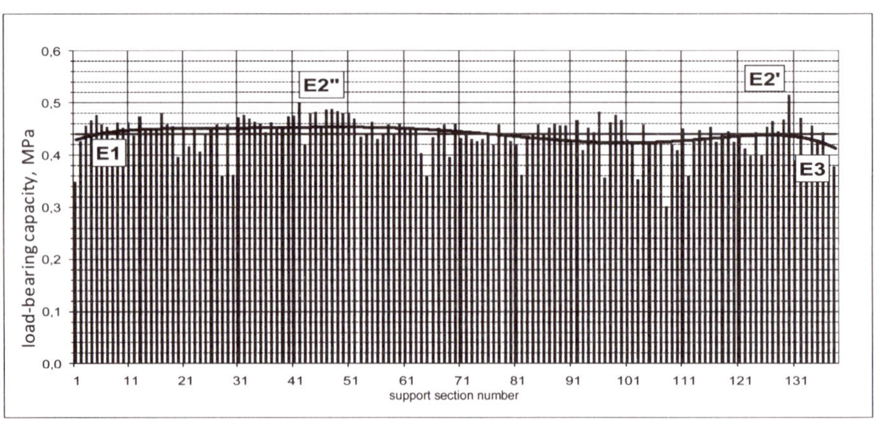

Figure 10. The average unit load-bearing capacity of section in plow longwall E

5.1 Bar charts of pressures distribution in section props in the tested longwalls

In the fields of the tested longwalls, mostly A, B and C, there was a significant variability of the unit load-bearing capacity values, which resulted from the registered fluctuations in the pressure from section props. Thus, the variability according to their pressures in the props was tested during the extraction in particular longwalls.

The prepared bar charts present the percentage of the field in longwall area, which include the maximum pressure value. The figures present a quantitative distribution of all registered pressure readings during the measurements with no relation to the selected maxims in the unit area, which have been drawn for comparison. In case of the plow longwall, only a single graph of quantitative distribution was possible regarding this part of data .

The bar charts have been organized and presented in Figures 11 and 12.

The case that relates to the reading of pressures in the space under prop pistons, thus, the load-bearing capacity (on account of slight difference in changes of the section height in the longwall) in shearer longwall D is distinctive. Its assessment was possible only after the proper organization of the pressure readings and drawing a quantitative distribution presented in Figure 11.

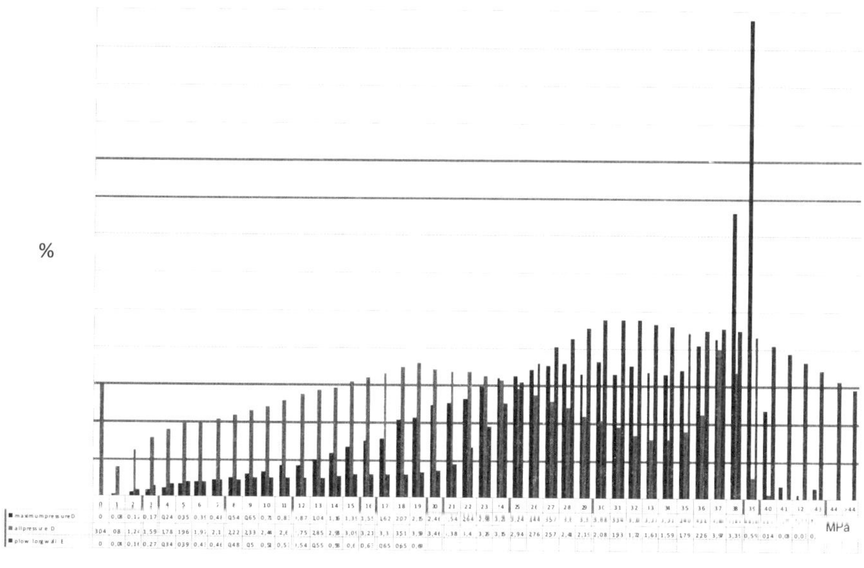

Figure 11. Percentage bar chart of pressure distribution in the section props in the shearer D and plow E longwalls

The average calculated for the whole set of data generally indicated low values of pressure which were situated close to the support setting load-bearing capacity value. Seemingly, it indicates that the load-bearing capacity was not fully used by sections. Only due to the relation of data to the area of extraction field, thus regarding the model of the phenomena of rock mass loading on the support, it was possible, potentially appropriate, to assess the development of the pressure view in the props – including load-bearing capacity of the section.

Due to the presented data selection, it was possible to draw conclusions basing on the diagrams depicted in Figure 11. The bar charts prepared from the set of all registered data shows that the raw data (green color) treated as the measurement set registered every 1 minute has relatively even intervals. Additionally, the amount of the registered measurements in the particular intervals of pressure values does not exceed the level of around 4% of the full set. Moreover, there are lots of measurements (62.8%) of values below the feeding pressure of the supports, 25 MPa. Thus, so many registered values of the setting load-bearing capacity have an impact on the course of the bar chart which was designed after organizing the maximum pressures (claret color) to the given area of longwall D field which has been measured. Its examination allowed to state the following:

- roughly 23.5% of the longwall field area, the support section reached values of pressures above 38 MPa, so close to the maximum yielding pressure,
- the maximum interval of quantitative distribution falls on roughly 20.4% of the longwall field area where the support sections reached the pressure values p within $38 \leq p \leq 40$ MPa,
- roughly 70% of the measurement-control field area, the support sections reached pressure values between setting pressure and yielding pressure, thus the proper scope of section operation. All the same, in the area of about 30%, the sections, for some reasons (e.g. failure, wrong maintenance) did not even exhibit the setting load-bearing capacity.

It can be stated that the pressures adjacent to the maximum working pressure, in the zone reaching ¼ of longwall field area, as well as envelope shape of the bar chart obtaining the maximum in the interval of $38 \leq p \leq 40$ MPa and falling directly under 42 MPa, are connected with symptoms of rock mass pressure as the aftermath of the small setting load-bearing capacity.

The low setting load-bearing capacity does not reduce sufficiently the rock mass movements in the direction of excavation, causing a relatively rapid increase of loading. It also favors the accumulation of periodical pressures from the rock mass connected with the delayed breakdown of superstrata behind the powered support.

The correlated diagram of plow longwall E (violet color) shows a more beneficial pressure distribution of which the highest percentage is situated in the higher pressure intervals than the setting one.

Undoubtedly, the method of running the powered support applied in the automatic system of setting load-bearing capacity control contributes that greatly.

Alternative pressure distributions for longwall A, B and C, presented in Figure 12, show a greatly 'flattened' course characteristics with maxims situated, in general, slightly above the area of setting load-bearing capacity. It seems that the phenomena largely results from the way the longwall monitoring system works. Thus, the data gathered is averaged by a proper equipment during acquiring in packages, transferred every 15 minutes. It would be more beneficial, regarding the possibility of performing analyses, to apply a system based on purely measuring principles without the necessity of initial data processing.

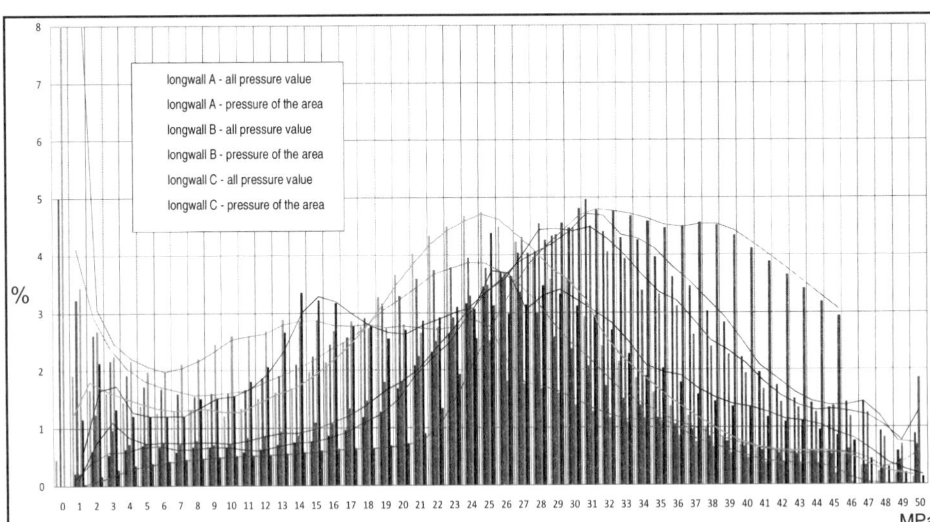

Figure 12. Percentage bar chart of pressure distribution in the section props in shearer A, B, C and plow E longwalls

6 SUMMARY

Upon analysis of the results of tests and measurement in the five longwalls with different kinds of supports driven by use of the plow and shearer mine systems, it is possible to spot some regularities which concern the view of the tested loadings in longwall. The measured pressure values in the space under props pistons are marked by high fluctuations, often between the adjacent sections. Nevertheless, even the partial measurement data from the plow longwall allows to observe the possibility of substantial reduction of this phenomena, for instance in case of the use of pressure regulation system in support props. It may also be seen a significant impact of the way the monitoring system works and data is gathered on the possibility of their interpretation. The access to raw data which is not processed by a system is very beneficial.

Looking at the average pressure values in the props, it can be stated that the powered supports often operate close to the values of setting load-bearing capacity. Additionally, in case of longwall D there are relatively low values of the developing unit load-bearing capacity. Different reasons may exist for the above-mentioned, which result from the following: insufficient section sprag, applying a section of high load-bearing capacity in, generally, favorable mining-geological conditions, the damage of hydraulic systems, driving the support with the wrong geometry etc. Answers to these questions will be clear after more measurements, analyses and experiences were performed.

In each longwall presented, a correspondingly high load-bearing capacity reserve, in relation to predicted rock mass loadings was being indicated, assuming that the section sets with pressure of 25 MPa.

Thanks to the presented observations and measurements, it is possible to draw the following conclusions:

1. The use of measurement and monitoring systems in longwall complexes allows to gather lots of important information, not only about the control function in the conditions of current extraction but also the following analyses.
2. In order to develop an effective measurement database, it is necessary to apply systems which allow to obtain a proper access to the completed initial data, without its processing by any program or equipment module.
3. Based on measurements, the presented calculations and dissertations revealed a significant impact of parameters and method of monitoring system work on the possibilities of gathering data and quality of the performed analyses.
4. By recording the pressure changes in the space under prop pistons, it is possible to indicate areas of different values of rock mass pressure in the longwall field caused by the influence of edges, residual pillars, gob surroundings, dislocations etc.
5. The measured level of the developed sections load-bearing capacity, related to their setting load-bearing capacity, proves that there are many areas in which the powered support sections have not been appropriately spragged.
6. Maximum pressure values assigned to the units of longwall field area are marked by significant variability along the longwall, which may not be beneficial for the cooperation between the support section and rock mass. The increased loadings can be observed in the central longwall part.
7. Load-bearing capacity values and their distribution turned out to be much less diversified in the plow longwall, which was undoubtedly influenced by automation of the section set in powered supports.
8. The loads influencing the marginal sections of powered support in the shearer longwalls turned out to be roughly 14÷30% lower than the average loadings of all sections in the longwall while in the plow longwall – about 7%. The difference, however, may result not only from the choice of mine system, but also different working heights of the excavation.
9. In order to perform a comprehensive analysis of loading and load-bearing capacity of the support, it is necessary to proceed with the measurements and tests, particularly in the diversified geological-mining conditions, to gather more data.

REFERENCES

Barczak T. 2006. A retrospective assessment of longwall roof support with a focus on challenging accepted roof support concepts and design premises. 25th Int. Conf on Ground Control in Mining, Morgantown, WV, USA, pp. 232-244.

Juárez-Ferreras R., González-Nicieza C., Menéndez-Díaz A., Álvarez-Vigil A.E., Álvarez-Fernández M.I. 2009. Forensic analysis of hydraulic props in longwall workings. Engineering Failure Analysis 16 (2009) 2357–2370, Elsevier.

Peng S.S. 1998. What Can a Shield Leg Pressure Tell Us? Coal Age, pp 54 57.

Stoiński K. 2000. Obudowy górnicze w warunkach zagrożenia wstrząsami górotworu. Główny Instytut Górnictwa, Katowice. (In Polish)

Szweda S. 2004. Identyfikacja parametrów charakteryzujących obciążenie sekcji obudowy zmechanizowanej spowodowane dynamicznym oddziaływaniem górotworu. Zeszyty naukowe Politechniki Śląskiej nr1648. Gliwice. (In Polish)

Trueman R., Lyman G., Cocker A. 2009. Longwall roof control through a fundamental understanding of shield–strata interaction. International Journal of Rock Mechanics & Mining Sciences 46 (2009) 371–380, Elsevier.

Trueman R., Lyman G. and Cocker A. 2008. Managing Roof Control Problems on a Longwall Face, in Aziz, N (eds.), Coal 2008: Coal Operators' Conference, University of Wollongong & the Australasian Institute of Mining and Metallurgy, 10-21.

Trueman R., Lyman G., Callan M., Robertson B. 2005. Assessing longwall support-roof interaction from shield leg pressure data. Mining Technology (Trans. Inst. Min. Metall. A), Vol. 114.

Płonka M. 2009. Rozkład obciążenia obudowy zmechanizowanej w ścianie zawałowej i jego powiązania z prognozą utrzymania stropu. Dokumentacja z realizacji projektu badawczego własnego, dokumentacja GIG nr 04040316-152, niepublikowana. (In Polish)

Płonka M. 2009. Measurements of load of power supports on caving longwalls. Школа подземной разработки. III Международная Научно-Практическая Конференция, Днепропетровск-Ялта.

Płonka M., Rajwa S. 2009. Podporność i rozkłady sił w węzłach sekcji obudowy zmechanizowanej. Wiadomości Górnicze, nr 10. (In Polish)

Płonka M. 2009. Statyczne i dynamiczne pomiary ciśnienia w stojakach obudowy zmechanizowanej w ścianach zawałowych. Wiadomości Górnicze nr 12. (In Polish)

Płonka M. 2010. Ocena obciążenia obudowy zmechanizowanej w ścianach zawałowych w oparciu o pomiary ciśnienia w stojakach. Polski Kongres Górniczy 2010. Katedra Elektryfikacji i Automatyzacji Górnictwa Politechniki Śląskiej, Gliwice. (In Polish)

Rajwa S. 2004. Wpływ konstrukcji stojaka hydraulicznego w ścianowej obudowie zmechanizowanej na utrzymanie stropu. Praca Doktorska. Katowice. (In Polish)

Rajwa S., Masny W., Prusek S. 2009. Powered Support in Conditions of Rock Mass Tremor Occurrence in the Polish Coal Mining. Aachen International Mining Symposia, 5th International Symposium - High Performance Mining, RWTH Aachen.

Rajwa, S., Płonka M., Lubosik Z., Walentek A., Masny W. 2008. Principles of safe use of powered supports. School of UNDERGROUND MINING, UKRAINA, Jałta.

International Mining Forum 2011, Dyczko et al. (eds) © 2011 Taylor & Francis Group, London, UK. ISBN 978-0-415-68938-0

Maintaining of the Headgate Behind the Exploitation Face in Difficult Geological and Mining Conditions in LW "Bogdanka" Mine

Z. Rak, J. Stasica
AGH University of Science and Technology, Krakow, Poland

ABSTRACT: The increasing costs of driving longwall headings result in systematic research concerning their maintenance behind the exploitation face in Polish coal mines. This paper describes the results of such attempts which were undertaken in relatively difficult conditions in LW "Bogdanka" mine in the 80s. of 20th century and – after a long break – in 2010. The article gives detailed description of the technology designed to protect the headgate 1/VI in the seam 385/2 maintained behind the exploitation face in the conditions of LW "Bogdanka" mine. The method of securing the headgate from the beginning of its drive until the last reinforcements built-in behind the longwall face. Previous results of the attempts to maintain the heading in one-sided surrounding of gobs were also described.

KEYWORDS: hard coal mine, longwall headings, costs of driving

1 THE EXPERIENCES CONCERNING THE MAINTENANCE OF LONGWALL HEADINGS BEHIND THE EXPLOITATION FACE IN THE CONDITIONS OF LW "BOGDANKA" MINE

The exploitation of hard coal in "Bogdanka" mine started in November 1982. Until 1990 longwall system with roof collapse "retreat" was used exclusively. Both longwall headings were driven together with the advance of the longwall and were maintained behind the exploitation face. In this system the problem of securing and maintaining the longwall headings in one-sided surrounding of gobs had a fundamental role. This was a serious issue which caused a lot of trouble since the beginning of exploitation. Due to this fact, numerous experiments and research were carried out. In the late 80s they led to the introduction of costly and labour consuming, but relatively effective solutions of driving and maintaining longwall headings behind the face in the so-called "advance" system. Between 1982 and 1986 a system of driving the longwalls "to the borders" was developed in "Bogdanka" mine. The following parameters of longwall headings support were commonly used:

- The headings were driven with a minimal advance in front of the longwall face (8 m),
- The support of the longwall headings was a mixed support using a V36 section – the roof-bars were two-element ŁP-10s and the side-walls arches had the ŁP-8 sizes,
- The spacing between the sets was usually 0.5 m,
- The fill zone which protected the heading from the gobs' side was made from a crib built from old rails and filled with stone, anhydrite and additionally supported with a series of V-shaped props built on straight elements from the same section.

The considerable lateral area of the longwall headings comparing to contemporary standards, labour consuming technology of driving and constructing the support where both loading of the output and construction of the support were manual as well as the difficulties connected with huge amount of work at the crossing of the longwall with the heading resulted in only about 2.5-meter-per-day advance of the heading. The basic elements of the system used to secure and maintain the longwall headings behind the face were so-called joint support and protective belt along the headings. To eliminate the destruction of support sets due to the side pressures in the headings in the zone of strong influences of exploitation pressures up to 40 m behind the exploitation face (from the coal mass side or from the old gobs' side), special sets of heading support were used. They had side-wall arches jointly connected with the roof element (so-called joint support). This kind of support allowed to cut the heaved side-wall (usually twice) and divert the side-wall arch back to the original position. In 1987 an experiment was carried out in a research longwall (no. 2/I in 382 seam) in order to optimize the construction of the protective belt of the heading from the caving side. Four experimental segments were made: with a narrow belt (without wooden crib) and with a wooden crib; both variants with an anhydrite fill and without anhydrite. The constructions of the protective belts in those experiments have been shown in Figure 1.

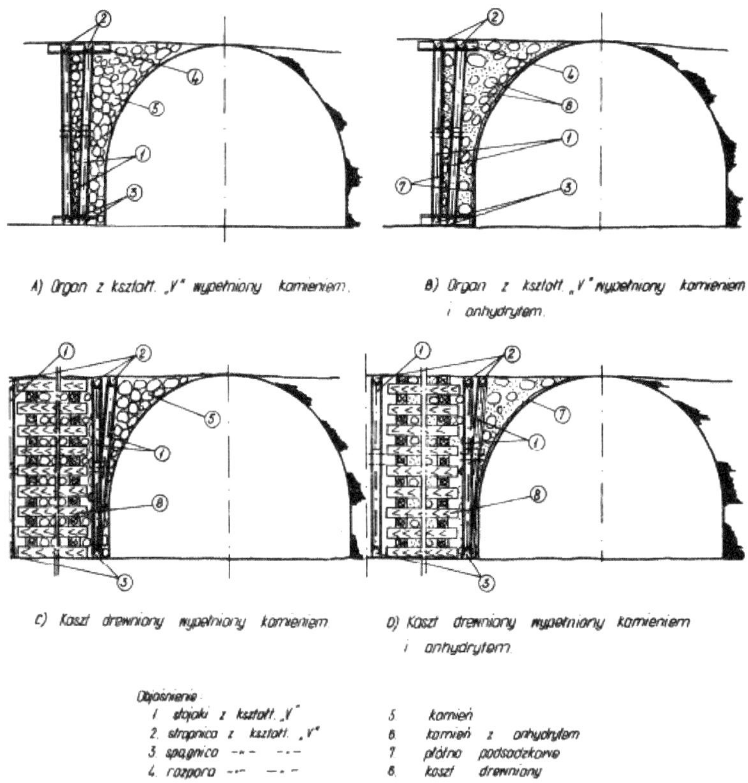

Figure 1. Different methods of securing the longwall heading used in "Bogdanka" mine

Notations: A) Organ z kształt. V wypełniony kamieniem – *A) V-shaped body filled with stone;* B) Organ z kształt. V wypełniony kamieniemi anhydrytem – *B) V-shaped body filled with stone and anhydrite;* C) Kaszt drewiniany wypełniony kamieniem – *C) Wooden crib filled with stone;* D) Kaszt drewiniany wypełniony kamieniem i anhydrytem – *D) Wooden crib filled with stone and anhydrite;* Stojaki z kształt. V – *V-shaped props;* Stropnica z kształt. V – *V-shaped roof-bar;* Spągnica z kształt. V – *V-shaped floor-bar;* Rozpora z kształt. V – *V-shaped sprag;* Kamień – *Stone;* Kamień z anhydrytem – *Stone with anhydrite;* Płótno podsadzkowe – *Filling linen;* Kaszt drewniany – *Wooden crib*

Following conclusions may be drawn from these and previous experiments:

- The use of wooden cribs filled with stone and reinforced with a series of V-shaped props lead to a definite decrease in convergence of the roof and an improvement in the maintenance of the longwall headings.
- The use of joint side-wall arches from the coal mass side eliminated the destructive influence of the creeping side-wall and lead to a significant decrease in horizontal tightening. It also efficiently prevented deformations in roof and side-wall parts of the support.
- Further improvement in the conditions of maintaining the longwall headings was possible due to the use of anhydrite binding agents to seal the cribs and the space between the crib and the support. It especially improved the work of the support from the caving side while ensuring full insulation from the gobs.

Only the system to maintain longwall headings behind the face based on three aforementioned elements ensured relative efficiency in protecting the headings in the used "advance" exploitation system.

The final construction of the longwall headings protective belt used in the late 80s is shown in Figure 2. This complicated method of protecting the headings has become one of the main reasons the "advance" exploitation system was rejected. It was a restraint in the advances of the longwalls. This system was characterized by very costly and labour consuming maintenance of longwall headings after the passing of the face and a need to use the support with joint side-wall arches, with a construction of the protective belt with a few rows of horseheads and steel props, with a wooden crib filled with anhydrite binding agent. Furthermore, there was a frequent need to cut the coal side-wall and the floor which accounted for volume often greater than the volume of the driven heading. Sometimes even a total reconstruction of the heading was necessary (Phot. 1).

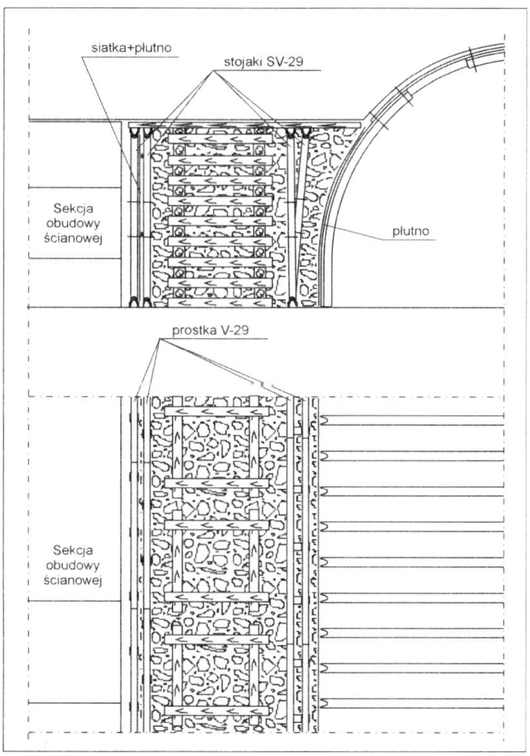

Figure 2. The final construction of the protective belt in the late 80s

Notations: Siatka+płótno – *Net+linen;* Stojaki SV-29 – *SV-29 props;* Sekcja obudowy ścianowej – *A set of mechanised longwall support;* Płótno – *Linen;* Prostka V-29 – *V-29 straight element*

From 1990 a gradual change in the exploitation system was made – the "advance" system with the driving of longwall headings and their maintenance after the passing of the face, was replaced with the "retreat" system where the longwall headings were driven in advance and liquidated together with the advance of the longwall face. This change in exploitation system has become a turning point in the capabilities of improvement in longwall advances and increasing efficiencies of work in LW "Bogdanka" mine. This system with only single time use of the longwall heading has been used for almost 20 years.

Phot. 1. The condition of the longwall heading before the reconstruction1/I– 1984

Despite this fact, from March to November 2010 an attempt has been made to maintain a longwall heading behind the exploitation face and use it for the next longwall. Several arguments support such action, but the fact that we now have new, more effective technologies and technical means to maintain longwall headings is the most important one. Some of them have been successfully implemented in LW "Bogdanka" for many years and the following are most important:

- New constructions of highly-supportive arch support based on steel with improved durability parameters,
- Binding the rock mass,
- Low and high bolting,
- Mechanical lining behind the support.

Furthermore, in mining all over the world many efficient solutions of securing the headings behind the exploitation face have been developed. These solutions use a variety of constructions (props, cribs and belts), supported with highly effective binding agents.

Except the aforementioned technical arguments, advantageous organizational conditionings occurred. In the 1st quarter of 2010 a first plow longwall 1/VI has stated its operation in 385/2 seam. This longwall has research character and its main goal is to implement the plowing technique. Therefore it was possible to simultaneously conduct a trial to maintain the heading behind the exploitation face in order to use it for the planned 2/VI longwall. The experiments described below were realized basing on a design prepared by a team of authors from AGH's Department of Underground Mining and engineering staff from LW "Bogdanka" with managerial role of the authors of this article.

2 GEOLOGICAL AND MINING CONDITIONS OF 385/2 SEAM IN THE HEADGATE 1/VI AREA

1/VI headgate was driven in seam 385/2 at the depth of 950 m. In the area of the heading, the seam 385/2 has been classified to 1st degree of water hazard. The inclination of the seam and surrounding layers is about two degrees. The thickness of the seam 385/2 at the panel length (about 1800 m) ranges from 1.3 to 1.6 m. Lithology of roof and floor layers shows significant variability. In general it may be assumed that directly above the seam a 0.8-to-2.0-meter-thick layer of siltstone occurs. Directly above the siltstone a layer of mudstone is deposited. Its thickness ranges from 0.2 to 7.0 meters and the greatest thickness may definitely be observed at the middle of panel length. Over the mudstone a layer of sandstone is deposited. It is 0 to 7 meters thick, locally interbedded with single pads of mudstone. Above the sandstone alternating layers of mudstone, siltstone and coal may be found as well as locally appearing layers of sandstone and mudstone soil. A layer of siltstone with thickness ranging from 0.3 to 1.5 meters exists at almost full length of seam's floor. Locally, directly under the seam siltstone soil may be found. Sphaerosiderite nodules may be observed relatively often, especially in siltstones and mudstones. No tectonic or sedimentary dislocations may be found at the heading's length. The excavation is also not affected by past exploitation events.

The parameters of the seam and surrounding rocks are relatively well identified. During the drive of the heading seven exploratory boreholes in the roof and floor were drilled to conduct penetrometer measurements, collect the cores and analyze them in a laboratory. Durability and deformation parameters of the rocks and coal are relatively low (Table 1).

Table 1. Properties of coal in 385/2 seam together with surrounding rocks

Layer	Young module E [MPa]	Poisson coefficient [υ] [-]	Internal-friction angle φ [°]	Cohesion c [MPa]	Tensile strength R_r [MPa]
Mudstone	11,760	0.25	44.27	1.875	1.15
Siltstone	4600	0.25	44.13	1.860	1.15
Coal	1040	0.39	45.34	0.699	0.4
Siltstone	4600	0.20	43.95	0.320	0.2
Mudstone soil	2000	0.20	37.38	0.229	0.2

To summarize one may state that the geological and mining conditions and physical and chemical parameters of the rocks in longwall 1/VI In seam 385/2 are typical of "Bogdanka" mine. In general one may classify them as relatively difficult where maintaining the headings in one-sided surrounding of gobs requires undertaking of complex precautions. The maintaining of the headings is especially complicated by the layers of siltstones and soils which exist directly over and under the seam. These are relatively weak layers, with tendencies to crack intensively even in the maintenance phase without the influence of the exploitation.

3 THE METHOD TO MAINTAIN THE 1/VI HEADGATE IN THE SEAM 385/2 IN ONE-SIDED SURROUNDING OF THE GOBS

Longwall 1/VI was the first one in "Bogdanka" mine equipped with a plow complex. Due to the necessity to cut the plow into the coal mass in the longwall headings and significant overall dimensions of the plow's and longwall conveyor's drives, the headgate 1/VI was designed with ŁPS/C 12/S support which has relatively large dimensions (Figure 3). Due to this fact the heading was driven with a significant dinting and topping. ŁPS/C set is a kind of set designed for the specific conditions of "Bogdanka" mine. Differently than in other types of sets, the side-wall arches in the near-floor part are diverted a little towards the side-walls. This kind of construction allows to decrease the effect of losing the active usable face area because of horizontal tightening of the heading (also from floor elevation). This system has proven itself in the most difficult conditions of this mine. In the analyzed heading V36 sets, made of steel characterized by increased durability parameters (S480W) which are a standard in the mine, were used. The set spacing was 0.9 meters.

The protection of the excavation maintained in one-sided surrounding of gobs where the durability and deformation parameters of the rocks around the seam are relatively low required many precautions even in the driving phase. These precautions should minimize discontinuous deformations (cracks) in their surroundings. For a few years mechanical lining has been used for this purpose in "Bogdanka" mine (Phot. 2). The lining bags are filled with mineral binding agents and located between roof and side-wall arches and the heading's contour they efficiently decrease the rock mass deformations and prevent it to relieve stress too quickly. Early introduction of the support allows to decrease the relief in the roof and the side-walls significantly during the whole time of heading maintenance.

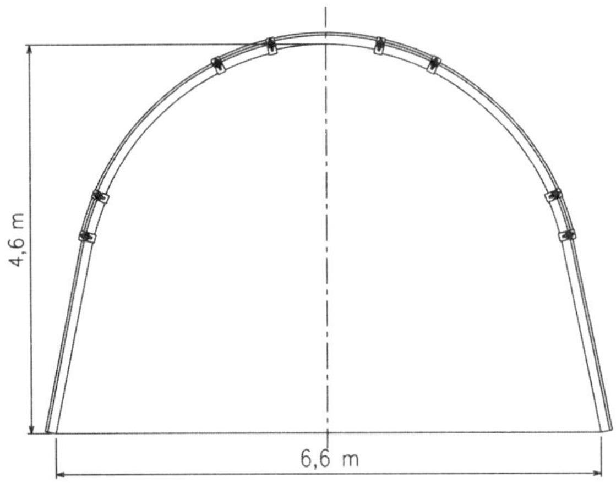

Figure 3. ŁPS/C 12 set

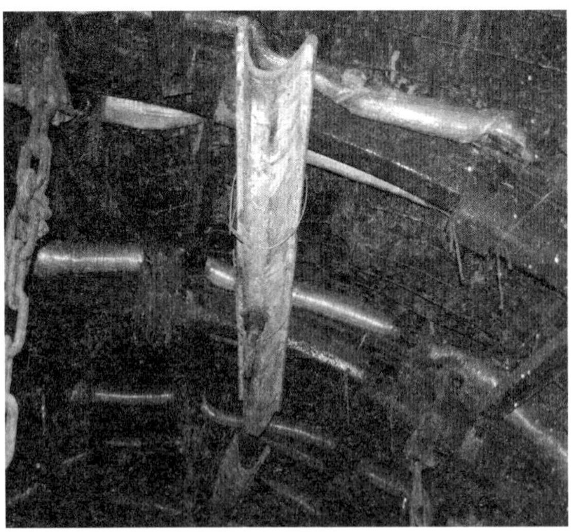

Phot. 2. Mechanical lining

Bolting is another action to improve the deformation parameters of the roof-bar over the heading. In 1/VI heading, directly by the face during its driving, rod bolts with 160 kN carrying capacity and total length of 2.7 meters were installed. The bolts were built-in symmetrically alternating 6 and 7 pieces in the spaces between the ŁPS/C support sets. The bolts are not connected with the support sets because their task is to reinforce the rock mass and therefore they are built-in individually to hold the roof through backing strips and the astel net in the heading. The aforementioned method to protect the heading remains unchanged until the exploitation begins. Further reinforcements are conducted in four stages, namely about 100 m before the longwall face, directly before the face, at the crossing of the face with the heading and behind the face.

Before the exploitation front, namely before the expected exploitation pressure occurs, at minimum 100 m in front of the longwall face the excavation is protected with a system of bolts and horseheads installed in the following way:

- Two rows of short horseheads (about 1.3 m long) are tied with every set made of V32 elements at minimum and bolted in the middle with string bolts with minimum carrying capacity of 280 kN and total length of 6.0 m (Phot. 2 and Phot. 4),
- The bolts and horseheads are built-in alternating in every second space between the ŁPS/C support sets, namely one is in the vicinity of the axis (or negligibly diverted towards the unexploited side-wall) and the other is built-in about 1.5 m from the axis towards the 1/VI longwall side-wall,
- Two rows of horseheads V32 at minimum are built-in overlapping at minimum one set, tied with every set and situated symmetrically in the distance of 1.8 m from the axis of the heading.

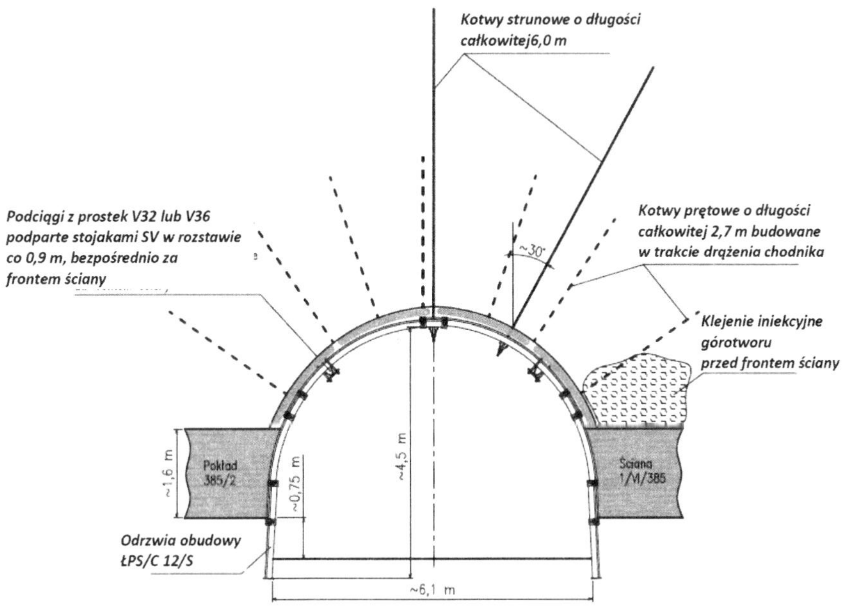

Figure 4. The precautions in the heading in front of 1/VI longwall

Notations: Kotwy strunowe o długości całkowitej 6,0 m – *String bolts of total length 6.0 m;* Podciągi z prostek V32 lub V36 podparte stojakami SV w rozstawie co 0,9 m, bezpośrednio za frontem ściany – *Horseheads made from V32 or V36 straight elements supported with SV props with 0.9-m-spacing, directly behind the longwall face;* Kotwy prętowe o długości całkowitej 2,7 m budowane w trakcie drążenia chodnika – *Rod bolts of total lentgth 2.7 m, built-in during the driving of the heading;* Klejenie iniekcyjne górotworu przed frontem ściany – *Injective binding of the rock mass in front of the longwall face;* Pokład 385/2 – *Seam 385/2;* Ściana 1/VI/385 – *Longwall 1/VI/385;* Odrzwia obudowy ŁPS/C 12/S – *ŁPS/C 12/S suport set*

The described precautions (Figure 4), together with the preparation of the excavation to be maintained behind the exploitation face, allow to maintain the crossing of the longwall with the heading without props or even to dismantle the side-walls arches in advance in order to make a small cavity and clean the corner of the longwall with a floor loader. Binding the roof is another treatment that is meant to prepare the heading to be maintained behind the exploitation face. The binding is only temporary and applied if cracks in the roof directly by the heading were noticed (Figure 4). These works were realized from a few to a dozen meters in front of the longwall face. Liquid two-component resins were used. The purpose of the injection was to bind the layers together to reinforce the stability of the roof-bar which was supported directly behind the face. At the crossing with the longwall, the heading was protected through astel, roof-bars, props and cribs. One of the most important elements of this technology is to build a wooden crib made of prefabricated units made of hard beech wood with carrying capacity at least 1500 kN and outside dimensions 1.2 x 1.2 meters (inside dimensions 0.9 x 0.9 m). The cribs were built directly behind the individual support of the cavity one by the other and created a relatively tight system that way (Figure 5, Phot. 3 and 4).

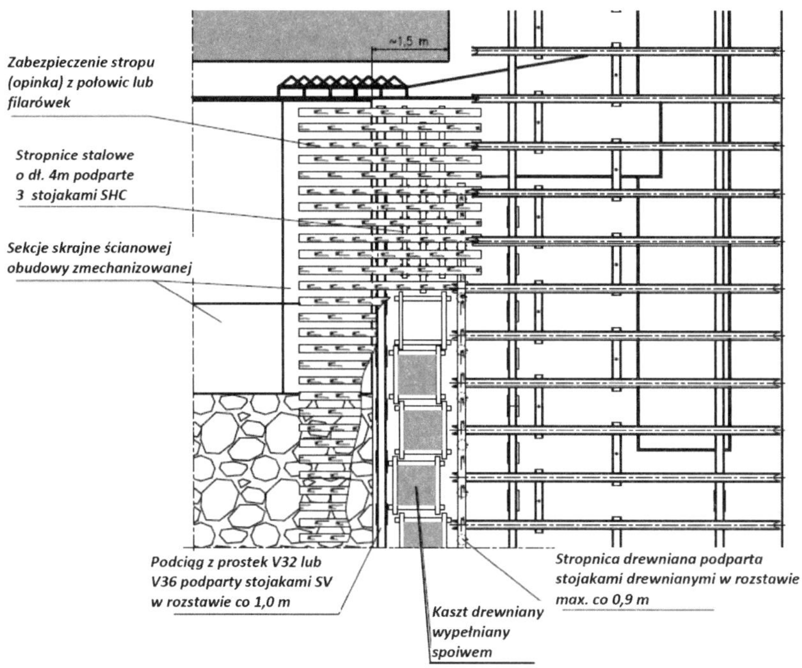

Figure 5. The method of building cribs behind the longwall face

Notations: Zabezpieczenie stropu (opinka) z połowic lub filarówek – *Roof protection (lagging) made from half-elements of pillar-elements;* Stropnice stalowe o dł. 4 m podparte trzema stojakami SHC – *4-meter-long steel roof-bars supported with three SHC props;* Sekcje skrajne ścianowej obudowy zmechanizowanej – *Outermost sets of mechanised longwall support;* Podciąg z prostek V32 lub V36 podparty stojakami SV w rozstawie co 1,0 m – *Horsehead made from V32 or V36 straight elements supported with SV props with 1.0-meter-spacing;* Kaszt drewniany wypełniony spoiwem – *Wooden crib filled with binding agent;* Stropnica drewniana podparta stojakami drewnianymi w rozstawie max. co 0,9 m – *Wooden roof-bar supported with wooden props with max. 0.9-meter-spacing*

After the construction, the crib was precisely bolted under the roof which was earlier protected with a wooden astel. Before the construction of the crib the floor was thoroughly cleaned to remove the loose material. Leaving the loose material would lead to the susceptibility of the crib and fast deflection of the roof-bar and in consequence to the additional weight put on the support. Apart from the crib, additional horseheads and SV steel props were built-in. In the last phase, namely directly behind the exploitation face some final reinforcements were made to maintain the heading in one-sided surrounding of the gobs. The final state of precautions undertaken in 1/VI heading is shown in Figure 6.

Phot. 3. Construction of the crib seen from the heading side

After the construction, the crib was precisely bolted under the roof which was earlier protected with a wooden astel. Before the construction of the crib the floor was thoroughly cleaned to remove the loose material. Leaving the loose material would lead to the susceptibility of the crib and fast deflection of the roof-bar and in consequence to the additional weight put on the support. Apart from the crib, additional horseheads and SV steel props were built-in. In the last phase, namely directly behind the exploitation face some final reinforcements were made to maintain the heading in one-sided surrounding of the gobs. The final state of precautions undertaken in 1/VI heading is shown in Figure 6.

Phot. 4. The final phase of the crib construction seen from the longwall side

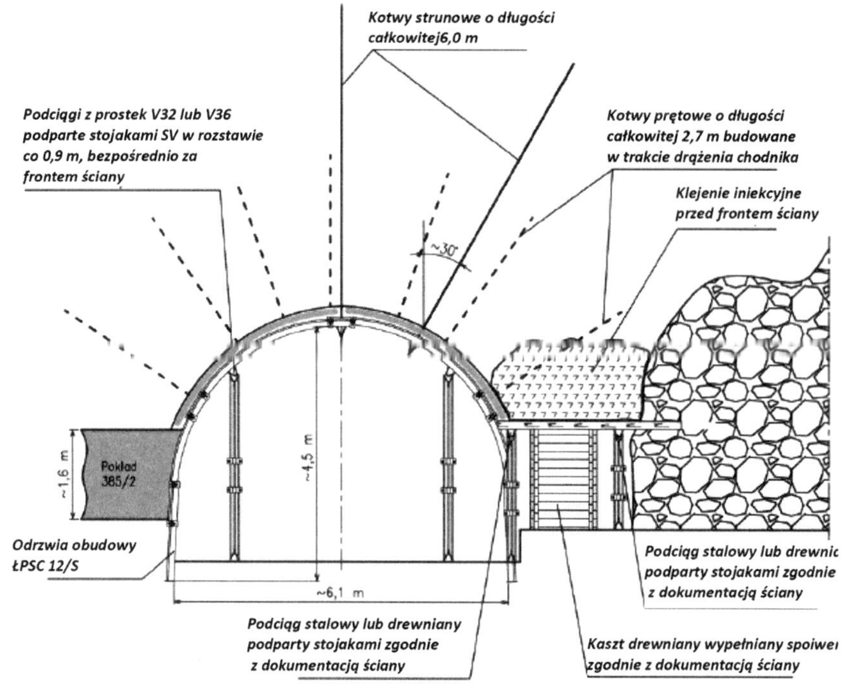

Figure 6. Final state of precautions in 1/VI heading behind the longwall face

Notations: Podciągi z prostek V32 lub V36 podparte stojakami SV w rozstawie co 0,9 m, bezpośrednio za frontem ściany – *Horseheads made from V32 or V36 straight elements supported with SV props with 0.9-m-spacing, directly behind the longwall face;* Kotwy strunowe o długości całkowitej 6,0 m – *String bolts of total length 6.0 m;* Kotwy prętowe o długości całkowitej 2,7 m budowane w trakcie drążenia chodnika – *Rod bolts of total lentgth 2.7 m, built-in during the driving of the heading;* Klejenie iniekcyjne górotworu przed frontem ściany – *Injective binding of the rock mass in front of the longwall face;* Odrzwia obudowy ŁPS/C 12/S – *ŁPS/C 12/S suport set;* Podciąg stalowy lub frewniany podparty stojakami zgodnie z dokumentacją ściany – *Wooden or steel horsehead supported with props accordingly to the longwall documentation;* Kaszt drewniany wypełniony spoiwem zgodnie z dokumentacją ściany – *Wooden crib filled with binding agent accordingly to the longwall documentation*

At this stage, the most important thing when it comes to the stability of the heading is to fill the crib constructed earlier with fast-binding and highly efficient mineral-cement binding agent called Reapack (Phot. 5 and 6). If it was possible, the filling was realized together with the advance of the longwall face. Because if the nature of the crib's work and the filling technology it was assumed that the filling should be completed with a delay of no more than two cribs if counting from the longwall side. The binding agent was delivered using the dry method with a help of a cement gun. This technology allowed to maximally use the parameters of the binding agent which are shown in Table 2. In consequence in just one hour from the filling of the crib, the durability of the material was no lower than 5 MPa. Unusually low hydration heat was one of the most important reasons why Reapack binding agent was chosen (apart from the physical and chemical parameters). The temperature inside the column during the binding and hardening of the binding agent is not greater than 40°C. As a result, no structural changes were observed and therefore the durability parameters of the binding material also have not changed. These kind of changes caused by excessive temperature (in typical materials above 60°C) inside the props or belts are often a cause of significant loss of carrying capacity and – in consequence – the loss of stability in the headings in one-sided surrounding of the gobs.

Phot. 5. Built-in cribs before filling with the binding agent

Phot. 6. Heading side-wall after the filling of the cribs

Table 2. Reapack binding agent parameters (according to the technical documentation)

Compression strength Rc [MPa]					
1 h	5 h	24 h	48 h	7 days	28 days
6.7	8.8	18.1	23.5	29.8	48.5
Bending strength Rc [MPa]					
1 h	5 h	24 h	48 h	7 days	28 days
1.1	1.9	3.9	4.5	5.5	6.1
Binding time					
Beginning			End		
55 sec.			4 min. 50 sec.		

The vectors' directions and the total dislocations values, the range of the rock mass plasticization area as well as the values of axis strengths in the supporting elements of the support and bolts were determined during numerical calculations performed using the finite element method. FLAC 3D programme produced by the company Istasca was used.

For the computing purposes a three-dimensional model of the rock mass containing 641,400 areas has been built. The rock mass was modeled as a Coulomb-Mohr medium. Due to the significant depth it was assumed that the state of pressure is characterized with equal values of normal components of the pressure tensor in all directions. During the calculations the degree of effort of all the elements that build the heading's precautions was obtained together with the rock mass' range of plasticization area around it (Figure 7).

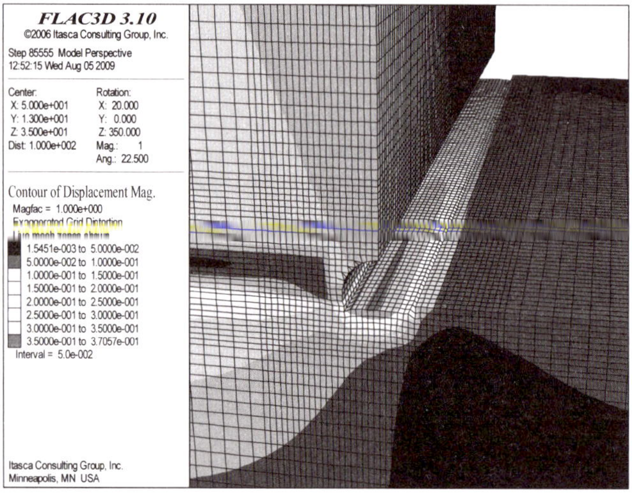

Figure 7. Total dislocations

The calculations were conducted in three stages, namely for the heading maintained in front of the longwall face, at the crossing of the longwall with the heading and behind the longwall face. The basic conclusions for the final stage are shown below. The use of rod bolts and string bolts allows to reduce the damaged area to about 5 m. Maximal axis strengths in the string bolts are around 285 kN behind the longwall face. It is worth mentioning that the most intensive increase in the load put on those bolts was observed at the crossing with the longwall where the bolts have already been loaded with almost critical strength. Maximal axis strengths in the rod bolts were around 184 kN and were noted in the longwall face area. Behind the face one may practically not observe any increase in the load values, but the highest values were observed in both side bolts both from the caving side and the undisturbed side. SV props built at the crossing of the longwall and the heading were supporting the horseheads and had the maximum load equal to 200 kN only 10 m behind the longwall face. The results of load capacity in the cribs filled with the binding agent deserve special attention. Similarly as in the case of SV props, the cribs received the nearly-maximal load (around 1150 T) very quickly, just 10 m behind the longwall face.

Two very important conclusions may be drawn as a summary of the numerical calculations:

- In the analyzed rock mass conditions the most intensive increase in the load put on the elements securing the heading appears relatively quickly (up to a dozen meters behind the longwall face),
- The rod and string bolts system reduces the damaged area by about 20% what results in a significant reduction in the load put on the props, including the cribs behind the longwall.

It may be expected that in real conditions the increase in load put on the support and reinforcements occurs a little slower. This is caused by some susceptibility of all of the props built at the crossing and behind the longwall face and which result from partial plasticization of the floor. The elevation of the floor at the longwall-heading crossing reaches a few dozens of centimeters while simultaneously leveling (almost to zero) the step between the heading and the longwall. Despite the cleaning of the floor before the construction of props and cribs, some degree of susceptibility in unavoidable in real mining conditions. However even this small delay in the load increase speed, we must be prepared that around 10–20 m behind the longwall face the loads on crib or the fill belt may reach maximal levels.

4 PREVIOUS EXPERIENCES CONCERNING THE MAINTENANCE OF HEADGATE 1/VI IN ONE-SIDED SURROUNDING OF GOBS

During the exploitation and after it finishes wide support stability and rock mass deformations measurements were carried out. In 19 measurement posts the following parameters were observed periodically:

- The change of height and width of the heading behind the exploitation face,
- Support dislocations behind and in front of the longwall face,
- Fissures of the roof rocks in the bolted area behind the longwall face,
- The distribution of cracks in the rock mass over the cribs reinforcement,
- The height of the roof over cribs reinforcements in relation to the roof in the longwall 1/VI field.

Before some of the measurement results and observations are briefly characterized, it must be mentioned that the method of securing the heading was slightly modified. The main goal of these modifications was to verify the possibility to reduce the lateral dimensions of the cribs built directly behind the longwall. The first sector of 1/VI heading about 760 m long (if counting from the starting cross heading) was constructed using the technology described in chapter 2. The cribs with lateral external dimensions of 1.2 x 1.2 m (internal dimensions 0.9 x 0.9 m) were used. The condition of the heading in this sector was assessed as very good (Phot. 7). No overload of the support sets or damages to the rod or string bolts have been noticed. The support sets maintain their arch shape without any deformations that would demonstrate their asymmetrical load. The average vertical convergence in the heading was less than 1800 mm. Elevation of the floor was noticed, locally reaching even more than 1.5 m. But with the height of the excavation equal to 4.6 m this fact does not generate any operating problems. Basing on previous experiences in "Bogdanka" mine much higher floor elevations were expected. It has also been noticed that the elevation process was gradually atrophic.

From circa 760 meter of the heading length the attempt was made to maintain the heading using smaller cribs. The previously used 1.2 x 1.2 m cribs were replaced with cribs of external dimensions of 0.9 x 0.9 m (internal dimensions 0.6 x 0.6 m). The lateral area of the binding agent filling the column in the crib was reduced from 0.81 m^2 to 0.36 m^2 which means over 50% reduction. The trial sector was about 100 m long. Because in the first phase of the use of the reduced cribs no disturbing symptoms of overloading the heading support were observed, a decision to continue the experiment was made. Unfortunately, after another 400 m of longwall advance first disturbing warning signs appeared, namely the roof-bars of the support were being straightened. The deformation of the heading support now advanced relatively quickly and affected the whole length of the heading where the reduced cribs were used. Despite building-in horseheads supported in the axis of the heading it was not possible to stop the advancing deformation process and local cracking of the roof-bars. The visible effect of the asymmetrical load on the support was a deflection towards the undisturbed side-wall. This means that – in conformity to the theory – that the maximal load of the heading resulting from the deflecting roof-bar, occurs between the axis of the heading and the side-wall from the gobs side. The direct cause of this situation is the advancing process of squeezing in the cribs filled with the binding agent into the relatively low-compact and easily plasticizing siltstone floor. The subsidence of the cribs reached a few dozen of centimeters and in consequence caused (apart from deformations in the roof parts of the heading support) increasing elevation of the floor next to the side-wall by the cavity. Obviously it cannot be ruled out that the decrease in heading stability in the sector of the heading with reduced cribs is a fault of the worsening geological conditions. However, the conducted measurements and observations suggest that the reduction in cribs' dimensions was the main reason of then existing state of the heading.

Phot. 7. The heading maintained behind the longwall

As a result of the failure in this phase of the experiment, the mine decided to go back to the original idea of using the cribs with larger dimensions. At the last 300 m of the panel length the heading was secured using the same technology as in the first sector. As a result the support condition has improved significantly and could be characterized as good.

Although the experiences described above allow to draw some conclusions, the real effects of the undertaken attempt to maintain the heading 1/VI will be known only in the future. The real test of the effectiveness of the means used in this heading will be conducted when it will serve as a tailgate for the next longwall and will be subjected to more exploitation pressures. In reality the described experiment is not about maintaining the heading just for the sake of maintaining, but about the possibility to re-use it in the future.

5 SUMMARY

As it was mentioned earlier, it is too soon to give a definite summary concerning the attempts to maintain the longwall headings in the conditions of "Bogdanka" mine, but the experiences gained so far allow to draw some conclusions:

- The preparation of the heading for maintenance behind the longwall starts in the driving phase. Not only the support type, its size, the size of the elements, set spacing, but also additional reinforcement of the rock mass with bolts and the method and quality of the lining behind the support are extremely important from the point of view of stability behind the longwall face,
- Long string bolts are very valuable tools for the improvement in stability of the excavation in every phase of its drive. While supporting the sets with horseheads they reinforce the heading in front of the longwall, allow to maintain the crossing without props and significantly reduce the area of cracking zone behind the longwall while not decreasing the area of the heading,
- The choice of cribs, filling belt or SV props as a prop type behind the wall should be based on two criteria: the carrying capacity of the prop and the effort of the floor in the area of its construction. Too small lateral dimensions of the props may result in their "drowning" in the floor and – in consequence – in overloading the support of the excavation,
- Low coal and surrounding rocks parameters with relatively high face advances may generate significant loads on the heading support in close proximity to the longwall. Due to this fact, the materials used to fill the belts or cribs should be characterized by an intensive hardening

process in first dozen hours. In the conditions of mines like "Bogdanka" where daily advances of the longwalls reach 10 m this issue is especially important,

- A satisfying speed of binding and hardening of the materials used to fill the cribs and belts may be achieved by the use of certain mineral-cement binding agents delivered using the dry method with a help of concrete guns,
- When taking the significant filling masses into account (huge dimensions of cribs and belts) and the lack of possibility to cure the binding agents, one must search for agents which generate little amounts of heat during binding and hardening. The overheating of the agents leads to loss of their mechanical and structural parameters.

REFERENCES

Korzeniowski W. et al. 2011. Analiza stanu utrzymywanego chodnika podścianowego ściany 1/VI w pokładzie 385/2, pod kątem możliwości i warunków dalszego wdrażania technologii utrzymywania chodników przyścianowych za ścianą w warunkach Kopalni "Bogdanka". Praca niepublikowana. Kraków. (In Polish)

Rak Z. et al. 2009. Opracowanie wzmocnienia obudowy łukowej chodnika podścianowego ściany 1/VI dla jego utrzymania w jednostronnym otoczeniu zrobów w pokładzie 385/2 Kopalni Lubelski Węgiel "Bogdanka" S.A. Praca niepublikowana. Kraków. (In Polish)

International Mining Forum 2011, Dyczko et al. (eds) © 2011 Taylor & Francis Group, London, UK. ISBN 978-0-415-68938-0

Design of Bolting Support for Maintenance of Longwall Headings for Plough Longwalls

P. Głuch
Silesian University of Technology, Gliwice, Poland

ABSTRACT: The article discusses the design of bolting support for the maintenance of longwall headings for plough longwalls when such factors like: difficult geological and mining conditions, significant depth of exploitation, long panel lengths or the necessity to keep large cross-section area of the excavation have to be taken into account. In the proposed solutions attention was paid to use bolting systems with high preliminary tension and support systems with high preliminary supporting capacity. In LW „Bogdanka" mine the heading maintained behind the longwall exploitation face may be used for storage of gangue and for ventilation purposes as well as for driving new secondary heading for the next longwall. Independent bolting support may be built in longwall cross-cuts, liquidation or technological drifts

KEYWORDS: hard coal mining, thin seam, plow technology, longwall heading, bolting support, longwall cross-cuts

1 INTRODUCTION

For the effective maintenance of longwall headings one must use apropriate technical and technological measures in all phases:
- in the phase of driving the heading in the undisturbed rock mass,
- in the phase of longwall face influence,
- at the crossing of longwall and heading,
- behind the face in one-sided surrounding of the gobs,
- at the crossing of longwall and heading (for the second chosen longwall).

The current experiences show that during the maintenance of longwall heading it is necessary to take work organization into account, especially when performing works like:
- works behind the crossing of longwall and heading,
- securing proper advance of the longwalls,
- maintenance works and local reinforcements of the support.

These works significantly influence the safety and economic effectiveness of the exploitation.

Design of bolting support for the maintenance of longwall headings has not been sufficiently solved yet. Practical in-situ trials of maintaining longwall headings understood as re-using the heading for the next longwall up to now have not been broadly implemented.

In English mining where higly-efficient, technologically advanced bolting system with is used (AT advanced technology system (Bigby 1997, Bigby et al. 2010, Bigby et al. 2004,

Reynolds 2006), the longwalls are driven with leaving coal pillars about 150-meter-wide. A test to decrease the pillar width to 10 meters and treat it as yielding layers did not stand the test.

In Australian mines where the exploitation depths are low (about 200 m) double longwall headings with bolting support and separation of 35–40 m are driven. Maintaining parallel heading for the second longwall requires many additional supporting elements (Turek 2001).

In German mines the attempt to maintain the longwall heading with bolting support has not been commonly implemented despite the use of a special support solutions and technologies of supporting belts construction (Langhanki 2001).

In this article the main focus is on presentation of a system for maintaining longwall headings with the use of highly-efficient rock bolting for plough longwalls in the conditions of – "Bogdanka" mine.

2 SYMPTOMS OF HIGH ROCK MASS PRESSURES IN PREPARATORY HEADINGS OF "BOGDANKA" MINE

The fulfillment of the support stability condition (meaning no cavity hazard or roof rock slide) does not always allow to keep the usefulness of the excavation, especially its required dimensions directly before and in the area of the longwall-heading crossing for the means of haulage currently used.

Significant dimensions of the belt haulage devices, especially so-called joint station (the connection between the armored face conveyor and the haulage with belt conveyor) or – in the case of plough longwalls – the plough drive cause problems with keeping required dimensions of the heading in the area where exploitation pressure affects the heading support.

Analysis of many exploratory boreholes drilled in "Bogdanka" mine during the exploitation of seam 382 which can be successfully applied to seam 385/2 allowed to associate the degree of excavation (longwall heading) deformation with the geological structure of the stratified rock mass and its durability and deformation parameters (Głuch 1999).

Four different classes of rock systems may be distinguished (Figure 1).

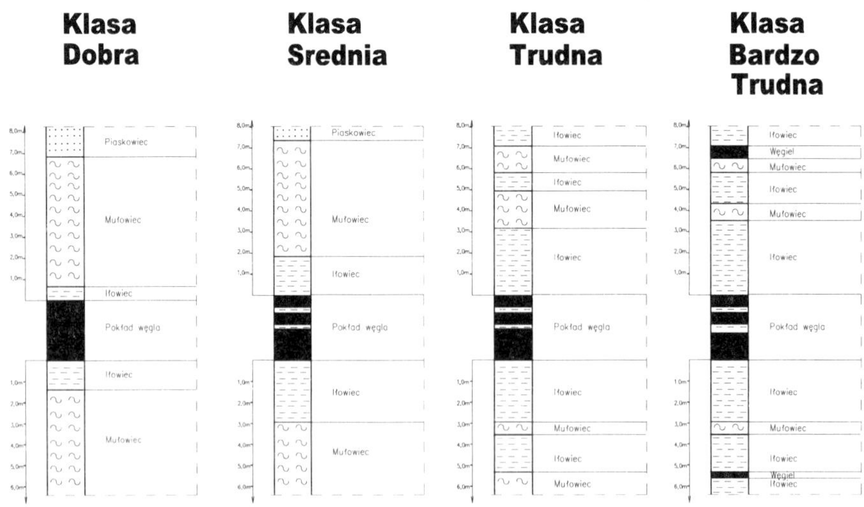

Figure 1. General classes of rock systems occurring around the longwall headings in the conditions of "Bogdanka" mine

Notations: Klasa dobra – *Good class;* Klasa średnia – *Average class;* Klasa trudna – *Difficult class;* Klasa bardzo trudna – *Very difficult class;* Piaskowiec – *Sandstone;* Mułowiec – *Mudstone;* Iłowiec – *Siltstone;* Pokład węgla – *Coal seam*

Particular classes of rock systems are characterised with stratified structure:

Good class – "D"

In general in these sections in the roof directly over the seam mudstone or sandstone without siltstone may be found. A low-thickness layer of siltstone (up to 0.2 m) may possibly be found adjacent to the seam roof and in the floor, low-thickness siltstone with predominant occurrence of sandstones and mudstones are acceptable.

Average class – "S"

In this class one may distinguish rock systems where siltstone layers are adjacent to the coal seam (their thickness in the roof exceeds 0.2 m and in the floor exceeds 1,5 m while some of them may be located in the cross-section of the excavation). The total thickness of the coal seam with adjacent layers of siltstone ranges from 4 to 6 meters.

When thick layers of mudstone and sandstone are deposited in the roof it is possible to maintain stability of the excavation in these conditions.

Difficult class – "T"

In this class the siltstone layers occur directly over the roof and the floor of the seam and have significant thickness (over 1.5 m) so that together with the coal seam and its partings they reach thickness exceeding 6 m. In greater distances from the excavation compact thick layer of mudstone occur and thin layers of sandstone are rare. For maintaining stability of this kind of rock system it is necessary to use special kinds of support.

Very difficult class – "BT"

In this class the rock layers around the excavation driven in the coal seam are built mainly from siltstones with additional interbedding of coal seam. The significant thickness of coal and siltstone layers (over 6m) with additional occurrence of thin coal seam and lack of thick layers of mudstones or sandstones creates very difficult conditions for the maintenance of longwall headings. In the area of exploitation pressure influence with low resistance parameters intensive tightening of the excavation takes place.

In these conditions when there is additional presence of water (which very often flows with coal seams) the siltstone is decomposed and is a subject of plastic flow towards the excavation. Significant movement of the side-walls and horizontal convergence in the excavation are characteristic features of deformations around the heading.

In the areas of exploitation pressure influence where significant concentrations of pressure occur the characteristic course of deformation goes as follows:

- elevation of the roof (on average 1.5 m in front of the longwall face and up to 3 m behind it),
- horizontal movement of the side-wall (up to 1.5 m on average),
- destruction of the rock structure around the excavation into cube structure and practical loss of cohesion of the floor rocks,
- horizontal movements between the rock layers occur together with movements along the layers' split areas,
- deformations of rocks result in significant crawling, especially in stratified rocks when moisture or water are present.

For example, in the roof and floor of seam 382 so-called "slippers" (layers of siltstone) exist which become a slide surface for the seam what results in its movement towards the excavation. In front of the longwall face it may reach 30 cm. The slippers are less common in the roof and floor of seam 385/2.

Some extreme symptoms of occurring headings deformations are shown in Figures 2 to 5.

Due to constant improvements of longwall headings support that have been developed in "Bogdanka" mine for many years both from construction side and the driving technology numerous new solutions were implemented:

- using large cross-section area of the heading S=6,8 m, W=4,5 m, area $A \approx 30$ m^2,
- using only V36 shapes from steel of increased strength for constructing support,

- constructing the support from an increased number of elements (6 or 7 pieces) with a spread of side-wall arches,
- spragging the support against the side-wall through spragging sacks filled with fast-binding construction material injected directly in the face of the excavation,
- using chain-and-knot net lining ensuring the continuity of the lagging,
- stabilization of the support with modular multi-element sprags,
- bolting of the rock mass between the sets of the arch support with minimum 5 or 6 bolts for every field of the support,
- bolting of the support sets through short steel horseheads with 6-meter-long string bolts with load capacity ca. 270 kN.

A view of a longwall heading driven most often with AM75 roadheader with average monthly advance exceeding 400 m with record advances reaching 600 to 800 m (Bigby et al. 2004) has been shown in Figure [illegible]

Figure 2. Extremely vertically tightened longwall heading behind the longwall face (seam 385/2)

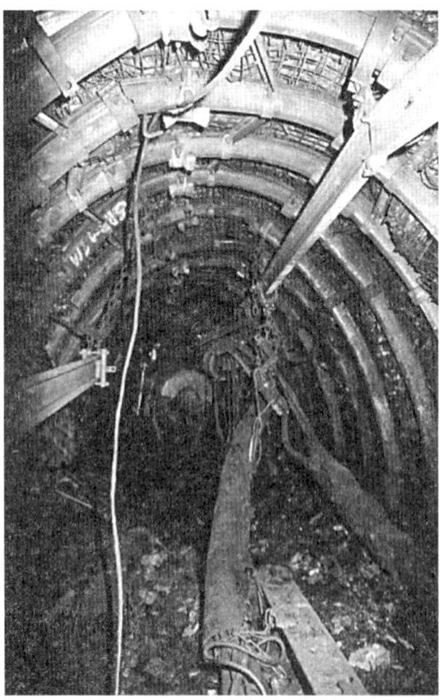

Figure 3. Deformation of longwall heading from the side-wall's side (seam 382)

Figure 4. Reconstruction of the heading after the influence of exploitation pressure (seam 385/2)

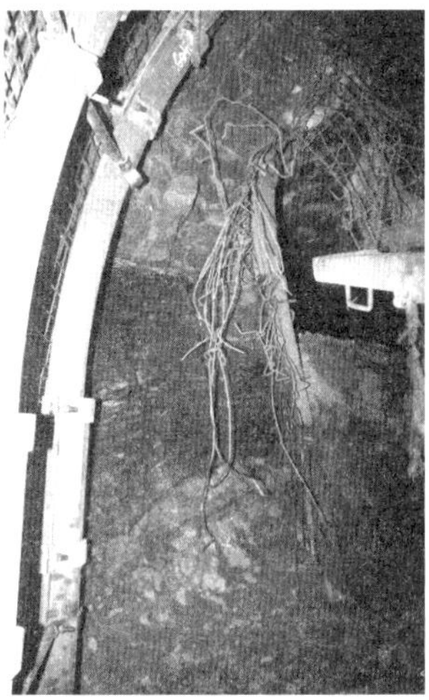

Figure 5. Deformation of coal side-wall in seam 385/2 behind the longwall face

Figure 6. Longwall heading with arch and bolting support after the driving phase

3 EXAMPLE OF A HEADING WITH INDIVIDUAL BOLTING SUPPORT IN "BOGDANKA" MINE

Individual bolting support in "Bogdanka" mine was used in 2001 in the cross-cut in the longwall no. 4/V in seam 382 (Głuch 2000).

The cross-cut was located about H=885 m below the surface.

The length of the cross-cut in longwall no. 4/V was about 250 m.

Nominal dimensions of the excavation with individual bolting support were as follows:

- width of the excavation measured along the floor – S_w = 5,4 m, (max 5,6 m),
- width of the excavation measured along the roof – S_w = 5,8 m, (max 6,0 m),
- height of the excavation – W = min. 3,2 m, max. 3,6 m.

The cross-cut in longwall no. 4/V was driven with roof ripping and floor dinting. The heading had the shape of a trapezoid with inclination of the side-wall from the longwall's side. General dimensions, the shape of the excavation and support solution are depicted in Figure 7 and the general view in Figures 8 and 9. The bolting support was realized while ensuring sufficiently high binding forces in the bolts (Drzęźla et al. 1997, Głuch et al. 1998) and building them in with a SWKB bolting cart.

The design of the cross-cut was based on the results of research from exploratory borehole BR-7/97 drilled in the headgate 4/V. During the driving of the cross-cut additional geotechnical research was conducted for verification of the design. A statement containing basic results of these research is shown in Table 1.

Table 1. Geotechnical parameters of the roof rocks in the cross-cut in longwall 4/V

No.	Location in the excavation	No. of borehole	Roof – 3 m packet			Roof – 5 m packet		
			R_c [MPa]	RQD [%]	Slakeability r	R_c [MPa]	RQD [%]	Slakeability r
1	Tailgate of longwall 4/V, 2033m	BR – 35/00	37.72	34.3	0.9	28.58	28.5	0.8
2	Cross-cut 4/V, 80m	BR – 36/00	32.50	70.7	1.0	33.19	80.6	0.96
3	Cross-cut 4/V, 190m	BR – 1/01	31.63	54,7	0.8	36.25	66.2	0.88
4	Headgate 4/V, 2023m	BR – 7/97	23.06	68.7	1.0	23.93	70.4	1.0

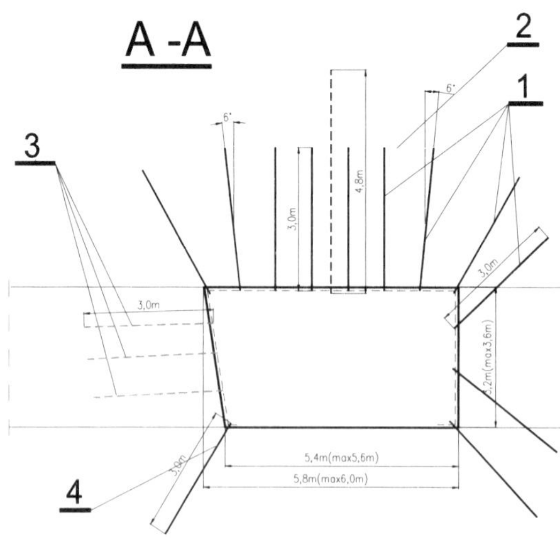

Figure 7. The construction of individual bolting support for the cross-cut in longwall 4/V in seam 382: 1 – 3,0-meter-long steel bolts, 2 – string bolt, 3 – wooden bolts binded with cement, 4 – 2,5-meter-long corner steel bolt, 5 – lagging from steel plates, 6 – welded net

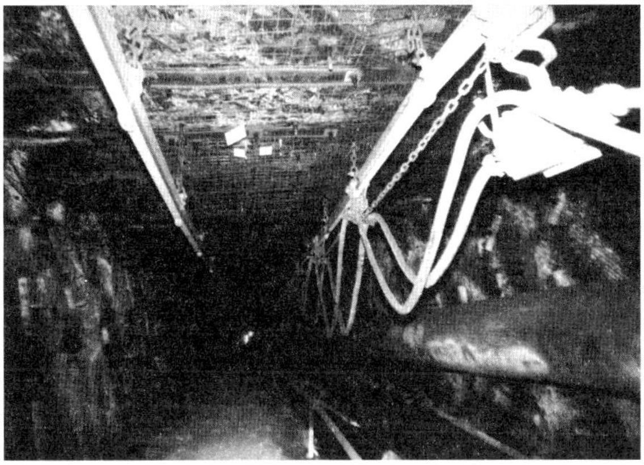

Figure 8. General view of the cross-cut 4/V during the driving phase

Figure 9. The inclined coal side-wall from the future longwall's side with built-in wooden bolts

4 NEW REQUIREMENTS FOR BOLTING SYSTEMS OF THE ROCK MASS FOR MORE ADVANTAGEOUS MAINTENANCE OF LONGWALL HEADINGS

The deformation of rocks around the excavation is especially intensive in the rock mass where the mining support does not actively influence the surrounding rock mass. The commonly used late-support system with steel sets made even from the heaviest shapes is not able to resist the high pressures occurring together with the destruction of rock structures around the excavation. Driving the excavation with arch support in highly-stratified rock mass causes significant deformations with deep range.

Especially adverse situation takes place when the surfaces of rock layers contact experience significant horizontal slips (slippers occur). Then deep sources of destruction (deformation) occur around the excavation which are quickly activated when pressure is added and cause the course of deformation and tightening of the excavation to be very intensive. The speed of rock deformation around the longwall heading is then a few times faster than in the case when the heading is not subjected to exploitation pressure. If the advance speed of the longwall is low, intensive tightening of the excavation occurs. It may be reduced by increasing the exploitation face advance.

Taking strong stratification of Carbon rock mass into account and its variable composition and structure it is necessary to actively influence the rocks around the excavation during the driving phase in order to prevent them from relieving stress and loosening or even to cause their preliminary compression at the borders of the excavation. In this solution the rock layers should adhere tightly to one another at their contact surfaces which will reduce lateral cracking and therefore the split of the rock mass into cube structure which is practically a rock breccia and does not carry any load. Because of its increased volume the breccia causes significant pressure on the surrounding rock mass strata (Drzęźla et al. 2000, Głuch 2000).

The stratified and naturally cracked rock mass around the excavation may be reinforced through adhering preliminary pressure before its deformation to its surface and to the deeper rock layers. This results in:

- creation of advantageous pressure state at the borders of the excavation. This advantageous (triaxial) pressure state leads to a qualitative change of rock work around the excavation. Compression strength and its deformability are also increased,
- increase of friction between the layers and between cracks as well. Pulling the rocks together at the borders of the excavation causes the shrinking of all natural fissures what results in increased compression strength of the rock layers against mutual horizontal movements and delamination,
- compression of the rock layers occurring at the beginning of deformation (with small distortions of the rock mass) favourably influences the weaker layers to transfer increased loads without excessive effort put on the layers with higher strength.

To achieve the effect of compression in strongly stratified rock layers one needs to use significant (200 to 600 kN) pressures put on the rock layers directly after their exposure to prevent their preliminary delamination.

Technical solution of this method of rock compressing consists of:

- high-capacity rod or string bolts with preliminary pull ranging from 200 to 600 kN. The pull is achieved directly after building-in the bolt (after about 10 minutes) with interval connection of the bolt at the bottom of the borehole with injection along the whole length directly after the pull is realized. Full building-in of the bolt should be realized (interval binding, pull and injection) directly in the face, so the next web is driven in fully supported rock mass. Final bolting, which for the maintained heading often requires bolts with a length of 8 m or even 12 m, should be fully mechanized.
- Supporting elements with preliminary capacity ca. 200 to 500 kN. This capacity is obtained through their spragging between the rock surfaces in the excavation. One of the simplest ways which allow to effectively achieve such capacity is to sprag steel props or trough shapes in the excavation.

Research conducted to date (Giza et al. 2011) has shown that with the right construction of the supporting prop it is possible to obtain capacity ranging from 600 to even 1000 kN with maximal capacity after stiffening reaching 1500 or even 2000 kN.

The proposed solutions concerning the support's elements with preliminary capacity are described as AT-AS (advanced technology – advanced support) (Głuch 2000) and have been successfully introduced in Knurów mine (Matuszewski et al. 2006). If used properly these solutions may allow to:

- Store gangue in old gobs directly behind the longwall face in the case of significant floor elevation in order to increase the length of roof support and therefore reduce its pressure on the floor (this solution may be used when using plough technique to mine thin seams). At present, excessive amounts of gangue mined from the headings driven in thin seams cause significant costs related to their transport, separation from the winning, utilization or storage.
- In thicker seams and stronger roof rocks control the length of the rock prop through its fissuring in front of the longwall face, so the weight of the bar and therefore its pressure is significantly reduced.

5 THE DESIGNED SYSTEM OF MAINTAINING PLOUGH LONGWALL HEADINGS

An exploitation system for plough longwalls with independent longwall headings driven with arch and bolting support which would take Bogdanka's mine geological and mining conditions into account is currently designed.

To obtain high effectiveness of bolting it is proposed to use support with straight roof while arch shape of the side elements is maintained.

The heading will be driven in two stages:

- Stage I – directly in the face bolting support will be used. Rod bolts with capacity ranging from 100 to 150 kN and string bolts with pull ranging from 300 to 400 kN will be connected with the rock mass along the whole length of the borehole.
- Stage II – directly behind the face in distance of about 40 m arch support with full mechanical filling of empty space with mechanical lining (especially behind the side-walls of the excavation).

The length of rod bolts due to significant width of the excavation along the floor (ca. 6,8 m) should not be lower than 3,0 m and the length of the string bolts is a result of their necessary mounting in the area above roof cavity of the mined longwall. When the roof cavity is determined from the thickness of the seam, the minimal length of the string bolt is as follows:

$L_{kstr} = 5 \cdot g + 2.5$, what for seam thickness $g = 1.5$ m gives the length of the string bolt equal to:

$$L_{kstr} = 5 \cdot 1.5 + 2.5 = 10 \text{ m}$$

The amount of string bolts and their capacity should be determined using the transfer of weight of a rock block of the relieved layers behind the face with safety coefficient 1.5 using the equation:

$$n_{kstr} = \frac{1.5 \cdot 5 \cdot g \cdot S \cdot \gamma}{N_{kstr}} = \frac{1.5 \cdot 5 \cdot 1.5 \cdot 6.7 \cdot 25}{700} = 2.69 \text{ pcs.}$$

where: $S = 6.7$ m width of the excavation; $\gamma = 25$ kN/m^3 average bulk density of the rocks, $N_{kstr} = 700$ kN capacity of string bolts.

Three string bolts with minimum capacity of 700 kN for every 1 m of the excavation is a minimal number of bolts to ensure the safety condition is fulfilled and men may safely work in the heading behind the longwall face with additional arch support built in the excavation. All bolts should be built-in directly to the roof through steel washers and tied with the borehole along their whole length.

Maintaining the continuity of seam exploitation and mining the next longwall requires having a longwall heading with full, not deformed cross-section. In the design it is proposed to drive a parallel secondary heading for the next longwall in the distance of ca. 400 m behind the longwall face. The maintained longwall heading is systematically connected along a 400-meter interval with the secondary heading using short cross-cuts.

When panel lengths are significant this solution allows to:

- Improve the ventilation of the mined longwall,
- Use the maintained heading along a 400-meter interval for ventilation of the new secondary heading the other longwall,
- Store gangue in the heading of the previous longwall.

In this solution one purposefully resigned from labour consuming construction of supporting belt behind the longwall face and instead proposed to use supporting breaker and to fill the side-wall from the exploited longwall's side with gangue from floor dinting. A special section of powered roof support should be considered which would work in the longwall heading and increase the stability and safety at the longwall-heading crossing. It would also facilitate filling part of the cross-section towards the exploited longwall.

6 FINAL CONCLUSIONS

1. Safety, technical, operational and economic reasons point at the necessity to use technology of maintaining longwall headings in which advanced solution should allow to compress the rock layers and counteract their stratification and destruction at the driving phase.
2. The improvement in maintenance of longwall headings may be achieved through using:
   - Rod bolts or string bolts with preliminary pull that would become final directly in the face after a web has been driven,
   - Support with preliminary capacity ranging from 200 to 500 kN built-in directly after exposing the roof gives the possibility to work as a yielding construction with capacity from 800 to 1000 kN and as a stiff construction with capacity from 1500 to 2000 kN.
3. W prospective projects for maintaining the longwall headings one should predict the possibility of storing gangue from floor dinting of a chosen longwall as well as improving ventilation and work safety, especially when longwall advances are significant and geological and mining conditions are variable.
4. In the conditions of "Bogdanka" mine an individual bolting support with high preliminary pull should be used in longwall cross-cuts, liquidation cross-cuts and technological cross-cuts together with arch and bolting support for longwall headings when the heading is maintained behind the longwall face at the interval of about 400 m for ventilation and gangue-storage purposes with simultaneous driving of new secondary heading for the next longwall while securing its high capacity and required dimensions.
5. The proper structure of longwall headings and mechanization of driving and bolting using bolting support (in cross-cuts) or mixed arch and bolting support (in headings) should ensure both high-quality bolting and advance of driving that would allow to fully reproduce the exploitation front.

REFERENCES

Bigby D.N. 1997. Developments in British Rockbolting Technology Coal International.

Bigby D., Hurt K., Reynolds Ch, Brown R. 2010. Evaluation of tensioned and non-tensioned long reinforcement in UK deep mining conditions. Research report 831, Prepared by Rock Mechanics Technology Ltd for the Health and Safety Executive.

Bigby D., Kent L., Hurt K. 2004. Safe application of mine roadway support systems Research report 229. Prepared by Rock Mechanics Technology Ltd for the Health and Safety Executive.

Chmielewski J., Kozek B., Masiakiewicz M. 2006. Drążenie chodników w kopalni „Bogdanka". Szkoła Eksploatacji Podziemnej. Szczyrk. (In Polish)

Drzęźla B., Głuch P. 1997. Siła wiązania kotwi wklejanych. Bezpieczeństwo Pracy i Ochrona Środowiska w Górnictwie, nr 7. (In Polish)

Drzęźla B., Głuch P. 2000. Obudowa kotwiowa dla warunków KWK „Bogdanka" Konferencja naukowa pt. XXV lat Lubelskiego Zagłębia Górniczego Lublin. (In Polish)

Giza D., Głuch P. 2011. Stojaki podporowe o wysokiej nośności. Konferencja Geomechanika. Ustroń. (In Polish)

Głuch P. 1999. Problemy doboru obudowy i zabezpieczenia przekroju chodników przyścianowych w trudnych warunkach geologiczno górniczych głębokich kopalń. Materiały konferencyjne pt. Wybieranie złóż na dużych głębokościach oraz w trudnych warunkach geotechnicznych. Świeradów. (In Polish)

Głuch P., Masiakiewicz M., Stachowicz St. 1998. Badania parametrów transferu obciążenia kotew – klej – skała dla oceny przydatności systemu kotwienia. Materiały konferencyjne, XXI Zimowa Szkoła Mechaniki Górotworu. Zakopane. (In Polish)

Głuch P. 2000. Aktywna ochrona chodników przyścianowych dla utrzymania ich w jednostronnym sąsiedztwie zrobów. Miesięczniki WUG, nr 8. (In Polish)

Głuch P., Kozek B., Koza H., Masiakiewicz M., Słomiany P. 2001. Obudowa kotwiowa przecinki ścianowej 4/V w kopalni „BOGDANKA" Konferencja pt. Nowoczesne technologie Górnicze. KOTWIENIE. (In Polish)
Kozek B., Ruchel A. 2011. Wykonywanie i utrzymywanie chodników przyścianowych w kopani Lubelski Węgiel Bogdanka S.A. – dotychczasowe doświadczenia. Szkoła Eksploatacji Podziemnej. Kraków. (In Polish)
Langhanki B. 2001. Planungskozeption zur Doppelnutzung einer Rechteckankerstracke Glückauf 137, Nr 11.
Matuszewski J., Mąka B., Głuch P. 2006. Utrzymanie chodnika przyścianowego 23b w jednostronnym otoczeniu zrobów dla wybierania ściany 23 i 24 w pokładzie 405/1 w kopalni Knurów. Materiały konferencyjne, Nowoczesne Technologie Górnicze. (In Polish)
Opolony K. 2001. Einsatz von ankerausbau bei der DSK Konferencja pt. Nowoczesne technologie górnicze. KOTWIENIE.
Reynolds Ch. 2006. Testing and standards for rock reinforcement consumables. Research report 411, Prepared by Rock Mechanics Technology Ltd for the Health and Safety Executive.
Turek W. 2001. Drążenie wyrobisk korytarzowych w obudowie kotwiowej z dużym postępem w kopalniach australijskich. Konferencja pt. Nowoczesne technologie Górnicze 2001. KOTWIENIE. (In Polish)

International Mining Forum 2011, Dyczko et al. (eds) © 2011 Taylor & Francis Group, London, UK. ISBN 978-0-415-68938-0

Construction and Maintenance of Longwall Galleries for Plough Longwalls in the Mine of Lubelski Węgiel Bogdanka S.A.

K. Kwiatkowski
Lubelski Węgiel Bogdanka S.A., Poland

ABSTRACT: The article discusses the determinants associated with the selection of the support system for longwall galleries for longwalls mined with the plough technique. It contains an overview of the different alternative solutions analysed and the solution employed in the mine of Lubelski Węgiel Bogdanka S.A. in the case of the first plough longwall 1/VI/385, located in seam 385/2 in Nadrybie field. It presents further works undertaken to improve the support system frames for the second longwall equipped with the first plough system (longwall 7/VII/385) located in the same seam in Stefanów, but smaller in thickness (1.2–1.5 m).

KEYWORDS: deep mining, hard coal, roadway workings, longwall galleries, plough mining technique, Lubelskie Zagłębie Węglowe [Lublin Region Coal Basin]

1 INTRODUCTION

The standard solutions in longwall gallery support systems, which have been developed as a result of many years of experience and are currently used at Lubelski Węgiel Bogdanka S.A., can be described as follows:

- each longwall has two new galleries, built entirely before its launch,
- the galleries are built with an arch-shaped cross-section in a special frame support system made of grade S480W steel V36 section bar (before, it was partly made of V32, too), with the most often used frame spacing being 0.75–0.9 m,
- the cross-section of longwall workings has undergone numerous changes, the typical size for the shearer longwalls being presently 10,
- until 2000, the support system of longwall galleries was same as that for main workings, and was closed with floor elements while driving workings immediately behind the roadheader,
- for about 10 years, the support system for longwall galleries has been different than that for main workings. During the driving process, the standard support system in a heading is built as open and is closed only in selected cases and after dinting.

Requirements related to longwall galleries of a plough longwall are higher than in the case of shearer longwalls. It is necessary to employ the use of wider cross-sections of workings, especially in the case of intensive convergence of galleries, which is the case in Bogdanka mine. The maintenance conditions of the longwall-gallery crossing are also more difficult. Furthermore, the possibilities of conducting maintenance works of the galleries located in the area of crossing with the longwall are smaller as, in most cases, it requires putting the plough in a stand-by mode. The most preferable support system for longwall galleries of plough longwalls would be a bolting system with a rectangular cross section, which is difficult to use in the geological conditions of Bogdanka mine.

2 SEARCH FOR SUPPORT SYSTEM SOLUTIONS FOR LONGWALL GALLERIES OF FIRST PLOUGH LONGWALL

One of the numerous problems discussed during the preparatory works, which started in 2008, to launch the first plough longwall was the issue of selecting a support system suitable for longwall galleries. As a matter of fact, the experiences gained for shearer longwalls could not be simply transferred and applied to plough longwalls.

The simplest solution for longwall galleries of plough longwalls seemed to be an intermediate solution between the bolting system and the support system, and between a rectangular and an arch-shaped cross section, which is *an arch and straight shaped bolting support system*. This system has proved to be effective in the geological and mining conditions of Bogdanka mine and continues to be used as the standard solution for start-up cross-cuts of longwalls. It is usually of size 13 and 6.4 m wide at the floor level, i.e. offering a cross-section which is sufficient for plough longwalls.

In the case of plough longwalls, the spacing of frames should be as wide as possible (e.g. 0.9 m) to minimise the amount of work to be done at the crossing with the longwall. Considering the catalogue of the support systems used at LW Bogdanka, these could be frames of the ŁPrP type, the "Ł" or "K" version. The "K" version is preferable as it ensures a bigger width of the gallery in the roof part, which might be of importance in a situation of intensive horizontal convergence.

The ŁPrP support system features a big roof-bar bending radius and therefore offers very low support capacity against roof pressures. Therefore, it would have to be used along with a considerable amount of bolting of roof rocks. As far as the bolting system is concerned, it would require using both low bolting by means of steel rod bolts (e.g. 7–8 pcs per frame) and high bolting with string bolts (two rows of horseheads made of V32/36 section bar, bolted with string bolts every two frames each), all of this built up in the heading zone of the gallery being driven, i.e. having a direct impact on the gallery driving progress.

The following preliminary assumptions on the gallery cross-sections were made:

1. Tailgate within a size 10 system. In the case of the ŁPrP frame, the floor level width would be 5.5 m, whereas the full height would be 4.0 m.
2. The headgate should be bigger. Its width could be 5.8 m when using the ŁPrP size 11 frame, which would not be enough since the reserve left to compensate the horizontal convergence would amount only to 1.2 m. Using size 13 seemed more reasonable, with the gallery height adequately reduced.

The gallery driving progress forecast was the decisive factor in final rejection of the ŁPrP frame. The comparison analysis conducted by the mine clearly showed that using the arch-shaped support system will allow to achieve gallery driving progress about 50% higher than in the case of the ŁPrP frame.

Use of the system which was typically used at that time in longwall galleries of shearer longwalls was also considered: a 5-element arch-shaped support system of the ŁPS/C type, frame size 10 for the tailgate and size 12 or 13 for the headgate. In such case, it would also be necessary to use low and high bolting since the cross-section was very big, and it would be required to apply maximum spacing between the frames. In this case, the possibilities to lower the frames and the height of the galleries would be insignificant because the width of the working decreases with the decrease in the height of the working. On the other hand, however, it would be possible to use, at least in part, bolting behind the roadheader, which would allow to increase the progress of driving the gallery, provided that the arch-shaped cross-section is applied.

It should be emphasised that the opening section of the headgate of longwall 1/VI/385 was made in the ŁPS/C type frame, with frame size 12 and certain modification of the frame. Figure 1 shows the ŁPS/C type frame with frame size 12, used in a part of the bottom gallery.

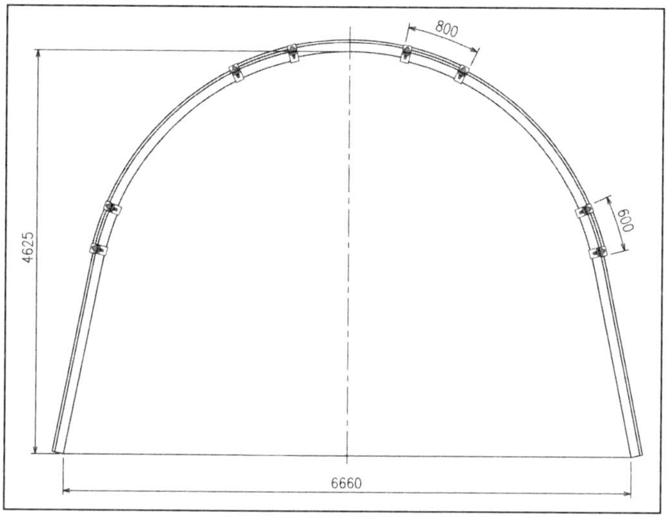

Figure 1. The ŁPS/C type frame, size 12, used in a part of the headgate of longwall 1/VI/385

The second crucial factor which prevented the use of the typical frames of the support system is the quantity and the arrangement of elements of the frames which does not reflect the need to maintain the longwall-gallery crossing. In shearer longwalls, it was most often necessary to cut the roof-bar elements of the frames. In the case of plough longwalls, it would be more difficult because it would require stopping the plough.

In this situation, the best solution would be to design a new frame, particularly for this purpose.

3 NEW 6-ELEMENT ŁPSC FRAME FOR LONGWALL GALLERIES OF SHEARER AND PLOUGH LONGWALLS

In the years 2008–2009, at Bogdanka mine, works were undertaken to develop a new system of frames which would be better adjusted for the purposes of longwall galleries. At that time, 100% of them were longwall galleries for shearer longwalls. Those works resulted in 6-element arch-shaped frames with transverse cross-section, designed to create and maintain a crossing with the longwall. Those frames were designed to be built in the heading of the driven gallery as open, with the possibility of closing the floor later after dinting (as in the case of the ŁPS/C frames). Originally, frames were designed for shearer longwalls, and later, when the preparatory works to introduce the plough technique in the mine were started, expanded also to comprise plough longwalls.

The characteristic features of the 6-element ŁPSC frame developed include:

- The number and length of frame elements have been adjusted in order to obtain a favourable position of the roof-bar arch against the roof of the seam being undercut in the heading of the driven gallery so as to provide a convenient longwall inlet after removing side-wall arches, without the need to cut the roof-bar.
- The frames use short side-wall bottom elements (floor elements), whose length is approximately equal to the floor upheaval value, increased by the overlap, so as to enable elimination of those elements when the formerly removed side-wall elements are built up again (after passing of the longwall face), creating non-symmetrical frames without unnecessary dinting of the floor.
- Overlaps of frame elements have a fixed value, except the connection between lower side-wall elements and upper side-wall elements of the frames. The overlap can be changed indi-

vidually within the range of 500–700 mm, depending on the actual size of the break made by the mining machine while driving the gallery. It enables setting the frame support feet on stable, solid floor, without having to provide a bed of loose rocks or sub-plates to correct the break height.

- Frames have been designed separately for longwall galleries for shearer longwalls of smaller frame size, marked "K", and for longwall galleries for plough longwalls of bigger frame size, marked "S".
- The "K" and "S" variants differed from each other in the frame height to width ratio. In the "S" frames, the height no longer increased in proportion to frame width.

The frame type offered six sizes – three for longwall galleries for shearer longwalls (variant "K" – size 9, 10, and 11), and three for longwall galleries for plough longwalls (variant "S" – size 11, 12, and 13).

The new 6-element support of the ŁPSC type was used for the first time in longwall galleries of longwall 1/VI/385 in seam 385/2, driven for the purpose of the first plough longwall by division GRP-4.

Figure 2 shows the 12/S size ŁPSC type frame. The photograph in Figure 3 presents a view of the gallery built within the new support system.

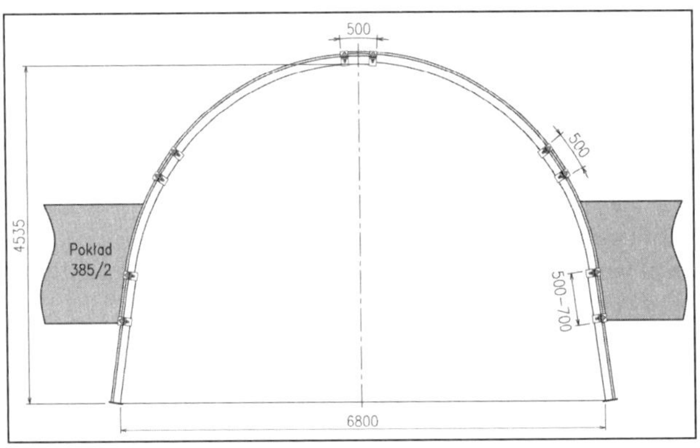

Figure 2. The 6-element size 12/S ŁPSC type frames used in the headgate of longwall 1/VI/385
Notations: Pokład – *Seam*

Figure 3. The longwall gallery of longwall 1/VI/385 made in the new ŁPSC 12/S frame

4 EVALUATION OF THE DESIGNED SUPPORT SYSTEM AND THE MAINTENANCE OF LONGWALL GALLERIES UNDER THE IMPACT OF OPERATIONAL PRESSURES OF LONGWALL 1/VI/385

According to the approved specifications, the headgate was made in the bolting support system with ŁPSC type frames, frame size 12/S, spacing 0.9 m. The rock mass between the frames was reinforced by alternating seven and eight steel bolts, 2.7 m long, built up in the heading zone.

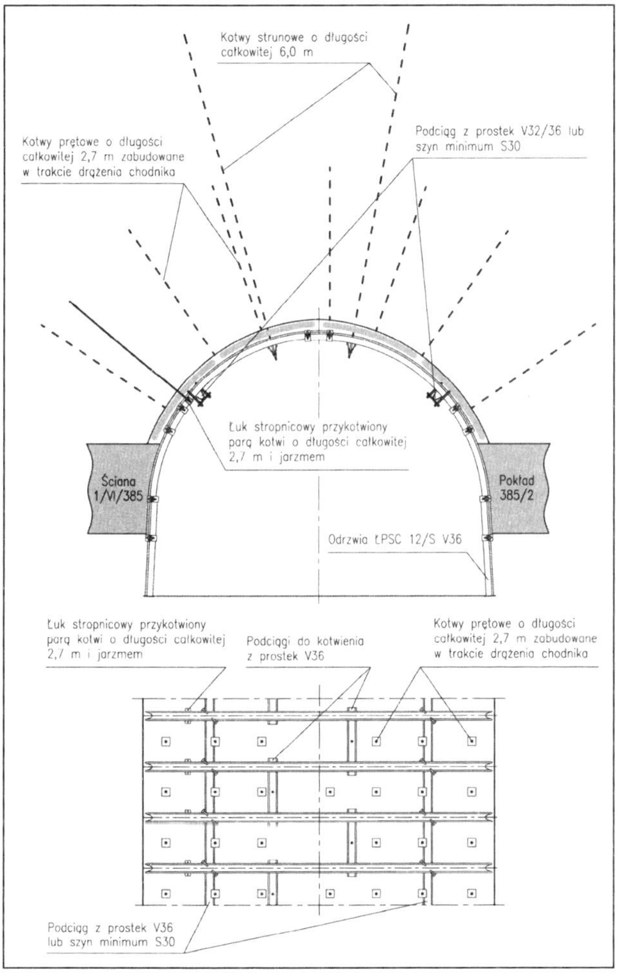

Figure 4. Reinforcement of the headgate of longwall 1/VI/385 for the purpose of maintaining the longwall-gallery crossing in a non-standing manner and keeping it behind the longwall

Notations: Kotwy strunowe o długości 6,0 m – *String bolts, total length: 6.0 m;* Kotwy prętowe o długości całkowitej 2,7 m zabudowane w trakcie drążenia chodnika – *Rod bolts, total length: 2.7 m, fitted during the process of driving the gallery;* Podciąg z prostek V32/36 lub szyn minimum S30 – *Horsehead made of straight elements V32/36 or rails, minimum S30;* Łuk stropnicowy przykotwiony parą kotwi o długości całkowitej 2,7 m i jarzmem – *Roof-bar arch, bolted with a pair of bolts of the total length of 2.7 m and yoke;* Ściana 1/VI/385 – *Longwall 1/VI/385;* Pokład 385/2 – *Seam 385/2;* Odrzwia ŁPSC 12/S V36 – *Frame, ŁPSC 12/S V36;* Podciągi do kotwienia z prostek V36 – *Horsehead for bolting made of V36 straight elements;* Podciąg z prostek V36 lub szyn minimum S30 – *Horsehead made of V36 straight elements or rails, minimum S30*

The tailgate of longwall 1/VI/385 was made in an identical support system; however, based on frame size 11/S and the rock mass was reinforced by means of alternating three and four bolts built in between each frame.

In order to reinforce the support system within the operational pressure zone and on the crossings of longwall galleries with the longwall for its non-standing maintenance, the roof-bar elements of the support system frame were bolted. The standard high bolting with string bolts and straight elements of V36 section bar with a hole was applied. The reinforcement in both galleries was made with the use of bolted horseheads, built in two rows with bolts every second frame in each row. Furthermore, the support system was reinforced by fixing horseheads made of V section bar or rails fixed to frames with hook bolts, on the right and left side of the longwall.

Due to the operation of the floor mining loader at the crossing of the gallery and the longwall, it was necessary to disassemble the side wall elements of the frames first, which significantly lengthened the section of the gallery maintained supportlessly within the crossing with the longwall. For that reason, the frame roof-bar element was bolted with an additional pair of steel bolts through a J-shaped profile yoke.

Figure 4 shows the support system of the headgate of longwall 1/VI/385 which was used along with reinforcement for the purpose of maintaining the longwall-gallery crossing.

5 LONGWALL GALLERIES FOR PLOUGH LONGWALLS IN STEFANÓW FIELD – THE ŁPSC FRAME IN A 7-ELEMENT VERSION

The new ŁPSC frames for longwall galleries, introduced to be used at Bogdanka mine, described in Section 3 and 4, were designed for seams which were more than 1.6 m thick. In the case of plough longwalls and mining seams of small thickness (from 1.2 m to 1.6 m), which is the case with the longwalls in field VIII and partly VII in seam 385/2 in Stefanów, ensuring a proper width of galleries and a proper position of frames against the seam at the crossing of the gallery and the longwall required further modification of the frame design. In thin seams, it is reasonable to drive a longwall gallery with little floor dinting, conventionally specified to be about 0.8 m. In that situation, building up the roof part of a working situated above the seam requires three in place where two elements of the frame roof-bar had been used so far in order to simultaneously maintain all the favourable solutions used in the 6-element ŁPSC frames. This gave birth to the concept of a 7-element frame. In addition, the moving (changeable) overlap on the connection of side-wall arch with the straight floor element was expanded in the 7-element frame. This overlap can be matched individually within the range of 500-900 mm, depending on the actual thickness of the seam and the size of break made by the mining machine while driving the gallery.

The "S" version of the 6-element ŁPSC type frame that had been designed earlier was additionally expanded by three sizes (11, 12, and 13) in the "S7" version 7-element frame. This frame was introduced for use in the headgate of longwall 7/VII/385 in seam 385/2 (size 12) in November 2009, and slightly later, in the tailgate of longwall 7/VII/385 (size 11). In December 2010, longwall galleries began to be driven in that support system for longwalls 1/VIII/385 and 2/VIII/385 in Stefanów. Figure 5 shows the exemplary support system with the 7-element ŁPSC type 12/S7 size frame which was used in the headgate of longwall 7/VII/385 in seam 385/2 in Stefanów field.

The ŁPSC type frame made for production purposes, comprising both the 6-element frame (version I) and 7-element (version II), was designed by the Central Mining Institute in Katowice.

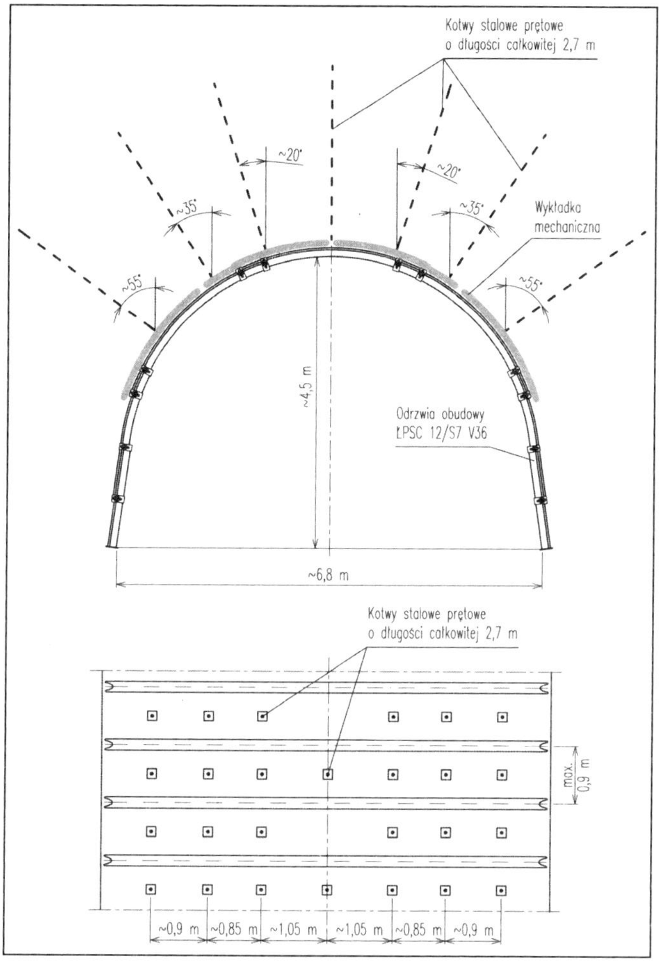

Figure 5. Support system with the 7-element ŁPSC type 12/S7 size frame used in the headgate of longwall 7/VII/385

Notations: Kotwy stalowe prętowe o długości całkowitej 2,7 m – *Steel rod bolts, total length: 2.7 m;* Wykładka mechaniczna – *Mechanical lining;* Kotwy prętowe o długości całkowitej 2,7 m zabudowane w trakcie drążenia chodnika – *Rod bolts, total length: 2.7 m, fitted during the process of driving the gallery;* Odrzwia obudowy ŁPSC 12/S7 V36 – *Support system frame, ŁPSC 12/S7 V36;* Kotwy stalowe prętowe o długości całkowitej 2,7 m – *Steel rod bolts, total length: 2.7 m*

6 LONGWALL GALLERIES OF LONGWALL 2/VI/385 FOR SECOND PLOUGH SYSTEM IN NADRYBIE FIELD

According to the schedule of mining operations development at Bogdanka mine, there is a second longwall planned for coal excavation through shaft 2.1 in Stefanów, to be launched in the second half of 2012. It will also be a plough longwall. Therefore, it was necessary to purchase a new plough system for LW Bogdanka. A contract to purchase such a system was signed in May 2011, with delivery within 14 months. The first longwall, 2/VI/385, where the second plough system will be installed, will be situated next to exploratory longwall 1/VI/385 in Nadrybie. Therefore, the geological and mining conditions of longwall 2/VI/385 in seam 385/2 will be similar to those of the exploratory longwall, with the planned panel length increased to

2200 m and the longwall run direction reversed. The longwall field for longwall 2/VI/385 was already parted. In fact, the same solutions were applied as for the exploratory longwall.

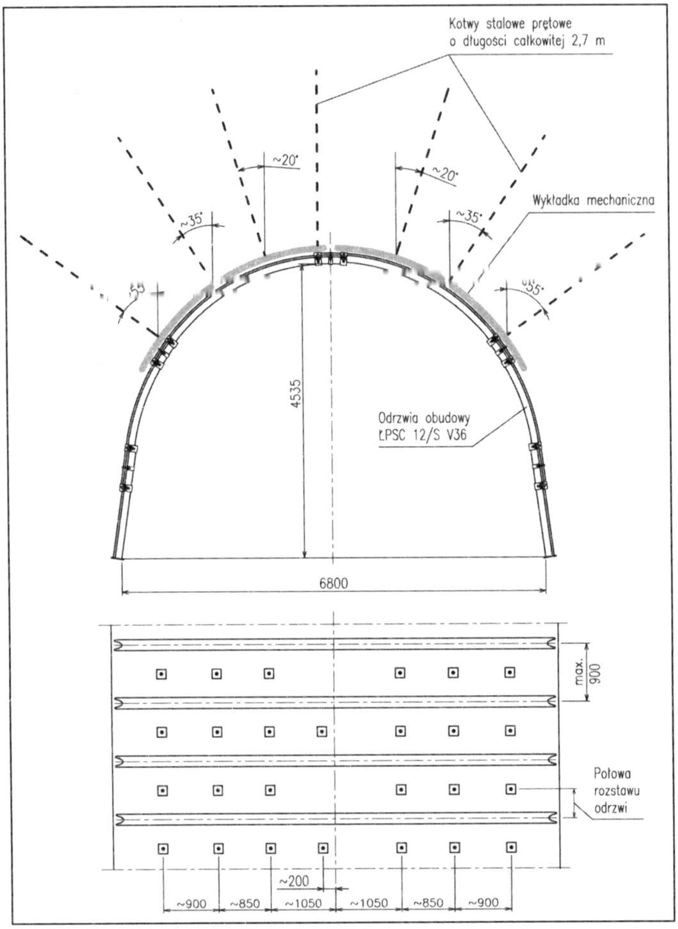

Figure 6. Support system of headgate2 of longwall 2/VI/385

Notations: Kotwy stalowe prętowe o długości całkowitej 2,7 m – *Steel rod bolts, total length: 2.7 m;* Wykładka mechaniczna – *Mechanical lining;* Kotwy prętowe o długości całkowitej 2,7 m zabudowane w trakcie drążenia chodnika – *Rod bolts, total length: 2.7 m, fitted during the process of driving the gallery;* Odrzwia obudowy ŁPSC 12/S V36 – *Support system frame, ŁPSC 12/S V36;* Połowa rozstawu odrzwi – *Half of the frame spacing*

The range of preparatory works for longwall 2/VI/385, completed by the mine's own GRP-4 department, comprised:

- headgate 2 of longwall 2/VI/385, about 2300 m long, built in the ŁPSC 12/S support system made of grade S480W steel V36 section bar, with frame spacing of 0.9 m, rock mass bolting between frames with steel bolts, 2.7 m long, six and seven units per each frame,
- a tailgate of a similar length which, over a section of 1750 m, is a behind-longwall headgate for longwall 1/VI/385, and its remaining part is a reconstructed section of the headgate of longwall 2/VI/385, which has been opened for many years,
- start-up cross-cut of longwall 2/VI/385, made by reconstructing a section of the existing eastern gallery 2/385.

Figure 6 shows the support system of the headgate 2 of longwall 2/VI/385.

The status of the headgate maintained behind longwall 1/VI/385 is subject to other studies and it will not be discussed here. It is worth emphasising that, after a waiting period of approximately six months, intended to let the movements of the rock mass within the caving zone of longwall 1/VI/385 become extinct, the patency of the gallery was restored by reconstruction of the most deformed section and by replacing the frame longwall side elements from the side of the coal body over the entire maintained length. Furthermore, the floor underwent dinting all over the whole panel length of the gallery. The detailed scope of works completed by GRP-4 to make the tailgate operational comprised:

- reconstruction of the old headgate of longwall 2/VI/385 because of the cross-section being too small, 377 m long, completed between 1 June and 8 July 2011,
- reconstruction of nearly 500 m of the gallery maintained behind the longwall – from 1000 m of the panel length to 1487 m of the panel length, completed between 30 July and 30 September 2011,
- floor dinting and replacement of longwall side elements from the side of the seam over the remaining length, i.e. 1263 m.

The works carried out to restore patency of the gallery will be completed at the end of November / beginning of December 2011. The gallery maintained behind the longwall is 5.6–5.9 m wide.

The start-up cross-cut of longwall 2/VI/385 was completed in April and May 2011. The geological and mining conditions were very good, despite reconstruction of the existing gallery. Figure 7 shows the typical solution of the support system that was used.

7 SUMMARY AND FINAL CONCLUSIONS

1. The requirements regarding the geometrical dimension of longwall galleries for plough longwalls are stricter than for shearer longwalls; therefore, the support systems used so far could not be directly used while implementing the new technique.
2. In the case of the first plough longwall at the Bogdanka mine, namely longwall 1/VI/385 in Nadrybie, the support system used for longwall galleries was of the arch-shaped ŁPSC type, newly designed both for the shearer longwalls ("K" version) as well as for the plough longwalls ("S" version). The headgate of longwall 1/VI/385 was made with frame size 12/S, whereas the tailgate – size 11/S. Both galleries had frames made of grade S480W steel V36 section bar, frame spacing of 0.9 m, with rock mass bolting between the frames.
3. In order to provide a supportless maintenance of the crossing of the gallery and the longwall, high bolting along with string bolts and straight elements of the V36 section bar was used in both galleries. Two rows of horseheads with string bolts were fixed every second frame in each row. Furthermore, the support system was reinforced by fixing horseheads made of V section bar or rails fixed to frames with hook bolts, on the right and the left side of the longwall, which was partly related to the maintenance of the headgate behind the longwall.
4. Due to the operation of the floor mining loader at the crossing of the gallery and the longwall, it was necessary to disassemble the side-wall elements of the frames first, which significantly lengthened the section of the gallery maintained supportlessly within the crossing with the longwall. For that reason, the frame roof-bar was bolted with an additional pair of steel bolts through a J-shaped profile yoke.
5. In the case of plough longwalls and mining seams of small thickness from 1.2 m to 1.6 m, which is the case with longwalls in field VIII and partly VII in seam 385/2 in Stefanów, ensuring a proper width of galleries and a proper position of frames against the seam at the crossing of the gallery and the longwall required further modification of the 7-element version of the frame. Such support system was used in the longwall galleries of longwall 7/VII/385 in seam 385/2 in Stefanów, the second longwall equipped with the first plough system launched at the end of September / beginning of October 2011.
6. In longwall 7/VII/385 in Stefanów, a new coal self-loading machine was used, which eliminated the need to employ the use of the floor loading machine at the crossing behind the

longwall. This will enable simplification of the maintenance of the crossing. This solution will be tested in the headgate but, if positively evaluated, it will be also applied to the tailgate, and later for the second plough system.

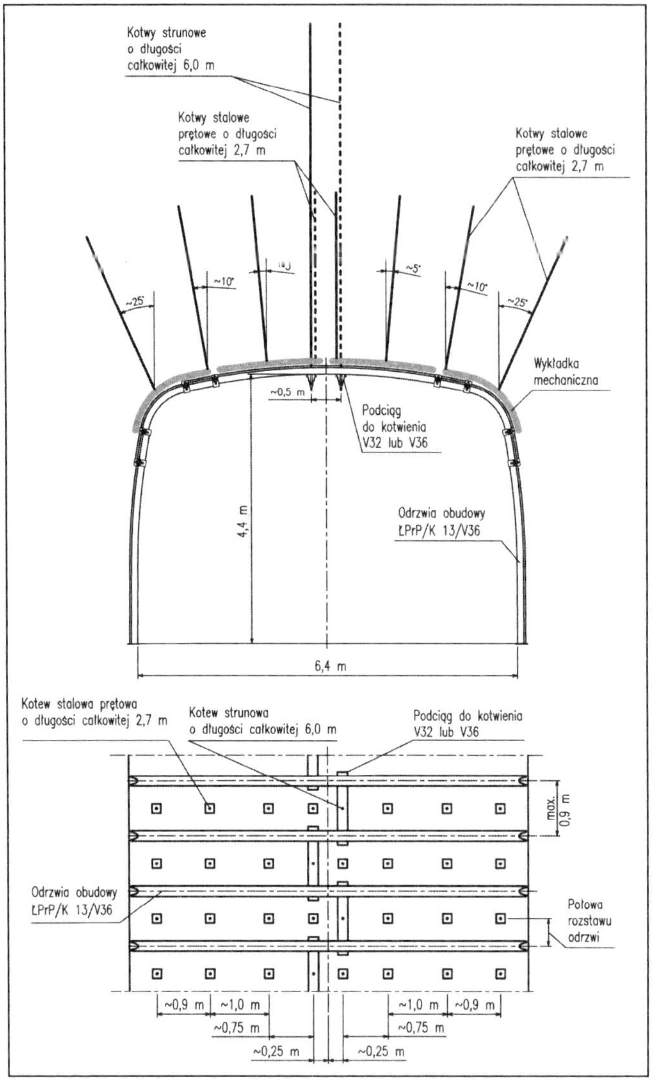

Figure 7. Support system of the cross-cut of longwall 2/VI/385

Notations: Kotwy strunowe o długości całkowitej 6,0 m – *String bolts, total length: 2.7 m;* Kotwy stalowe prętowe o długości całkowitej 2,7 m – *Steel rod bolts, total length: 2.7 m;* Wykładka mechaniczna – *Mechanical lining;* Podciąg do kotwienia V32 lub V36 – *V32 or V36 bolting horsehead;* Odrzwia obudowy ŁPrP/K 13/V36 – *Support system frame, ŁPrP/K 13/V36;* Kotew stalowa prętowa o długości całkowitej 2,7 m – *Steel rod bolt, total length: 2.7 m;* Kotew strunowa o długości całkowitej 6,0 m – *String bolt, total length: 6.0 m;* Podciąg do kotwienia V32 lub V36 – *V32 or V36 bolting horsehead;* Odrzwia obudowy ŁPrP/K 13/V36 – *Support system frame, ŁPrP/K 13/V36;* Połowa rozstawu odrzwi – *Half of the frame spacing*

7. The parting of the longwall field for longwall 2/VI/385, in which the second plough system contracted by the Bogdanka mine will be used, employed the use of the same solutions as

those for exploratory longwall 1/VI/385. The tailgate for this longwall is mostly the headgate of longwall 1/VI/385, maintained behind the longwall.

8. New challenges for longwall galleries for plough longwalls will appear during the construction of galleries at abandoned workings. Considering the big panel lengths, such galleries will be driven in two directions from the engineering cross-cut. The technology, their construction, and maintenance can be problematic.

REFERENCES

An unpublished collective work by the Mine of Lubelski Węgiel Bogdanka S.A. 2009. Poszukiwanie nowych rozwiązań w zakresie obudowy chodnikowej dla warunków Lubelskiego Węgla „Bogdanka" S.A. Etap VII. Odrzwia eliptyczne. Analiza wyników pomiarów w odcinkach doświadczalnych i opracowanie wniosków końcowych [*In search of new solutions for gallery support system in the conditions of Lubelski Węgiel Bogdanka S.A. Stage VII. Elliptical frame. Analysis of the measurement results in experimental sections and formulation of final.* (In Polish)

Chmielewski J., Kozek B. 2009. Poprawa postępów drążenia wyrobisk przygotowawczych w świetle doświadczeń kopalni LW „Bogdanka" S.A. [*Improvement of the preparatory working tunnelling progress in the light of experiences of the mine of LW Bogdanka S.A.*] Materials of the School of Underground Mining, Szczyrk. (In Polish)

Chmielewski J., Kozek B., Stopa Z., Masiakiewicz M. 2006. Doświadczenia Lubelskiego Węgla „Bogdanka" S. A. w zakresie drążenia wyrobisk korytarzowych [*Experiences of Lubelski Węgiel Bogdanka S. A. in driving roadways*] Przegląd Górniczy No. 10. (In Polish)

Kwiatkowski K., Leś A. 2010. Wdrożenie pierwszej zautomatyzowanej ściany strugowej w Lubelskim Węglu Bogdanka S.A. [*Launching the first automated plough longwall in Lubelski Węgiel Bogdanka S.A.*] Zarząd Główny SITG (Main Board of the Association of Mining Engineers and Technicians), Katowice. (In Polish)

Prusek S., Rotkegel M., Zabój K., Kozek B. 2011. Obudowa wyrobisk przyścianowych dla ścian strugowych – wymogi konstrukcyjne oraz doświadczenia praktyczne [*Longwall working support system for plough longwalls – design requirements and practical experiences*] Materials of the XX School of Underground Mining. (In Polish)

Stopa Z., Masiakiewicz M., Kozek B. 2010. Doświadczenia Lubelskiego Węgla „Bogdanka" S.A. w zakresie drążenia obudowy kotwiowej. Perspektywy stosowania obudowy kotwiowej w polskich kopalniach węgla kamiennego [*Experiences of Lubelski Węgiel Bogdanka S.A. in using bolt support system. Prospects of using bolt support system in Polish black coal mines*] Prace naukowe GIG Górnictwo i Środowisko. Quarterly edition No. 2/1/2010. (In Polish)

International Mining Forum 2011, Dyczko et al. (eds) © 2011 Taylor & Francis Group, London, UK. ISBN 978-0-415-68938-0

MR620 Roadheader Offers Opportunity for Improvement of Drift Driving Efficiency

A. Jagiełło, P. Nowak
Sandvik Mining and Construction Sp. z o.o. in Tychy, Poland

ABSTRACT: This paper presents a roadheader of 600 series suitable for mining of rocks of UCS of up-to 140 MPa. In addition, optional equipment for the roadheader, in particular bolting equipment, is discussed. Impact of changes in technology of gangway driving on daily advance rate is discussed on the basis of simulations.

KEYWORDS: hard coal mining, roadheader, bolting equipment

1 INTRODUCTION

Sandvik, performing one of the elements of the group strategy, takes due care of all safety aspects related with the manufactured products. In case of roadheaders, it is expressed by thorough analysis and monitoring of machine operation over its whole lifetime.

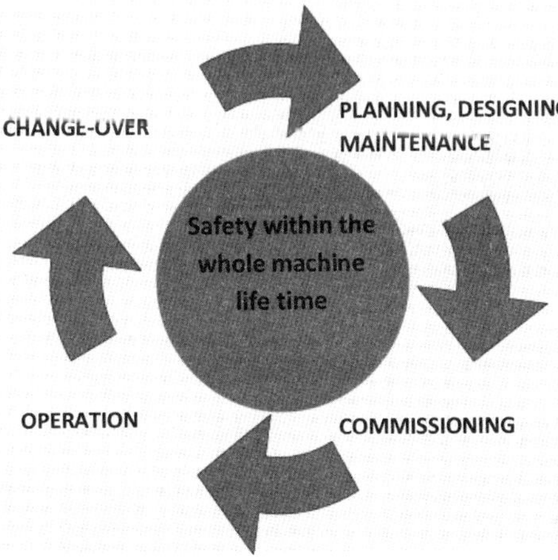

Figure 1. Concept of machine lifetime control

Effective national and EU laws are considered already at the stage of engineering works. Comprehensive EU legislation system requires machine designers and manufacturers to keep due care and carefulness when applying any solution responsible for safe machine operation.

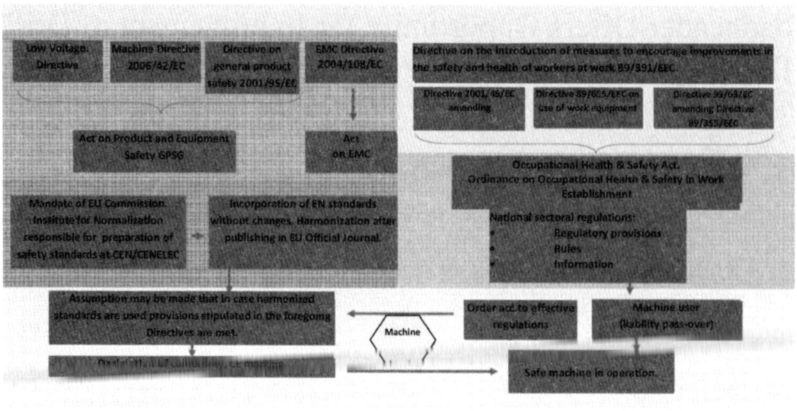

Figure 2. System of product safety assurance

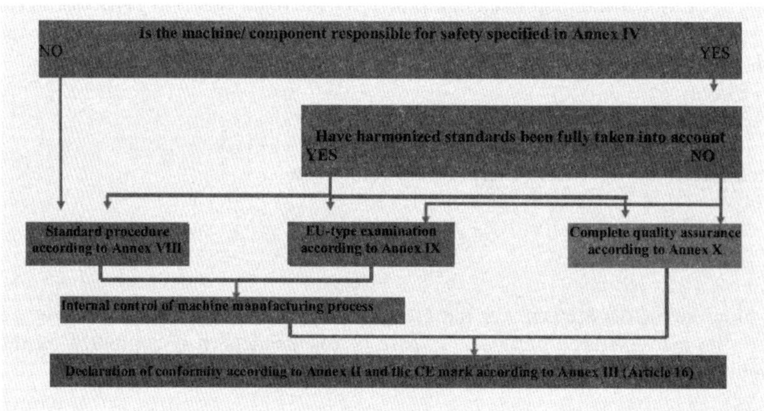

Figure 3. Evaluation Procedure of Product Compliance with the requirements comprised in the Machine Directory

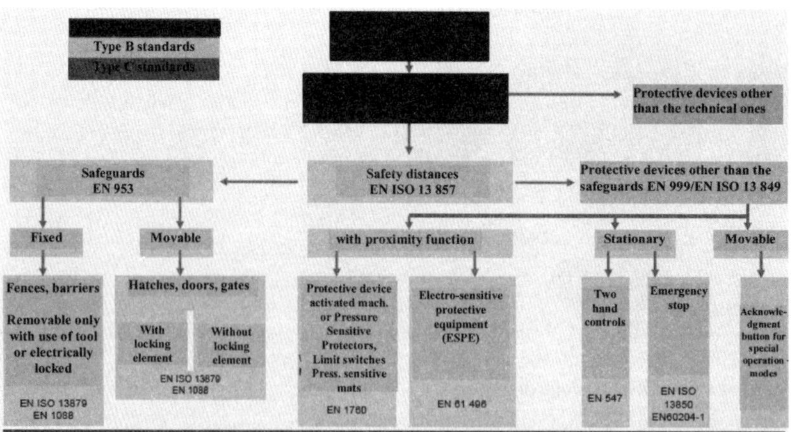

Figure 4. Standards related with risk analysis

One of the manners by which the compliance with effective laws is achieved is execution of the Evaluation Procedure of Product Compliance with basic safety requirements comprised in the Directives.

One of the requirements of effective law is the obligation to perform analysis of risks related with the product application already at the stage of engineering works. The risk analysis should also comprise product operation in emergency conditions and the cases of its use not in line with Operating Manual provided by the supplier along with the product.

As a result of the foregoing considerations and hazards related with roadheader operation such like:

- Hazards arising from work environment:
  - firedamp hazards,
  - coal dust hazards,
  - rock burst hazards,
- Hazards arising from roadheader operation:
  - hazards related with moving machine parts,
  - hazards related with combined operation of roadheader with other face equipment (winning evacuation, dedusting and ventilation, etc.),
  - fire hazard related with the roadheader itself,
  - electrical shock hazards,
- Hazards associated with performance of service and maintenance works
  - hazards caused by unintentional machine start-up during performance of diagnostic works,
  - hazards caused by uncontrolled machine motions,
  - hazards caused by special tools.

Whole spectrum of engineering solutions was integrated with the roadheaders in order to make them suitable to work in mining gangway conditions.

Key component of each roadheader is a cutter properly designed and adapted to specific operating conditions and equipped with high quality picks enabling rock cutting with the maximum efficiency at the minimum power consumption.

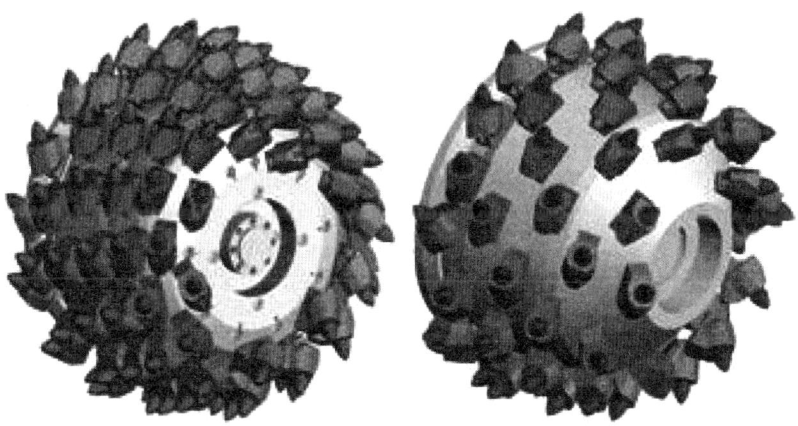

Figure 5. Various versions of cutter for different applications

Even the best design of the cutter will not eliminate generation of huge heat and dust amounts during cutting – therefore, the use of efficient spraying system is required.

At the present time the most effective solution is a double spray system composed of water spray of the cutting track combined with external water spray of the cutting space in various configurations.

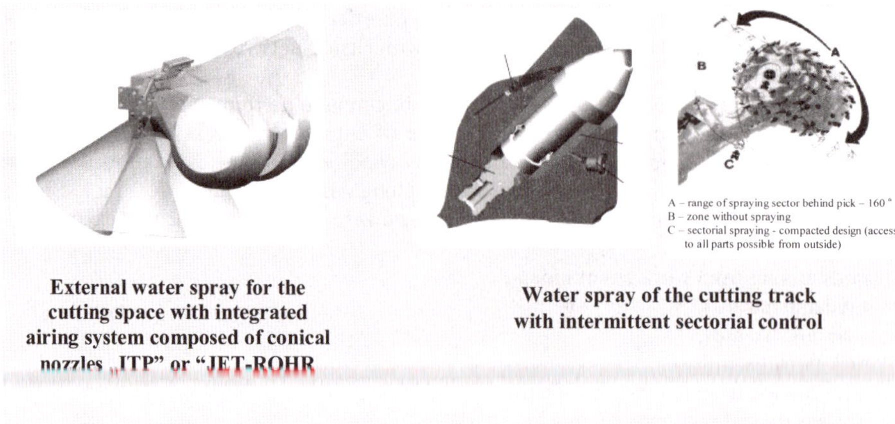

Figure 6. Double water spray of cutting head

For proper functioning of spraying system, appropriate control and monitoring unit of SIL 2 safety integrity level has to be provided. Spray control system is equipped with a dedicated PLC, which communicates with main PLC unit supervising operation of whole roadheader.

Figure 7. Control and supervisory devices

Control and supervisory devices comprise emergency switches and interlocks such like:

- four classical emergency switches, which switch off the roadheader,
- portable emergency switch for performance of maintenance and operation works,
- interlock preventing from machine switch-on from the place other than the operator stand,
- working platform position sensors.

All foregoing measures and additional safety equipment as well as power supply and roadheader control systems have been manufactured in accordance with the most recent European standards and upon consideration of:

- electric shock prevention rules,
- explosion safety rules,
- rules for safe control of face equipment (e.g. winning evacuation equipment, dedusting, etc.),

and they guarantee safe machine operation even in the most difficult operating conditions.

Figure 8. Switches and interlocks and optional safety equipment

2 ROADHEADER SERIES 600

There are many versions of the 600 series-roadheader. One of the most advanced solutions is MR620 with bolting platform. Machines of this type have been working for German mining for many years and they proved their suitability. Drifts of cross-section of up-to $40m^2$ may be driven with use of roadheaders of this type.

Technical date*):

Weight *(total)*	[t]	approx. 131
Length total *(total)*		approx. 18200
Length *(incl.ABSE, without conveyor)*	[mm]	approx. 12900
Height *(total)*		approx. 4060
Height *(without ABSE)*		approx. 2240
Width *(of crawler mechanism)*	[mm]	2600
Width of loading table (min. / max.)	[mm]	4800–5500
Position of loading table above lower edge of crawler max.	[mm]	600
Position of loading table below lower edge of crawler max.	[mm]	250
Width of crawler	[mm]	720
Ground pressure	[MPa]	0.24
Ground clearance	[mm]	230

Cutting profile with use of telescopic cutting arm:

Max. cutting height above lower edge of crawler	[mm]	5300 / 5800
Max. cutting width at swinging angle +/- 37° in horizontal direction	[mm]	8000 / 8500
Max. undercut below lower edge of crawler	[mm]	300 / 500
Max. limit of transverse cutting measured to the lower edge of crawler *(with/ without a boom)*	[m^2]	48 / 40

Supply voltage: 1000 [V], 50 [Hz]

Total electric power installed	[kW]	522
Cutter motor - water cooled	[kW]	300
Rotary speed	[rpm]	1475
Rated torque	[Nm]	1950
Hydraulic drive - water cooled	[kW]	150
Rotary speed	[rpm]	1485
Rated torque	[Nm]	975
Drive of loading stars	[kW]	2 x 36
Conveyor drive – hydraulic	[kW]	55

*) Values may vary subject to roadheader version and equipment

Bolting platform is equipped with:

- Set of bolting equipment:
  - Two NVT bolters,
  - RHR52 rotary drill,
  - HSX percussion & rotary drill,
  - Adapter for One Step Bolts,
- Manipulator for rolled insert,
- Jib crane for lifting materials to upper platform.

During cutting, the platform is retracted to home position and bolters are folded. Once the cutting has been completed and cutter boom lowered, the control system allows the platform lifting to working position and bolter unfolding. Each unit is operated by two operators.

The pictures below shows home, tramming and working positions of the platform.

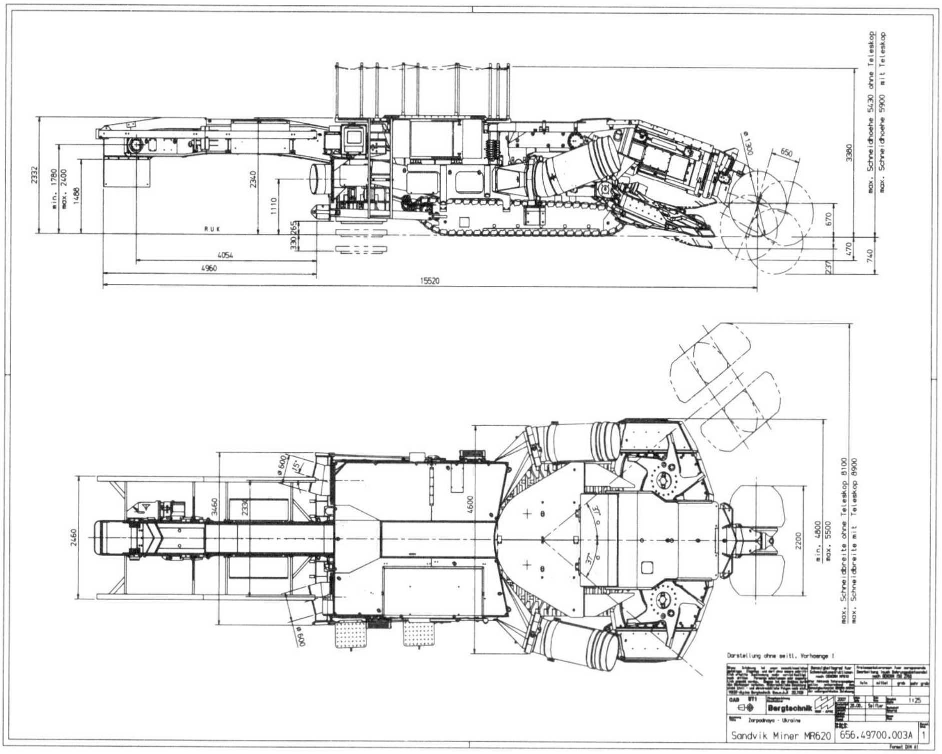

Figure 9. Overall dimensions of roadheader series 600

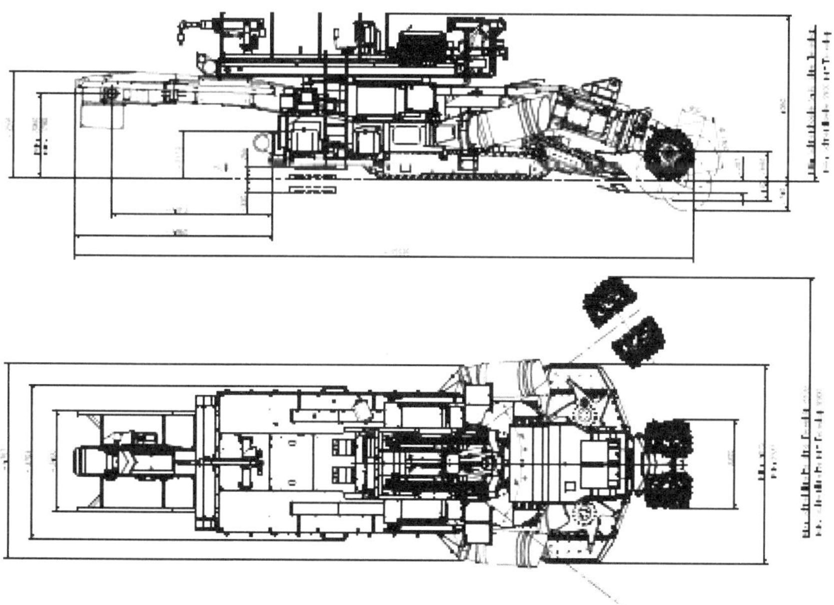

Figure 10. Overall dimensions of MR620 roadheader with bolting platform

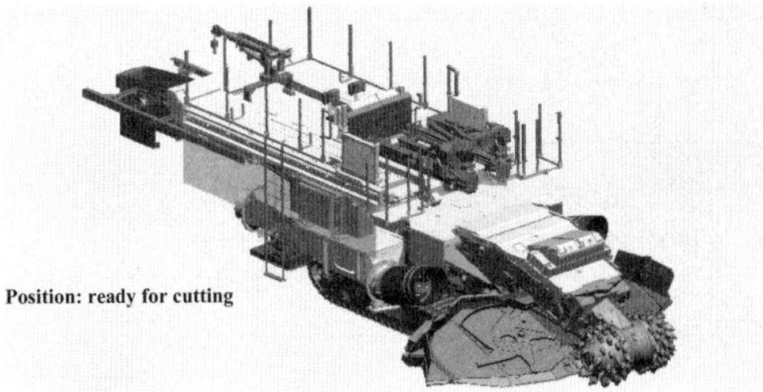

Position: ready for cutting

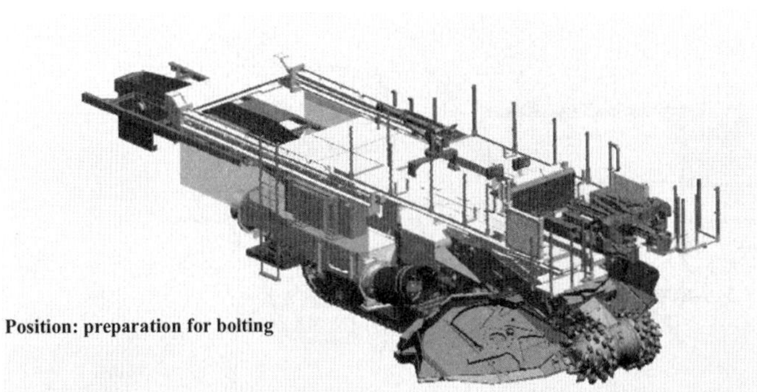

Position: preparation for bolting

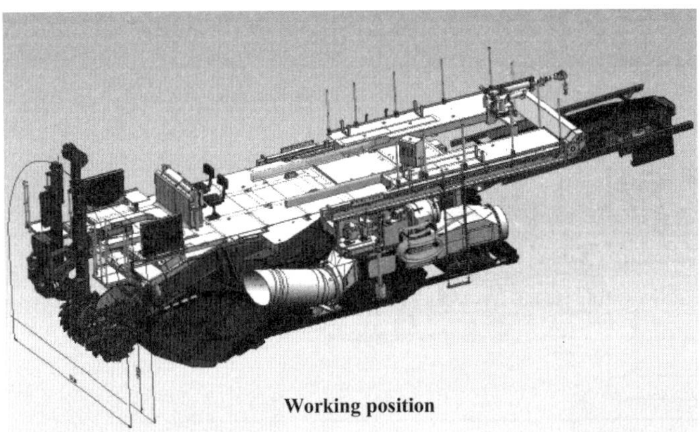

Working position

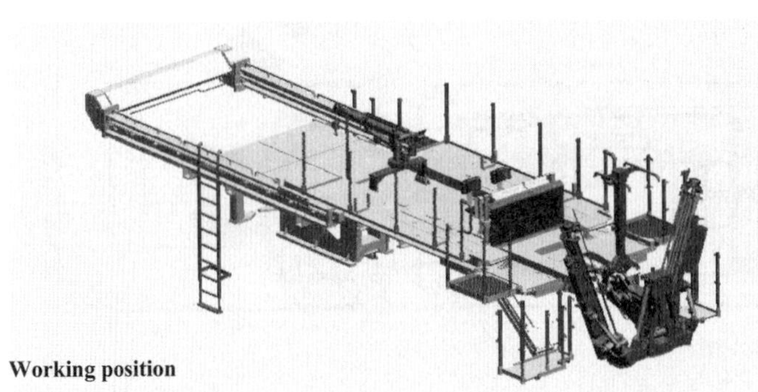

Working position

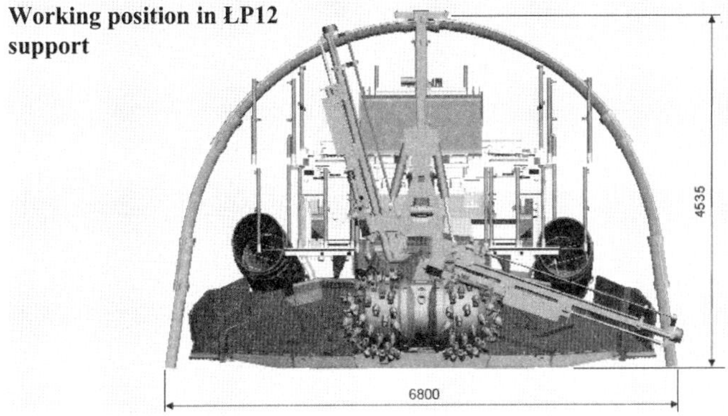

Figure 12. Operation stages of bolting platform

Another technically interesting solution is MH620/024 roadheader installed in copper mine of ZG Lubin KGHM Polska Miedź SA. It is a fully remote controlled unit without operator cab. Originally, the machine was equipped with a bolting platform very similar to the one described above, but because of varying geological conditions such design did not meet the customer's expectations and bolting platform was finally replaced with a bolter mounted on manipulator. The machine in current version is shown on Figure13.

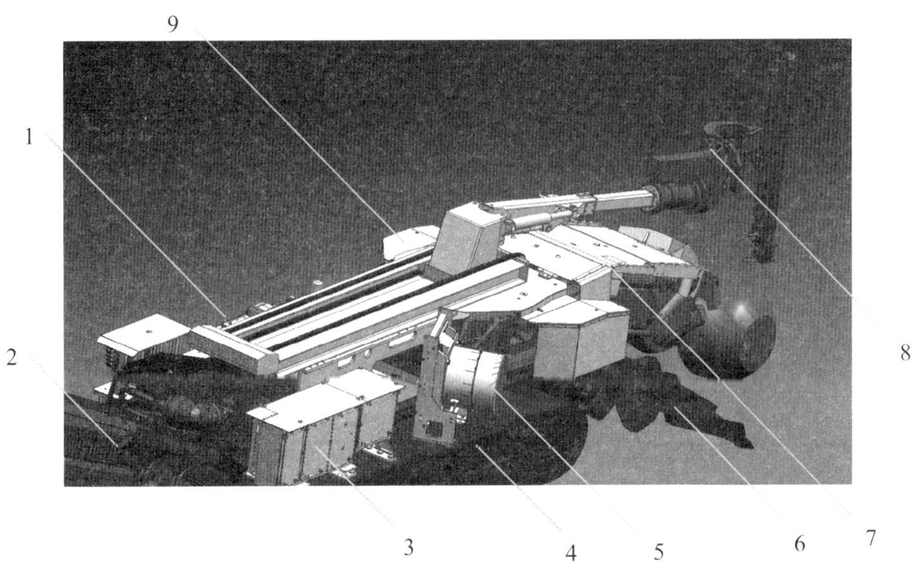

Figure13. General view of MH620 roadheader: 1 – hydraulic unit; 2 – tilting belt conveyor ; 3 - switch gear; 4 – crawler mechanism; 5 – additional cooling system; 6 – loading unit; 7 – boom arm with cutter; 8 – ABSE unit; 9 – turret

Characteristics of MH620 roadheader

- No operator cab installed,
- Remotely controlled semi-automatic bolter and compressor for lifting adhesive cartridges,
- Cooling system is extended by external air cooler,
- Control system is extended by AUTO-testing and data recording functions.

Figure 14. Example of bolting pattern for MH620 roadheader

Bolters mounted on boom-type manipulators of roadheaders also found application in coal mining. A dozen MR300 roadheaders equipped with such bolting units are currently working for the Polish and Czech coal mines. Roadheaders of 600 series may be also equipped with bolting unit of this type.

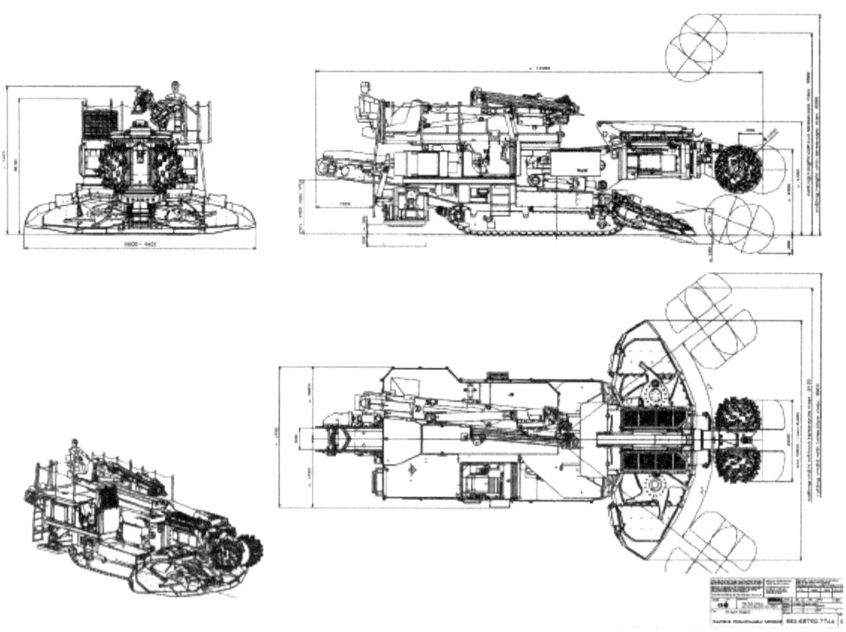

Figure 15. 600 series roadheader with bolter mounted on manipulator

In this case the manner of mounting and functional design of manipulator-mounted bolters are similar to the bolters already used in smaller machines. The unit is mounted on central console (10) with use of mounting plate (11) (see Figure 15.)

1. Operator stand
2. Hose layer
3. Boom bearing
4. Boom
5. Control block
6. Drill rig
7. Drill rig support
8. Manipulator
9. Bolter
10. Bolter unit console
11. Mounting plate
12. Bolter unit support
13. Boom lifting cylinder
14. Horizontal tilting cylinder
15. Position change drive

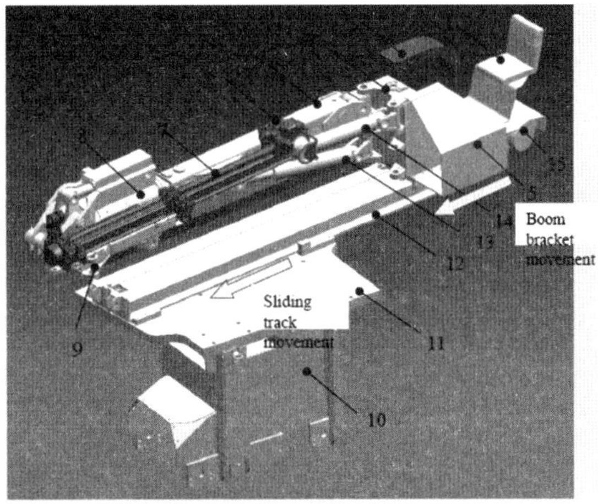

Figure 15. Fastening of bolter mounted on manipulator to roadheader

Note that this solution may be used only in drifts of large cross-section areas.

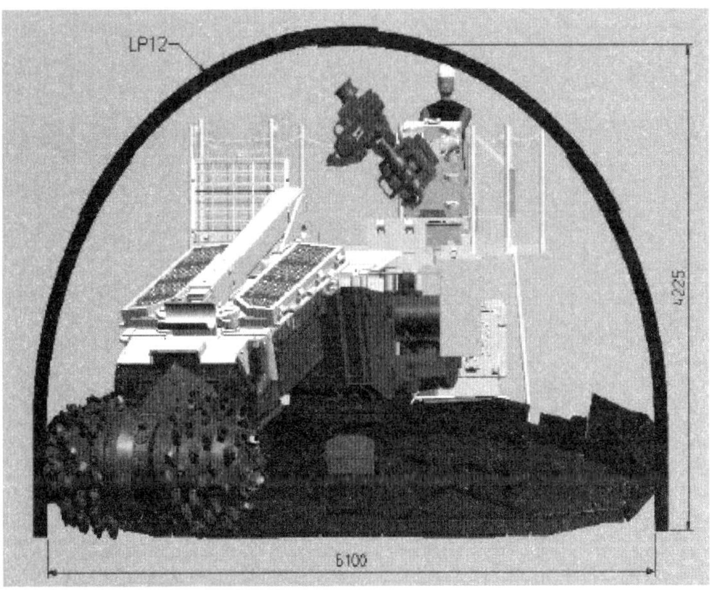

Figure 16. MR 620 roadheader with a bolter in ŁP 12 lining

3 AUTOMATIC TU-C BOLTER

Numerous solutions of bolting with use of hydraulically driven bolters are used in coal mining. However, two trends are clearly noted:

- Maximum simplification of bolter design and manual performance of auxiliary operations (e.g. tool replacement, bolt positioning in the drive, injection of adhesive cartridges) by the operator,
- Striving after minimising of man intervention in whole bolting process through enhanced mechanization and automation.

Figure 17. Top view of TU-C bolter's working components

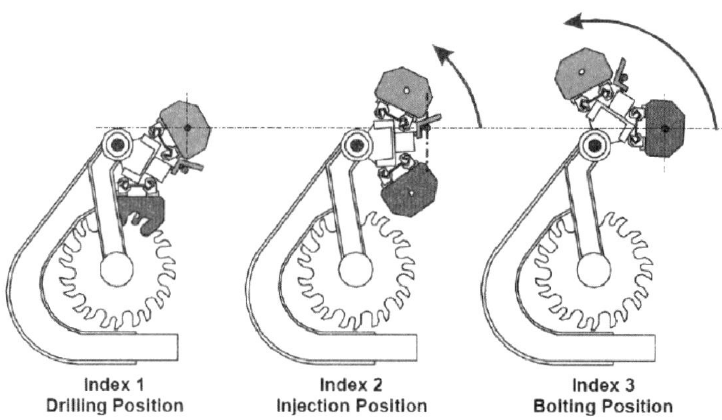

Figure 18. Mechanization of individual stages of bolting with use of TU-C bolter

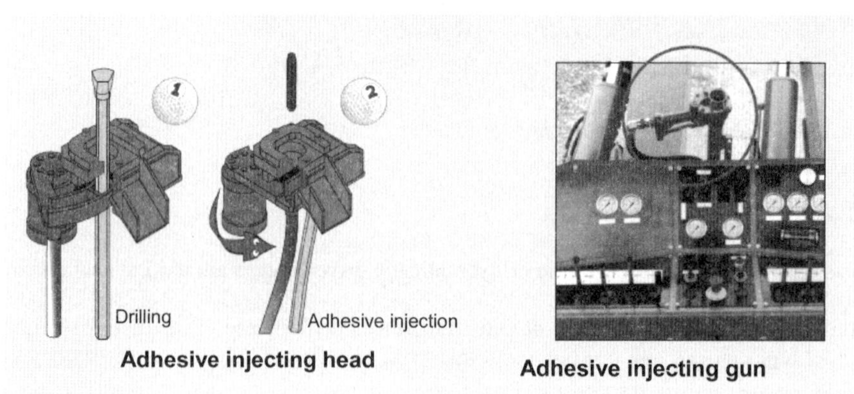

Figure 19. Realization of a concept of remote adhesive injection in TU-C bolter

Automatic TU-C bolter materializes the efforts aiming at improvement of bolting process efficiency through mechanization and automation of operations required for completion of bolting support.

Mechanization of individual bolting process stages has been achieved owing to their division by separate working components such like:

1. Drill rig drive,
2. Adhesive cartridge head,
3. Bolt insertion drive (adhesive stirring, nut tightening).

These components move pendulously around common axis, which runs through a strutting point to the roof.

TU-C bolter may operate without bolt feeder. In such case bolts are manually loaded or are loaded from revolving magazine holding up-to 15 bolts. In the latter case one row of bolts may be installed without a need for operator to leave the working stand. Such solution on one hand reduces bolting time and on the other hand improves safety at work by elimination of a need for operator to stay in vicinity of unsecured roof prior to bolting.

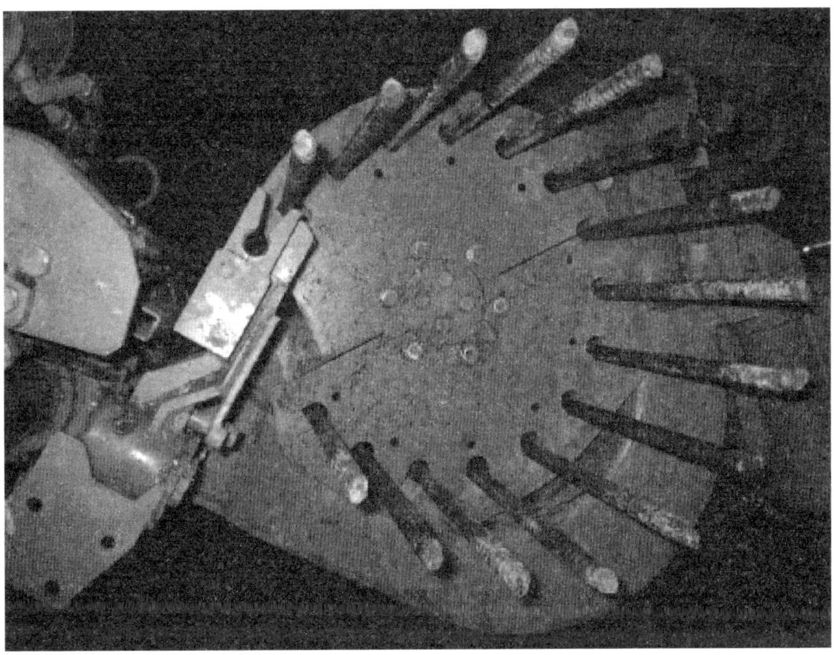

Figure 20. Revolving bolt magazine

4. CAPACITY

In case of roadheader, the only unbiased measure of machine capacity is cutting rate expressed as a volume of rock cut within a time unit. For making this index independent from the factors not related with the roadheader, the term of "net cutting time" has been introduced. "Net cutting time" is a time at which a cutting head remains in effective and active contact with a rock mass. Then, Net Cutting Rate (NCR) is a volume of cut rock divided by effective cutting time (expressed in net cutting hours).

For roadheader users, cutting capacity measured e.g. as a daily advance is more important. In this case, NCR is one of the parameters, which enable forecasting of expected advance in actual conditions. These analyses are one of the tools enabling optimization of future working face operation.

5. OPTIMIZATION

For simulating operation of hypothetical working face following assumptions were made:

Project Information		
Project name:	XXXXXX	
Tunnel shape:	ŁP	
Tunnel cross section:	28,00	[m²]
Number of shifts per day:	4	
Shift time:	7,00	[h]
Daily available working time:	20,00	[h]
Daily machine working time:	18,00	[h]

Machine type:	MR620	
Cutter head type:	R400W-TC60	
Pick type:	22 mm TC diameter insert	
Number of rock bolters:	2; 1; 0	
Availability:	90%	
Utilization:	83%	

Number of rock bolts per row:	0; 7; 14	
Rock bolt length:	2,70	[m]
Distance between rock bolt rows:	0,90	[m]

Quantity of coal	40%	
Average UCS_R of coal:	20	[MPa]
Quantity of sandstone:	60%	
Average UCS_R of sandstone:	40; 60; 80; 100; 120; 140	[MPa]

Figure 21. Pre-assumptions to working face simulation

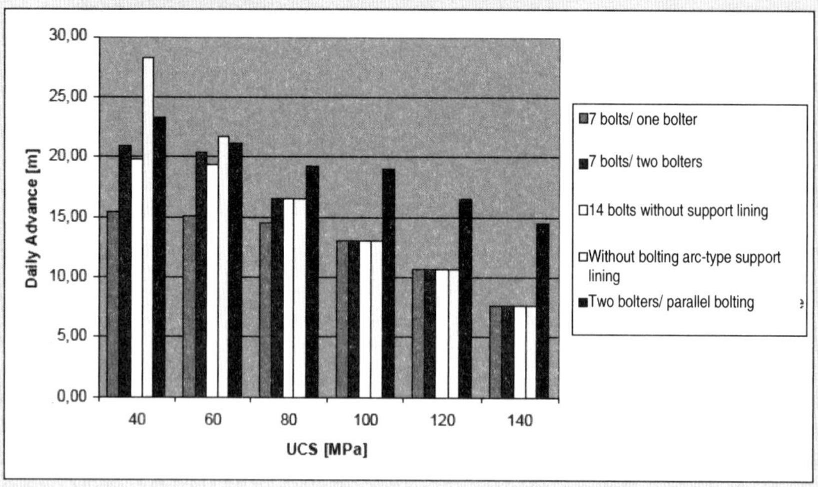

Fig. 22. Results of analysis (daily advances in function of cut rock Rc and support completion technique)

Analysis carried out with consideration of mining & geological conditions, work organization, type of gallery support, etc. provided the result, which shows impact of selected factors on achievable results i.e. daily advances.

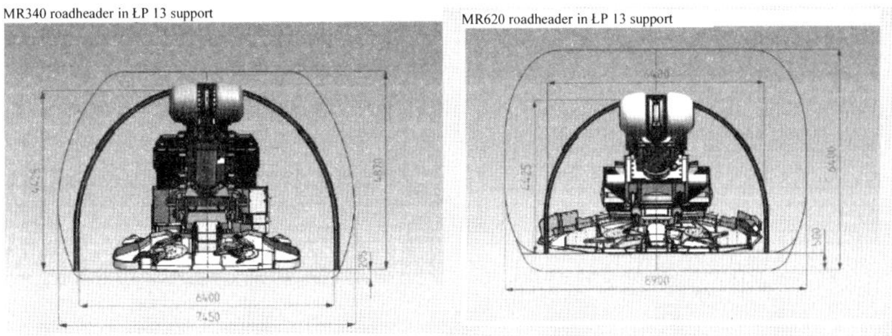

Figure 22A. Comparison of MR340 and MR620 roadheaders in ŁP 13 support

Figure 22B. Comparison of MR340 and MR620 roadheaders in ŁP 13 support

6 FINAL CONCLUSIONS

Roadheaders of 600 series offer new possibilities to adapt drift driving methods to the challenges set by plough technique (large heading cross-sections, high daily advances of plough complexes) for development works.

Roadheaders of 600 series are safe, efficient and reliable machines, which enable meeting such requirements as reliable operation in any mining conditions, achievement of high daily advances, and flexible adaptation to varying mining & geological conditions.

The roadheaders series 300 (AM75, MR340), which are commonly used at the present time, in case of development works for plough complexes have already arrived at their performance limits (particularly, no more margin remains in the scope of cut cross-section area). Therefore, implementation of equipment and technical & organizational solutions taking into account the current needs is required.

Implementation of the machine has confirmed achievement of cutting capacity of the roadheader assumed by the engineers of Sandvik Mining and Construction. Feedback with the user was very helpful because it enabled modifications of the roadheader resulting in improvement of bolting performances.

Possibility of forecasting real machine performances, which is essential for the users for analysis of economic and organizational aspects of the project, is of key importance.

REFERENCES

Sandvik's own sources.

International Mining Forum 2011, Dyczko et al. (eds) © 2011 Taylor & Francis Group, London, UK. ISBN 978-0-415-68938-0

Prospects of Using the MR 620 Roadheader with Advanced Mechanical Bolting System in "Bogdanka" Mine

M. Masiakiewicz
Lubelski Węgiel "Bogdanka" S.A., Poland

ABSTRACT: The article presents a new unique solution for the system of driving roadways developed for the conditions existing in the mine of Lubelski Węgiel "Bogdanka" S.A. The purpose of the presented organisation and technical project is to increase the progress rate of driving roadways within arch-shaped bolt support. The system will comprise a high power roadheader, a mechanized platform for installing gallery support frames, and a double boom bolting machine. The new face system will enable simultaneous performance of certain basic operations of the engineering cycle, such as mining and bolting, which will bring a significant increase of the driving progress rate. The system involves building a support system in the face to ensure fast support of the rock mass by means of mechanical lining, using containers filled with a fast-setting binding material under pressure, and bolting, in order to reinforce the rock mass behind the roadheader.

KEYWORDS: hard coal mining, underground roadway, roadheader, mechanical bolting system

1 INTRODUCTION

Driving roadways is the key element of deep mine hard coal extraction, which significantly determines the output and economic results of the mine. Improvement of the roadway driving engineering systems is therefore an important factor in the improvement of mining efficiency, especially that the wall progress rates in beds featuring coal thickness of approx. 1.5 m increase due to the required high level of output generated by a single wall. Moreover, the range of works required to complete 1 m of roadway increases because the cross section of the gallery widens, and the support system becomes increasingly complicated and time-consuming to build. Presently, the standard shoring used in the "Bogdanka" mine is the 6- and 7-element frame support system, with additional rock mass reinforcement by means of bolt support installed between the frames of the support system. The purpose of the project is to increase the roadway driving efficiency, i.e. to increase the heading progress rates which will be related to a reduction in roadway driving costs. This is possible due to parallel performance of several operations which are part of the driving process, such as:

- mining,
- loading,
- preliminary installation of arch canopies,
- installation of lagging,
- filling containers for mechanical lining,
- bolting,

– extending the conveyor belt.

The essential purpose of the proposed solution is to obtain a "highly efficient heading system for operating large section area excavations". In the case of headings driven by means of roadheaders, the factor which significantly limits the driving progress is the serial performance of operations which make up the engineering process. The shoring completion time is often much longer than the excavation operation. This time can be shortened by using equipment in the heading which will work in a coordinated way and enable parallel performance of operations in the entire process of driving roadways. Such equipment includes: a roadheader, a self-propelled working platform for the installation and transportation of the gallery support system, a drilling and bolting device with its own propulsion and a return station of the conveyor belt with a separation system.

2 THE PRESENT TECHNOLOGY OF DRIVING ROADWAYS IN LUBELSKI WĘGIEL "BOGDANKA" S.A.

The primary equipment of the driven roadway heading, which has not been significantly modified for many years now, includes:
- AM-75 roadheader,
- PTB-BOA feeder,
- DT-SIGMA belt feeder,
- Gwarek 1000 belt conveyor,
- HBKM-1/400 dry dust extractor by Hëlter (Germany),
- DV-290 air cooler by WAT (Germany)[1],
- overhead heading apparatus train,
- Korfmann ES9-500/80 air-duct fan (Germany).

Figure 1 presents the layout of the heading zone tunnel equipment.

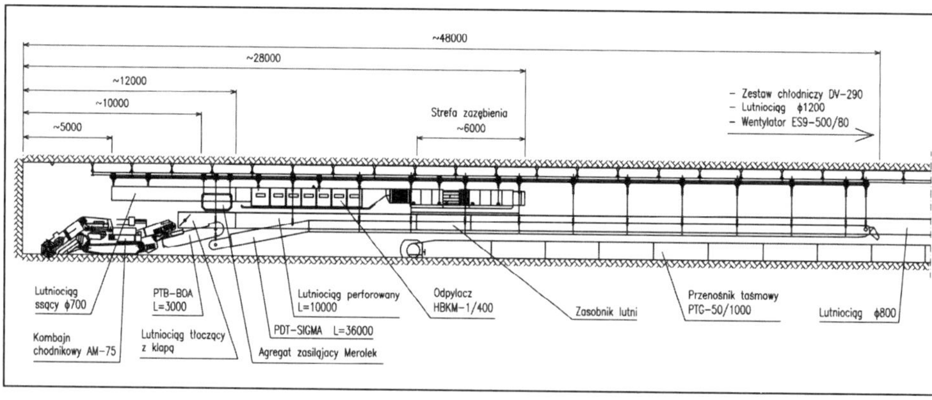

Figure 1. Typical equipment of an driven gallery heading

Notations: Strefa zazębienia ~6000 – *Overlap zone, ~6000;* Zestaw chłodniczy DV-290 – *DV-290 cooling set;* Lutniociąg Ø 1200 – *Air-duct, Ø 1200;* Wentylator ES9-500/80 – *ES9-500/80 fan;* Lutniociąg ssący Ø 700 – *Intake air-duct, Ø 700;* Kombajn chodnikowy AM-75 – *AM-75 roadheader;* Lutniociąg tłoczący z klapą – *Force air-duct with damper;* Lutniociąg perforowany L=10000 – *Perforated air-duct, L = 10000;* Agregat zasilający Merolek – *Merolek power generator;* Odpylacz HBKM – 1/400 – *HBKM – 1/400 de-duster;* Zasobnik lutni – *Air-duct tray;* Przenośnik taśmowy PTG – 50/1000 – *PTG – 50/1000 conveyor belt;* Lutniociąg Ø 800 – *Air-duct, Ø 800*

[1] Since May 2011, Stefanów mining field has been using central air conditioning

The most restrictive requirements with regard to driving roadways concern plough walls in longwall galleries that must feature wider dimensions than in case of shearer walls. The standard solution used for longwall galleries is the ŁPSC type 6- or 7-element frame support system made of S480W steel V36 section bar, where the frame size used for shearer walls in both galleries is 10/K, and, for plough walls, it is 11/S for the top gate and 12/S for the bottom gate. In galleries for shearer walls, the support system features a frame with the spacing of 0.75 or 0.9 m. According to the common standard, longwall galleries for plough walls are driven in the support system with rock mass bolting between frames. The frame spacing is usually 0.9 m. The number of bolts is usually 7–8 bolts per 1 frame.

Figure 2 presents a typical layout of the support and bolt system, characteristic of the bottom gate for a plough wall.

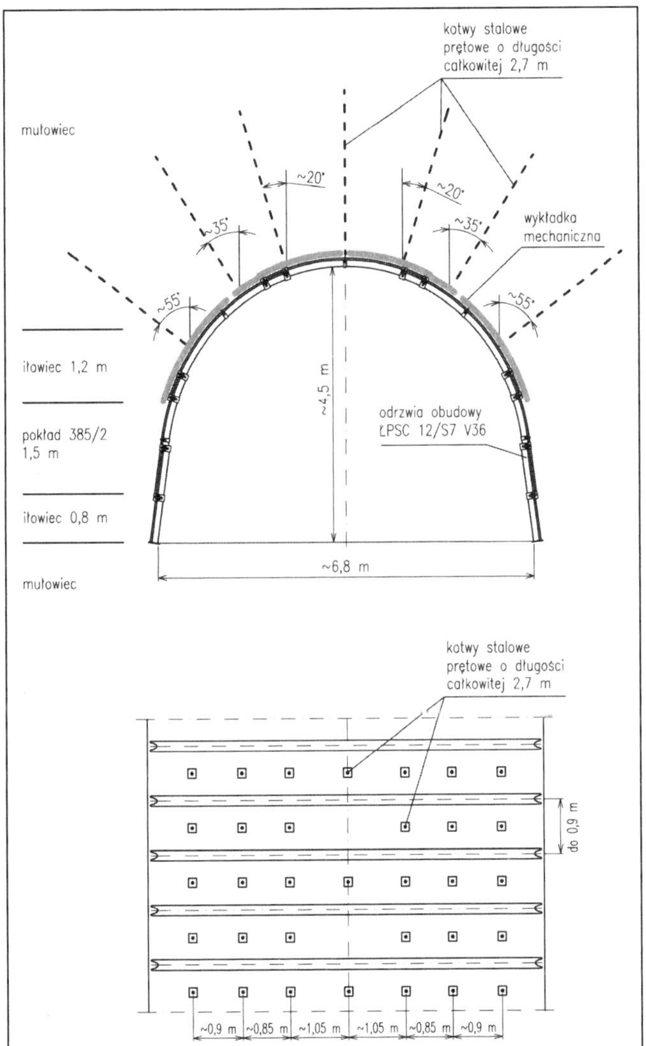

Figure 2. Typical layout of a support and bolt system of the bottom gate for a plough wall

Notations: mułowiec – *mudstone*; iłowiec – *siltstone;* pokład – *bed;* kotwy stalowe prętowe o długości całkowitej 2,7 m – *steel bar bolts, total length of 2.7 m;* wykładka mechaniczna – *mechanical lining;* odrzwia obudowy – *support system frame;* do 0,9 m – *to 0.9 m*

The basic machine for driving roadways is the roadheader, which is a collective operation machine, performing several different tasks at the same time: mining, loading and removing the yield behind the roadheader onto other means of transport. Operations within the roadway driving process that are independent include: making the lagging, construction of support system frames, and bolting between frames. Performing those works involves a roadheader arm with an installation grabber fixed on it, operational platforms, and a drilling and bolting device (installed on a roadheader or on a self-propelled chassis). After making a breach, the heading brigade performs scaling and lagging of the roadway roof, then backs the roadheader in order to install the support system canopies. Shoring canopies are installed under a built-up roof, at least in the third field from the heading.

After installing the set of canopies and building lining containers, the roadheader operator drives the roadheader into the heading, lifts the set of canopies up to the roof and connects them with stretchers. The next operation is to complete the lagging and install side arches. After assembling complete frames and connecting them to other frames, the heading brigade begins to fill the lining containers with cement binder. The described works are performed cyclically during three shifts, and the fourth shift completes bolting and makes an extension of the belt conveyor. Using such technology for a roadway excavation with cross section of approx. 30 m^2, it is possible to achieve daily progress of 13.5 m and monthly progress at around 350 m.

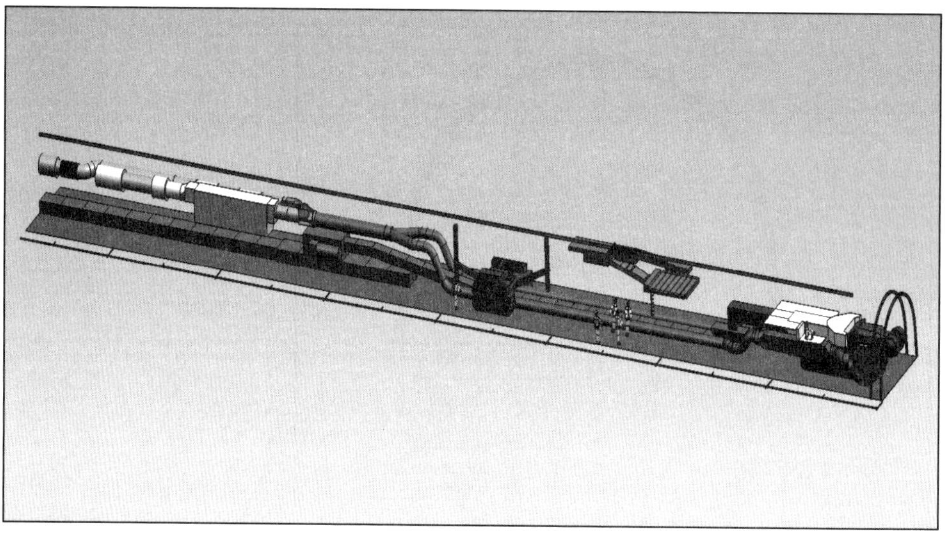

Figure 3. Concept of the new mechanical system

3 METHOD OF IMPLEMENTING THE PROJECT

Roadways will be driven inside the support system with bolting between the frames. The important difference compared with the present solution is that particular operations, such as mining, installation of the support system, and bolting, will be performed simultaneously. The key machine of this system is a heavy-duty roadheader, which enables mining of the entire section of the roadway from a single position of the machine, equipped with a mining profile automatic control system and radio control. It is a collective operational machine, performing several different tasks simultaneously: mining, loading, and transporting the yield away behind the roadheader onto a scraper conveyor, and further onto other means of transportation. Operations related to the building up the support system frames and bolting between frames will use other equipment, such as the operational platform with installation brackets, the self-propelled double-station drilling, and the bolting unit. Using that equipment will allow simultaneous performance of the individual operations related to driving a roadway – mining, preliminary assembly, and installation of steel bolts.

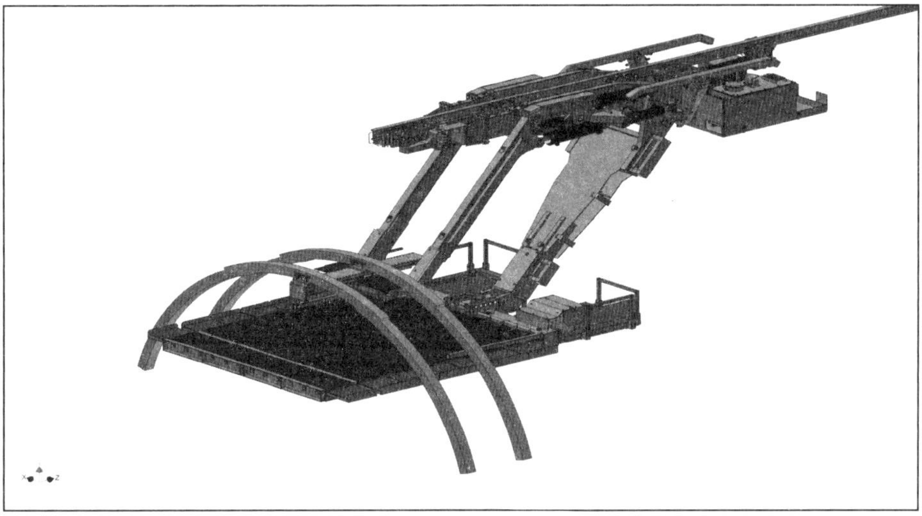

Figure 4. Design of a self-propelled platform for installation of the support system

Canopies are installed on an assembly bracket of the operational platform, located beyond the roadheader working zone. After installing two sets of canopies, building up lining containers and lagging net, the platform operator drives into the heading, lifts the set of canopies up to the roof, and connects them with stretchers. The next operation is to complete the lagging and to install side arches. After assembling complete frames and connecting them to other frames, the heading brigade begins to fill the lining containers with cement binder. Using mechanical lining, the roof is immediately supported, which allows the building up of the bolting system within the distance of 20 m from the heading face. This can be achieved by using the double-station bolting unit, located right behind the operational platform.

Figure 5. The double boom bolting vehicle for building up a bolting support system behind the roadheader

The belt conveyor will be extended as necessary during shifts, which will be possible due to innovative solutions – a track station or a self-propelled wheeled station.

Using the described technologies for the roadway with a area of approx. 30 m^2 will make it possible to achieve daily progress of 27 m.

4 SUMMARY AND FINAL CONCLUSIONS

1. The present technology of driving roadways in the Lubelski Węgiel "Bogdanka" mine by means of the AM-75 roadheaders allows to achieve heading driving progress of 20 m per 24 hours and uses a support system with size 10 frames.
2. Implementation of the plough technology in "Bogdanka" required increasing the cross section size of longwall galleries from 10 to 12, which, in turn, requires using a support and bolting system. As a result, the longwall gallery driving progress has declined in the case of plough walls down to 13.5 m per 24 hours.
3. In order to enhance the process of driving large section roadways (about 30 m^2 in the breach) in a support and bolt system, it is necessary to use new mechanized equipment for the heading which allows changing the presently used driving technology. This goal is achieved by the unique mechanized system, which is designed specifically for the conditions of the mine of Lubelski Węgiel "Bogdanka" S.A.
4. The new system will comprise a high power MR-620 roadheader, a mechanized platform for installing gallery support frames, and a self-propelled double boom bolting machine. The new face system will enable simultaneous performance of certain basic operations of the engineering cycle, e.g. mining and bolting, which will shorten the driving cycle time.
5. The system involves building a support system in the face to ensure fast support of the rock mass by means of mechanical lining, using containers filled with fast-setting binding material under pressure, and bolting, to reinforce the rock mass behind the roadheader.
6. Manufacturing of the first prototype of the heading system is planned for 2012 and it is intended to be used for the first time for the opening up of field VII and VIII in bed 385/2 in Stefanów.
7. In the conditions of the "Bogdanka" mine, successful implementation of the new roadway driving technology should enable a significant increase in the progress of driving wide section longwall galleries in a support and bolt system up to 30 m per 24 hours and more.

REFERENCES

Chmielewski J., Kozek B., Masiakiewicz M. 2006. Drążenie chodników w LW „Bogdanka" S.A. [*Driving roadways in LW "Bogdanka" S.A.*] Materiały Szkoły Eksploatacji Podziemnej, Szczyrk. (In Polish)

Chmielewski J., Kozek B., Stopa Z., Masiakiewicz M. 2006. Doświadczenia Lubelskiego Węgla „Bogdanka" S.A. w zakresie drążenia wyrobisk korytarzowych [*Experiences of Lubelskiego Węgla "Bogdanka" S. A. in driving roadways*] Przegląd Górniczy No. 10. (In Polish)

Chmielewski J., Kozek B. 2004. Poszukiwanie nowych technologii wykonywania i utrzymywania wyrobisk korytarzowych. Dwadzieścia lat w podziemnych technologiach górniczych [*Searching for new technologies of driving and maintaining roadways. Twenty years of the deep mining technologies*] Górnictwo i Geoinżynieria. AGH University of Science and Technology. Quarterly. Kraków 2004, Vol. 28, Journal 4/1. (In Polish)

Chmielewski J., Kozek B. 2006. Doświadczenia Lubelskiego Węgla „Bogdanka" S.A. w zakresie wykonywania obudowy chodnikowej z zastosowaniem wykładki mechanicznej [*Experiences of Lubelski Węgiel "Bogdanka" S.A. in building gallery support system with the use of mechanical lining*] Conference materials of the Central Mining Institute and Mittel Steel Poland. Katowice 2006. (In Polish)

Chmielewski J., Masiakiewicz M., Hajduk J., Szyszka A. 2011. Pomyślnie próby zastosowania samobieżnej stacji zwrotnej przenośnika taśmowego w kopalni „Bogdanka" [*Successful test operations of the self-propelled return station of the belt conveyor in the "Bogdanka" mine*] Wiadomości Górnicze, No. 9/2011. (In Polish)

Głuch P., Kozek B., Koza H., Masiakiewicz M., Słomiany P. 2001. Obudowa kotwiowa przecinki ścianowej 4/V w kopalni Bogdanka [*Bolt support system of 4/V wall entry in the "Bogdanka" mine*] Scientific and technical conference on Innovative Mining Technologies 2001 – Bolting, Gliwice, Ustroń. (In Polish)

Kozek B., Ruchel A. 2011. Wykonywanie i utrzymywanie chodników przyścianowych w kopalni Lubelski Węgiel „Bogdanka" S.A. – dotychczasowe doświadczenia [*Driving and maintaining longwall galleries in the mine of Lubelski Węgiel "Bogdanka" S.A. – Experience to date*], Materiały Szkoły Eksploatacji Podziemnej 2011, conference materials. (In Polish)

Stopa Z., Masiakiewicz M., Kozek B. 2010. Doświadczenia Lubelskiego Węgla "Bogdanka" S.A. w zakresię drążenia obudowy kotwiowej. Perspektywy stosowania obudowy kotwiowej w polskich kopalniach węgla kamiennego [*Experiences of Lubelski Węgiel "Bogdanka" S.A. in using bolt support system. Prospects of using bolt support system in Polish hard coal mines*] Prace naukowe GIG Górnictwo i Środowisko. Quarterly edition No. 2/1/2010. (In Polish)

International Mining Forum 2011, Dyczko et al. (eds) © 2011 Taylor & Francis Group, London, UK. ISBN 978-0-415-68938-0

"Lubin" Mine Experiences in Driving Heading Excavations with Sandvik Mining and Construction's MH-620 Heading Machine

M. Czechowski
KGHM Polish Copper, "Lubin" Mine, Poland

ABSTRACT: The experiences gained during the drive of heading excavations with Sandvik's MH-620 in two entirely different areas of "Lubin" mine are shown in the article. The influence of imperfect construction of the machine on the heading drive process has also been highlighted.

KEYWORDS: heading machine, heading drive, copper ore

1 INTRODUCTION

During the Underground Exploitation School in 2007 we have presented an article called "Technical and economical experiences of "Lubin" mine in driving inclined headings with Voest-Alpine AM-75EX-s heading machine". The article was concluded with a following statement:
"*The gathered experiences allow to think about wider use of such machines, not only in our mine. The procedures to implement such machines have already been developed. In particular, reinforced versions of machines such as Alpine Hardrock Miner AHM 105 could be used to drive first workings and development headings after the copper ore has been reached, namely in definitely harder, more compact and durable rocks.*"

In this paper we would like to present our experiences in driving heading excavations realized by the company PeBeKa S.A. in "Lubin" mine with Sandvik Mining and Construction's MH-620 heading machine. The described experiences will be based on the areas of the mine located far apart and distinctly different from each other. The first one may be considered as a testing ground which identified strengths and weaknesses in the heading machine's construction. The influence of improvements in the machine's construction on its work the other area have also been described.

2 DRIVING THE HEADINGS IN SOUTH-WESTERN AREA

2.1 The assembling of the heading machine and start-up trials in the South-Western area of the mine

The South-Western area together with the assembly chamber and the area where the drive started is shown in Figure 1. Whereas in Table 1 we show the following stages of works associated with descending the machine's elements until the start-up trials have started.

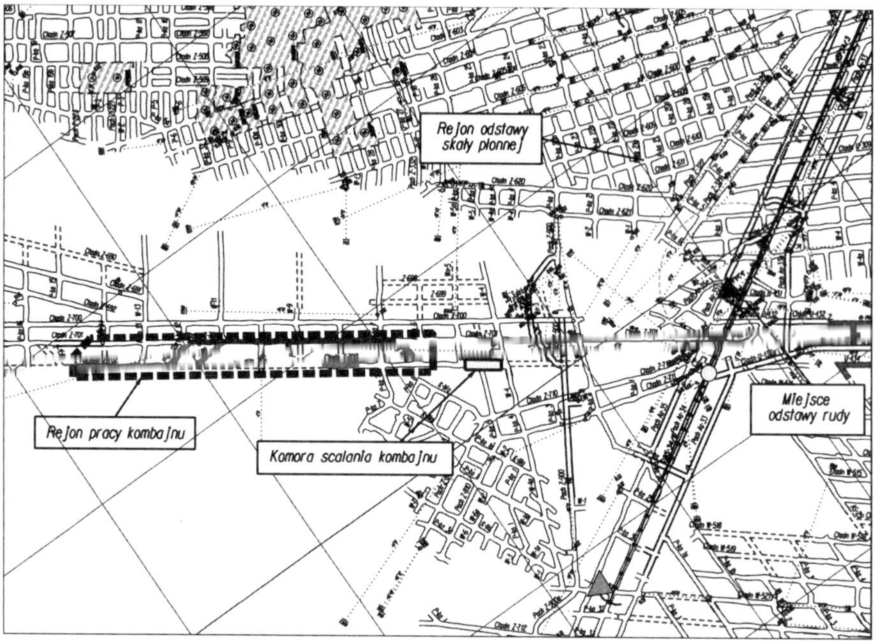

Figure 1. South-Western region of MH-620 works

Notations: Rejon odstawy skały płonnej – *Gangue haulage area;* Rejon pracy kombajnu – *Roadheader work area;* Komora scalania kombajnu – *Roadheader assembly chamber;* Miejsce odstawy rudy – *Ore haulage area*

18.12.2008 12.02.2009	Preparation of the assembly chamber and transport of the machine's components from the surface to the assembly spot
26.02.2009	Finish of the assembly works
27.02.2009 22.03.2009	Start-up trials in the assembly chamber
23.03.2009 31.03.2009	Relocation of the machine from the assembly chamber to the exploitation face in the planned development heading
01.04.2009 14.08.2009	Driving the planned development heading, the crossing headings and leveling the floor in the existing neighbouring excavation. Modernization works at the heading machine. Executing exploitation and bolting trials

2.2 Parameters of the copper ore deposit in the South-Western area

The deposit in this area is characterized by very difficult geological and mining conditions such as poor recognition with headings, multiple fault deformations, packet structure with numerous gypsum inserts, rock slides hazard or third degree of water hazard. The average Rc value for the surrounding rocks which create the exploitation face has been defined as an average value from the rocks creating the face and is equal to 116 MPa.

2.3 MH-620 heading machine

SANDVIK HARDROCK MINER MH 620 is a heading machine designed for exploitation of different types of rocks with compression strength of up to 130 MPa.

1	Exploitation arm	7	Electrical system
2	Turn-table	8	Hydraulic unit
3	Loading device	8.1	Greasing system

4	Chain conveyor	9	Cooling system
5	Crawl mechanism	10	Belt conveyor
6	Frame	11	Bolting system
6.1	Rear support		

Technical parameters:

Total weight *[t]*	approx. 131
Total length *[m]*	approx. 18.2
Total height *[m]*	approx. 3.3
Width of the machine *(undercarriage)* [m]	2.6
Width of the loading table min. / max. [m]	3.5–4.5
Max. exploitation reach over the lower contour of the crawl [m]	5.3/5.8
Max. exploitation width for deflection angle +/- 37° (horizontally) [m]	8.0/8.5
Max. exploitation reach under the lower contour of the crawl [m]	0.3/0.5
Max. lateral exploitation area *(with/without the telescope)* [m^2]	48/40
Supply voltage [V]	1000
Total electrical power [kW]	522
Exploitation engine – water cooled [kW]	300
Hydraulic engine – water cooled [kW]	150
Loading elements drive [kW]	2 x 36
Hydraulic drive of the conveyor [kW]	55
Driving speed [m/min]	0–15
Conveyor speed [m/s]	0.9

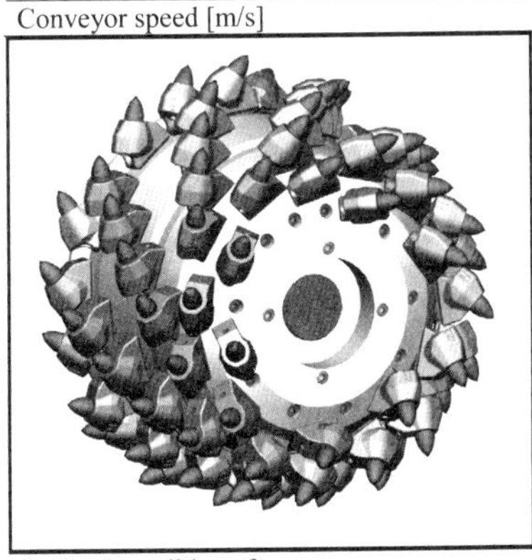

Optimal conditions for use	Rocks with compression strength up to 130 [MPa]
Equipment	2 x 72 pieces of nests for cutting tools with the mandrel diameter 38 mm
Diameter	1290 [mm]
Max. exploitation speed	1.36 [m/sec]
Driving speed	0.17–0.19 [m/sec]

2.4 The scope of the works performed with MH-620 machine and achieved work parameters

In the period between 01.04 and 14.08.2009 approximately 259 meters of heading with lateral area of 27.6 m^2 were made. During this period dinting has also been accomplished at the length of approx. 262 m. The scope of the performed works has been presented in Fig. 2. Following work parameters have been achieved: average efficiency 79 Mg/h which is equal to 28.5 m^3/h and cutting tools wear at 0.107 piece per m^3.

2.5 Summary of MH-620 works in the South-Western area

The geological and mining conditions in the South-Western area are considered to be exceptionally difficult and when it comes to the work of the heading machine even extreme. But the allowed to thoroughly check the machine in a perfect testing ground. This opportunity brought benefits for both the user and the producer, although all weaknesses of the miner have been revealed. The most important were: no possibility to build prop support from the machine and imperfect bolting system which allows to reinforce the roof of the heading only 0.6 m over the machine's height. The construction of the loading table is another disadvantage as it allows to load material only from 3.5 m (when the loading table's width is 4.5 m) what meant leaving a lot of unloaded material at the floor as the heading is 6.5-meter-wide. Works associated with adding a manipulator for prop support were started. The machine was working with constant supervision of the producer. The need to constantly move the machine along the area due to the imperfections in bolting and susceptible arch support construction was another downside.

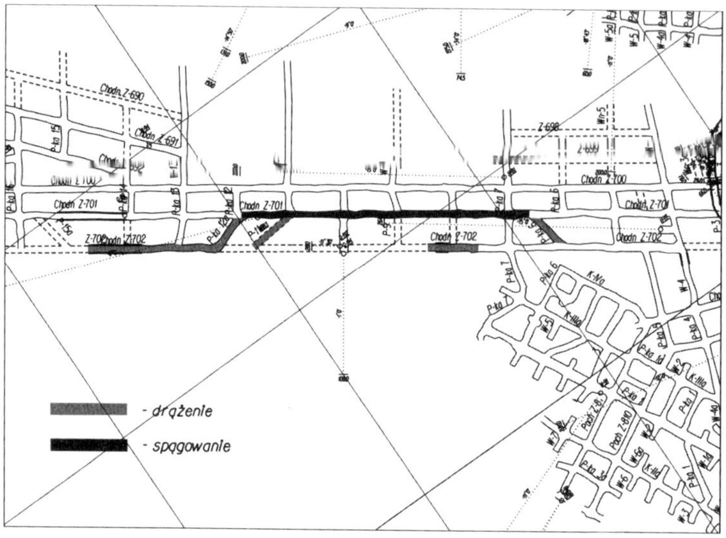

Figure 2. Scope of the works in the South-Western area

Notations: Drążenie – *Driving;* Spągowanie – *Dinting*

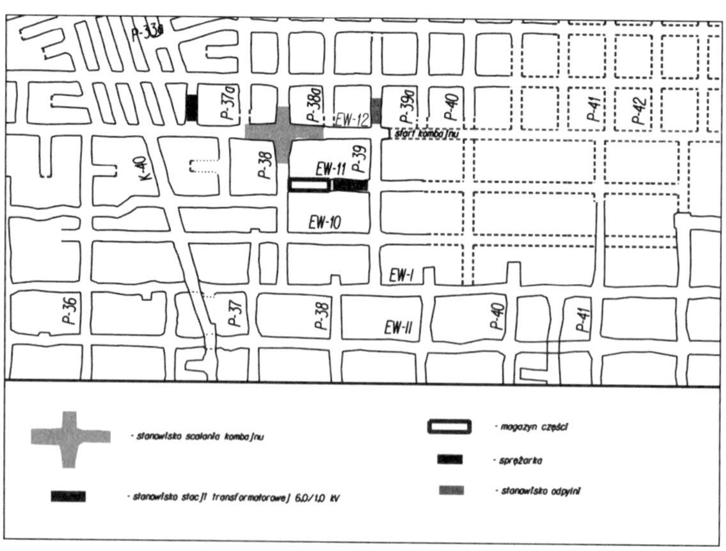

Figure 3. MH-620 assembly site in the North-Eastern area

Notations: Stanowisko scalania kombajnu – *Roadheader assembly spot;* Stanowisko stacji transformatorowej – *Transforming station;* Magazyn części – *Spare parts warehouse;* Sprężarka – *Compressor;* Stanowisko odpylni – *De-dusting station*

2.6 *Conclusions from the works in the South-Western area*

The experiences gained during the works conducted in the South-Western area lead to a decision to temporarily stop the driving and begin the modernization of the machine and relocation to a different area of the mine. A manipulator for the susceptible support was added during the works in the South-Western area. It was successfully tested during construction of thirty sets of ŁP12 support. It was agreed that the modernization will apply particularly to the bolting system and the loading table. It has to be emphasized that all of the works leading to improvements in the machine were carried out in cooperation of the staff from the mine, from PeBeKa and from Sandvik's team. All of the elements accounting for the modernization were implemented in the new area before the start of the works.

3 DRIVING THE HEADINGS IN THE NORTH-EASTERN AREA

3.1 *North-Eastern area*

The North-Eastern area with the assembly chamber and the location where the driving of the heading was started has been depicted in Figure 3 together with the scope of the works.

3.2 *Copper ore deposit parameters in the North-Eastern area*

The copper ore deposit in North-Eastern area is inclined towards the North by 3–5° and is characterized by high stability, low tectonic involvement and the possibility to lead the excavations in stable thickness of dolomitic shales and quartzite sandstones. The area is surrounded by existing excavations and post-exploitation gobs what makes it well recognized and well drained. The average compression strength of the rocks in this area is 52 MPa.

3.3 *Transport of the miner to the new work area*

Transportation of MH-620 miner to the area located approximately 8000 m away was performed in November of 2009. The operation took 23 hours and the machine was moving on its own caterpillars with an average speed of 300 m/h using the sixteen-tonnes LKP-0805 wheel loader most popular in KGHM mines as a drive. The loader was connected with the miner using a stiff towrope. A hydraulic connection between both machines was made allowing for transferring the drive of the loader to the miner. The described ride has to be considered as pioneer.

3.4 *Work of the machine in North-Eastern area*

The scope of the performed works and the parameters of work achieved by MH-620 are shown in Figure 4.

In the period between 14.06 and 31.12.2010 approximately 822 meters of headings with lateral area of 22.2 m^2 were made and over 5850 pieces of 1.8-meter bolts were built-in. 1683 cutting tools were used during the exploitation. In the first phase the heading was horizontal and in the second phase it was inclined accordingly to the deposit's inclination of 3–5°. The basic parameters of MH-620 miner achieved during the works in North-Eastern area are shown in Table 1.

3.5 *Summary of MH-620 works in the North-Eastern area*

Geological and mining conditions which were typical for this area of „Lubin" mine were not a serious difficulty during the driving of the heading. Self-mining of the Eastern side-wall took place what significantly increased the number of bolts that had to be build-in. This phenomenon did not occur in the neighbouring headings drived using explosives. In the new area the miner worked after the modernization has been finished which significantly improved its working parameters. The work is organized in four shifts of which the first one is assigned for maintenance and inspections and the remaining three are for driving the headings. Overlapping of I and II as well as III and IV shifts lead to the miner's dependability coefficient and technical availability reaching 90%. Another observation concerned the significant influence of high abrasiveness of excavated rocks on high wear of particular components of the machine: cutting tools' holders, the cylinder surface of the cutting organ, armored face conveyor's construction, the chain. It seems reasonable to resign from use of rope driving system of the bolting device because of its failure frequency and time consuming maintenance.

Table 1. Basic parameters of MH-620 works

Parameter	Unit	June	July	August	September	October	November	December
Length	m	36.6	80.5	102.9	109.8	136.2	163.5	193.7
Area	m^2	22.3	21.8	23.5	23.5	20.9	21.0	21.8
Advance	m/day	3.56	4.03	5.70	5.80	6.49	7.79	8.07
Bolts	pcs.	238	610	424	792	1152	1201	1389
Cutting tools	pcs.	171	177	178	182	191	332	452
Cost per meter	%	100	57.2	40.2	41.2	36.2	28.4	28.5
Total cost	%	100	129.4	116.1	127.3	138.7	130.3	155.1

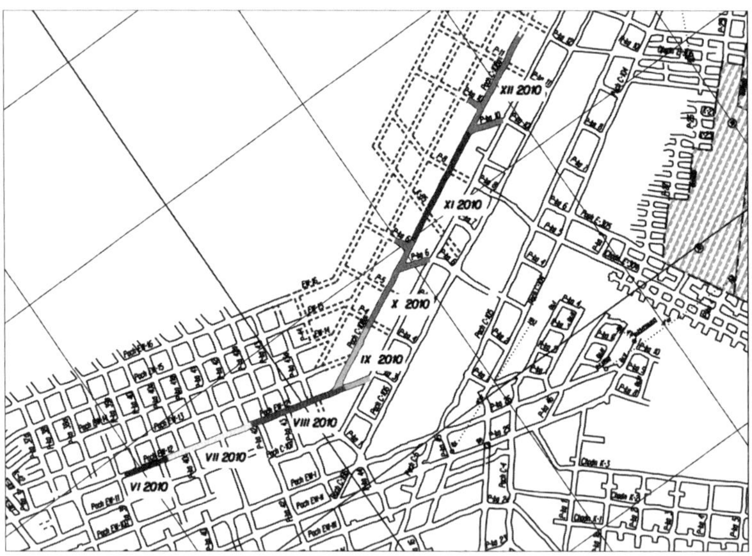

Figure 4. Scope of works conducted in the North-Eastern area

3.6 Conclusions

The experiences gained during the works in both described areas of "Lubin" mine allow to state that the right work organization during the day is a base for right and effective work of the heading machine. It allows to use the available time of machine work maximally, keep the maintenance and inspection regime, design an individual exploitation and bolting cycle, keep constant contact with the producer to analyze the parameters of miner's work, retain proper work culture and experience among the staff. What is the most important is design of the machine's construction. It should take geological and mining conditions into account in broadly understood workplace and allow for driving of the headings independently of local tectonic deformations.

REFERENCES

Internal KGHM Polish Copper plc and "Lubin" mine documents were used in the paper.

International Mining Forum 2011, Dyczko et al. (eds) © 2011 Taylor & Francis Group, London, UK. ISBN 978-0-415-68938-0

New Fire-Resistant Hydraulic Fluids Type HFA for Mining Use – Critical Analysis

H. Dwuletzki, B. Pfaender
CARL BECHEM GMBH, HAGEN, Germany

K. Niemczyk
LUBRICANT s.r.o., Czech Republic

ABSTRACT: Fire-resistant hydraulic fluids of type HFA form a very important subgroup within the hydraulic fluids H. Both water mixable species are used for underground mining; type HFA-S (solutions) and type HFA-E (emulsions) are discussed in detail. Test procedures according to various requirements need to be fulfilled: The 7th Luxemburg report (1994) describes most of the tests followed by the requirements of OE-Manufacturers and end users. Ecological properties and technical requirements are discussed and compared with the chemical formulation of different species of HFA-fluids

KEYWORDS: underground mining, fire-resistant hydraulic fluids, HFA-type fluids

1 INTRODUCTION

Hydraulic fluids are used to transfer force and pressure in closed systems. Such fluids can be divided into three main groups:

- Oils H acc. to standard DIN 51 502 [1], products based on mineral oil,
- Hydraulic Fluids Fire-resistant [2/2a] HF,
- Hydraulic Fluids Environmentally acceptable [3] HE,
- Other hydraulic fluids: for ex. ATF – oils for automatic gearboxes.

Table 1 shows other important way of classification.

Table 1.

Hydraulic fluids					
Based on mineral oil	Ecologically	Fire-resistant			
Oils H (DIN 51 502)	Environmentally acceptable: HE	Fire-resistant: F			
H	HE	HF			
HL	HETG	HFA			
HLP	HEES	HFA-E		HFA-S	
HVLP	HEPG	HFB			
		HFC			
		HFD			
HLPD	HEPR	HFD-R	HFD-T	HFD-S	HFD-U

Based on the fire-resistant criteria, HF Fluids are acc. to DIN ISO 12922 [2] separated by the chemical composition into following groups:

- Pressure fluids containing water: HFA (HFA-E or HFA-S), HFB, HFC
- Pressure fluids without water: HFD

2 HFA-FLUIDS

There are discussed in this article only water-containing hydraulic fluids type HFA intended for use underground in mines – fire-resistant hydraulic fluids soluble in water (HFA-S) and emulsified in water (HFA-E).

Below we will discuss also the concept of "synthetic", "partially synthetic" and "standard" HFA-fluids.

2.1 Requirements for the HFA-fluids used in mines

A large number of accidents caused by fires in the mines caused, that after the disaster in Marcinelle (Belgium, 1956) the relevant authorities initiated the development of new, fire-resistant hydraulic fluid.

To this day, it is effective 7th Luxembourg Report from 01.04.1994 [5] (for sure it will not be changed due to the well-known situation in Germany, resp. European mining industry). This report describes two test ranges applicable for this group of hydraulic fluids:

- Obligatory tests,
- Recommended tests.

Fulfills of the criteria for the 7th Report of Luxembourg is a necessary condition to obtain official permission to use HFA-fluids in the underground (in mines).

2.1.1 Terms included in the 7th Luxembourg Report (1994) regarding the use of HFA-fluids in the mines

The 7th Luxembourg Report defines a total 4 types of water quality. It follows that the water that we have available, has a significant impact on the use of fluids miscible with water.

Due to the extremely high consumption of water in the mine (even about 3500 m^3/day), there are used mostly natural waters. Properties of such waters over the annual cycle change substantially and therefore marked synthetic waters are used for the purpose of tests and measurements. Such waters have the following composition:

Table 2. Test water acc. to 7th Luxembourg Report

Type of test water	Sum of alkaline earth ions			Concentration of ions, mg/l					
	Equivalent of $CaCO_3$	mmol/l (Ca + Mg)	°d hardness	Ca^{2+}	Mg^{2+}	Na^+	Cl^-	SO_4^{2-}	NO_3^-
X	250 mg/l	2.5	14	50	30	130	200	240	0
Y	358 mg/l	3.58	20	61	50	113	201	145	20
V	500 mg/l	5	28	100	60	130	200	480	0
Z	750 mg/l	7.5	42	150	90	130	200	720	0
DIN	358 mg/l	3.58	20	121	13	0	215	52	0

Necessary tests should be carried out with the worst water quality type, in the ideal case should be used water of quality Z – see table.

Meeting the requirements arising from the use of water Z is associated with significant costs of developing HFA-fluids and is reflected in the high cost of production of these products.

The quality of water used to preparation of working fluid also exerts influence on all the indicators determining the effectiveness of the fluid, e.g. corrosion protection, stability of the fluid, microbiological load, production of zinc soap, etc. As you can see, water quality is critical, therefore, comes increased attention addressed to this problem.

Many years of test of water used in mines for preparation of working fluid show that its quality varies considerably.

Therefore, adequate HFA-fluids, particularly in terms of fluid's stability and the degree of corrosion protection in such extreme conditions, are very desirable.

For this reason it is essential to require the tests in the worst water quality – the quality of type Z.

2.1.1.1 *Obligatory tests and controls*

The following tests are carried out obligatory:

- flammability test of water-mixable HFA-fluid
- health hazards tests of water-mixable HFA-fluid
  - Determining of degree of toxicity
  - Oral toxicity (LD50 value)
  - Skin and eye irritation
  - Toxicity of product mists and aerosols
- Data on chemical composition
- Control of identity (repeatability)
- Technological tests:
  - Viscosity (20°C) of concentrate
  - Stability of emulsion (600h, 50°C or 168h, 70°C)
  - pH level: 6,7 – 11
  - Water content in the concentrate:
  - Corrosion inhibiting properties of cover materials:
    - Insulated trial metal
    - In electrolyte
  - Compatibility with sealing materials

2.1.1.2 *Recommended tests and controls*

Most recommended tests concerns further, mostly physical properties, for example:

- Density,
- Freezing point,
- Contents of oils (HFA-E), determination of concentration of the emulsion,
- Compatibility with tank liner,
- VKA tests at rolling and wearing,
- Solids contamination.

2.1.1.3 *Non-hazardous for environment of HFA-fluids*

Study of the environmental impact with determination of water hazard class by concentrate of HFA liquids:

- Biological degradability,
- Toxicity to fish,
- Toxicity to aquatic invertebrates,
- Toxicity to bacteria,
- Biological Accumulation.

In recent years, there were increased requirements (also by the authorities issuing the relevant permits) regarding the impact of HFA-fluids for environment. Determination of the water hazard classes showed that there are clear differences between the various types of HFA-fluids in terms of their biological degradability. In principle, there was established the following classification (Managing Director for Mining of the National Higher Mining Office in Dortmund, Mr. Moench Information from 05.02.1998):

Results of biological degradability of HFA-fluids (National Higher Mining Authority LOBA, 1998)	
HFA – S	HFA – E
< 55 %	> 75 %

Due to the increasing emphasis that is placed in the policy on environmental protection, we observe ongoing care for improving ecotoxicological properties of used fluids.

According to the basic law record for classifications of substances that are threaten to waters (so-called Water Hazard Class), we distinguish in principle between substances which do not represent hazards and substances presenting a danger for water (Water Hazard Class "WGK" 1, 2, 3).

Careful analysis showed that either HFA-E or HFA-S fluids in the concentrated or diluted form do not meet all the necessary criteria needed to ensure that they can be defined as "a substance non-hazardous to the water".

2.1.2 End-user requirements: for ex. DSK (German Coal Mines) Organization

In addition to the 7th Research Luxembourg Report tests requirements, there are other ones which should be fulfilled by HFA-fluids:

- Compatibility with seals made of polyurethane
  P 5000 (1000h) acc. to RAG N 900830,
- Corrosive protection in water type Y approximately 20°dh
  Fe / Zn <10 mg,
- Crevice corrosion in water type Z [9]:
  approximately 42°dH: RAG N 762830,
- Controlled experiment in mine min. 6 months.

Supervision: RW TÜV and the Central Research Department DSK (German Hard Coal Mining) and Mine, which will introduce test fluid.

2.1.3 The requirements of basic equipment suppliers (OEMs)

HFA-fluids are widely used in hydraulic roof supports around the world. These working almost automatically High-Tech support units equipped with electro hydraulic controls of various covers, put increased requirements for HFA-fluids in range of their purity and filtration techniques.

We also are dealing here with phenomena specific to this technique, such as extra-long pipe systems connecting the support units in the longwall with the feeding pumps. In this case it's required a high degree of compatibility of HFA-fluids with zinc, because these pipelines are usually made from galvanized pipes. At F&E Study of Ruhranalytik Laboratory / DSK (German Mining Coal) Organization [10] already in 2001 stated that "during studies involving emulsions HFA-E (P 87 AF) did not came no remarkable change, while use HFA-S fluid (GS 460 NH) showed a number of significant changes, also in a short time, additionally in the hydraulic fluid was made a zinc soap".

It is clear that significant differences in the composition of HFA fluids can cause differences in their technical application, because only the "function of memory" HFA-E emulsions allows the generation of efficient protective layer on the metal surface.

Leading manufacturers of enclosures report, in addition to many of the above-described conditions for tests, among other requirements, so called producer standards, which also must be met. For example:

- Caterpillar (Bucyrus):
  - 7th Luxembourg report
  - Crevice corrosion test in water work
  - Long-term studies of seals swelling
- Joy Company - Mining, United Kingdom:
  - Compatible with solenoid valves in the soft water
  - Test of lubricating properties

So we can briefly characterize described range of tests:

HFA-fluids are among the best studied lubricants. The development of HFA-fluids is the most difficult and most expensive research program for lubricants.

3 GENERAL STRUCTURE OF HFA-FLUIDS

The chemical composition of HFA-fluids is largely based on the formulas of the coolant and lubricant fluids miscible with water and used in machining metalworking; significant impact on their composition have the main functions, such as protection against corrosion and stability of HFA fluids miscible with water. There are not used in the vast majority any high-pressure additives (EP Extreme Pressure).

Each lubricant consists of base fluids and additives depending on the specific application. Generally, the composition of the HFA-fluids can be described as follows:

Main component		HFA Fluid	
		HFA – E	HFA – S
Base fluid			
	Water	up to 40 %	up to 70%
	Glycol	no	up to 20%
	Hydrocarbons: Natural: mineral oil, white oil; Synthetic: polyalphaolefins	up to 80 %	0 %
	Esters: Natural: rape oil; Synthetic	up to 80 %	generally 0%
Anti-corrosions additives			
	For ferrous metals	yes	yes
	For non-ferrous metals like: Cu, Ms, Al, etc.	yes	yes
Lubricating system			
	Hydrocarbons: mineral oil, white oil	yes	no
	Additives	yes	yes
	Polar lubricants	yes	yes
	Eventually AW (Anti-Wear) additives	rarely	rarely
Biostabilizers			
	Biostatics Boric acid compounds	rarely	yes
	Bactericides	yes	yes
	Fungicides	yes	yes
Emulsified system			
	Emulsifiers	yes	no
	Coemulsifiers	yes	no
	Solubilisators	yes	no
	Stabilizators	yes	no
	Wetting agent	yes	ev. yes
Anti-foaming additive	eventually	yes	yes

3.1 The general division of HFA-fluids on HFA-S and HFA-E type

With reference to the classification of coolants and lubricants in accordance with DIN 51 385 [4] HFA-fluids can be divided into two distinct groups: water-soluble HFA (HFA-S Solution) and the emulsified HFA (HFA-E Emulsion). However, this should be regarded only as an analogy.

For cooling and lubricating fluids, there are additional rules that are anchored in the following laws and regulations:

- REACh: Registration, Evaluation, Authorization of Chemicals [11],
- The Law on the chemical [12],
- Regulation on Hazardous Substances (GefStoffV) [13],
- Guidelines on the prohibition of chemical agents [14],
- TRGS Rules (TRGS Technical Rules regarding hazardous substances), especially TRGS 611 [15].

Problem is mainly about the various national laws and regulations, and often rather a lack of laws and regulations outside of Germany and the European Union, which would prohibit the use of chemicals which the EU has already prohibited. Here are some examples:

- In the U.S. and Australia there are used formulations of HFA-S fluids, which contain among others additives sodium nitrite as a very effective anti-corrosion agent.
- In the U.S., Australia, Russia and many other countries it is still a permitted use of secondary amines for production of anti-corrosion additives or additives to emulsifiers.

With just incomprehensible reasons, the health aspect of HFA-fluids is in the country and abroad strongly underestimated.

The table presented above shows the two basic types of liquid HFA:
The water-soluble HFA-S (Solution) fluid and water miscible HFA-E (Emulsion) fluid.
There should be correctly used terms: HFA solution eventually emulsion HFA.
Traders also coined for themselves the terms “fully synthetic”, resp. “standard emulsion”.

Image below shows the relationship:

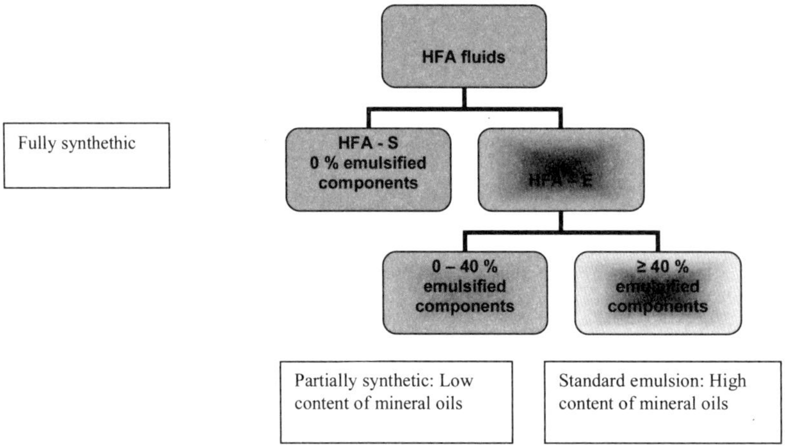

Systems with a low content of mineral oil contain max. 40% of mineral oil, and especially up to 40% water in the solution. For additives left 20% or so.

The decisive technical advantages have conventional systems HFA-E with a high content of mineral oils:

- High tolerance to fluctuations in service concentration in relation to the concentration required,
- Excellent “memory effect” on the working surfaces,
- Stable anti-corrosion layer on metal surfaces,
- Excellent protection against corrosion also in the case of lower pH.

4 EVALUATION OF HFA-FLUIDS

4.1 pH level of used fluids

In the case of HFA-fluids manufacturers apply the rule: higher pH level – more effective corrosion protection.

Disadvantages of this concept are primarily:

- Risk of serious damage to the skin in places that are often exposed to liquid,

- Clearly reduced compatibility with sealing materials, which are basically little resistant to the higher level of pH. The result tends to frequent leaks and failures of seals on the cylinders (rollers) and the control elements,
- In case of use galvanized pipe there is increased zinc releasing as a result of high alkalinity. For this problem, Ruhranalytik Laboratory / DSK (German Mining Coal) Organization [10]] conducted a study that showed significantly lower compatibility of synthetic fluid with high pH values in comparison to the standard emulsions with a low pH.

4.2 Concentrations problem – refractive correction calculator

Due to the criteria of fire-resistance of hydraulic fluids concentration above 5% is generally permissible only in extreme, exceptional situations. The real working concentration was accepted approximately concentration of 1–3%.

Due to the different refractive indices of HFA-S and HFA-E fluids in a water solution, for the safe determination of the actual concentration there is needed correction refractive correction calculator, in this case it works the principle:

Concentration of HFA-fluid = **The value read on the refractometer, [°Brix]** × **refractive correction calculator F = 0,9 to 3**

With the higher content of emulsified components (ester, mineral oils, etc.), the correction calculator F is closer to the value of 1.

With the higher water content, the higher is the correction calculator, which in the case of solutions can be within value of 2 to 3.

Rules are:

Refractive correction calculator for HFA-liquids		
HFA – S containing water	HFA - E containing water	HFA-E (almost) without water
F = 1.,8 – 3.3	F = 1.2 – 2.2	F = 0.9 – 1.0

4.3 Specific density of HFA-fluids concentrate problem

Production of working fluids form of water-miscible IIFA fluids (HFA solution, resp. HFA emulsions) is generally carried out by adding the concentrate to the basic component of a solution – water. This operation may take place within the hydraulic station or on the surface in the special blending station.

It is, in essence, addition into a certain amount of water, a certain amount of HFA-fluid concentrate, which is a volumetric mixing. It should be taken into account, due to the laws of thermodynamics, values of density of the concentrate used.

The following rules apply:

To produce 1000 dm^3 of HFA-fluid with a specific concentration, e.g. 2%, it needs 980 dm^3 of water and 20 liters of concentrate HFA. The density of HFA-S (synthetic) concentrate is about 1.01 g/ml, density of standard HFA-E concentrate is about 0,91 g/ml always at 20°C. We must therefore use 20.2 kg of HFA-S concentrate, resp. 18.2 kg of HFA-E concentrate to obtain 1000 dm^3 of HFA working fluid at concentration of 2%.

This means, that 100 tons of HFA concentrate gives 4951m^3 of HFA-S solution or 5495 m^3 of HFA-E emulsion. So density differences let get up to 10% more working emulsion of HFA-E with the same amount of concentrate!

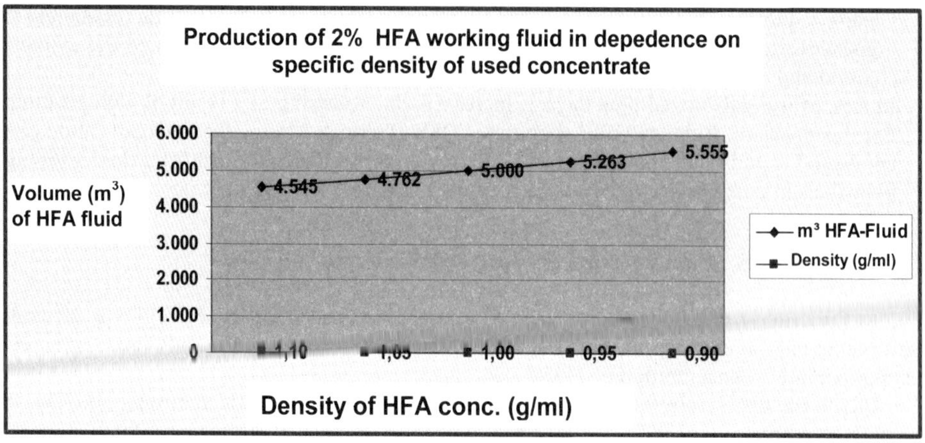

It should be noted that the HFA concentrates having the lowest specific density of specific, in other words concentrates of HFA-E with the highest possible content of mineral oil allow to preparing of the maximum amount of working fluid HFA, while the water-containing concentrates in this respect are less efficient. The difference can be up over 10%.

HFA-E fluids are, as it follows, much more economical.

4.4 Effect of temperature on HFA concentrate: the problem of freezing temperature

The following division was developed based on the experience:

Freezing temperature of HFA-liquids		
HFA-S containing water	HFA-E containing water	HFA-E (almost) without water
approx. -5°C	approx. – 15 °C	approx. -30°C

Water contained in concentrate frozen at lower temperature, which can result of damage of concentrate to the extent that eliminate product for use. In such situations it is worth, no matter the cost; strictly obey any conditions for the safe transport and storage of the product.

4.5 Stability of HFA-fluid in the worst quality waters

The advantage of HFA-S systems which are water-based solutions is fact that they have no emulsifying system that could be sensitive to substances presented in water. In the case of HFA-E emulsion it should be proved that they are suitable for use in the worst water quality. They must pass the required test in the trial waters.

The user requires a maximum degree of stability of the product also in "critical water". The term "critical water used to produce working fluid" it should be understood for water which exhibit:

- The high conductivity: > 1500 μS/cm,
- High content of chlorides: > 200 ppm,
- High content of cfu (colony forming units): > 105 cfu/ml,
- Low pH level: < 6,5,
- High carbonate hardness: > 20° of carbonate hardness.

Comparisons with standard quality test water type Z shows that was already met so many of the criteria for determining "critical water for mixing liquids" that should be insist on the general application of the water this quality for assessing the quality criteria at name "Stability of liquid HFA". Thanks to use this method, fluctuations in water quality will not lead to a situation where the working fluid turns out to be unreliable.

Very high quality emulsions HFA-E are characterized by very good stability also in very critical water used to prepare the working fluid, based on results of studies using the water quality test water type Z, so they should be rated as equal to HFA-S systems.

4.6 Property of HFA-fluid for emergency mode of work: "Memory effect" [16]

In connection with the search for causes of failure of the actuator cylinder of advancing mechanism DMT organization in Germany conducted a study which showed that HFA emulsions with a high content of mineral oils have a marked tendency to form a protective layer, important for the work in emergency mode of work and for protection against corrosion.

The best results were achieved with the hydraulic concentrate P 87 AF-TAD [17]. Previously damaged by corrosion cylinder roof supports failed after 1640 cycles of load. This result deserves special attention, because all the HFA-fluids, which have been tested, fared much worse. Observations indicate that, depending on the content of mineral oil in the test, HFA-fluids reach almost number of loads cycles achievable for new cylinders of roof supports (see Figure 7).

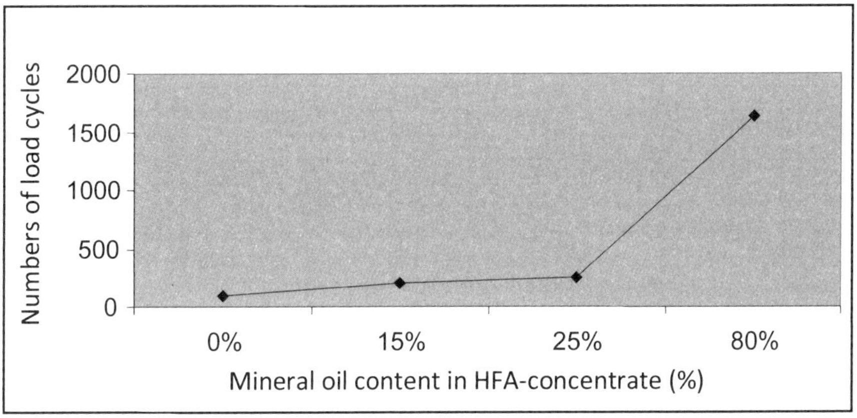

Figure 7. Depending of numbers of load cycles for actuator cylinder of advancing mechanism from mineral oil content in HFA-concentrate

Because all the tests took place under conditions of 2% dilution in water quality Z, it is therefore necessary minimum concentration of "active" emulsified mineral oil that may have occurred to produce a protective layer on the surface of hydraulic cylinders.

Result of large-scale extensive testing shows that:

Hydraulic Concentrate P 87 AF-TAD has a clear tendency to actively create a protective layer on the surfaces, which takes tribologic effect.

The layer of mineral oil which is in dynamic equilibrium of sorption and desorption results in excellent compatibility with seal materials, and exhibits the characteristics necessary to work in safe mode, so that corrosion attacked and very rough surfaces of hydraulic cylinder do not damage mechanical cylinder sealing components.

It all makes that short-term insufficient concentration of HFA fluid emulsion is tolerated by the system and it does not cause its immediate damage.

HFA-S fluids does not possess this property because they contain only substances completely soluble in water, which in turn makes it through desorption can be very quickly removed from the metal surface. Solubility in water under these conditions clearly undermines the effectiveness of "memory effect".

4.7 Attempt of evaluation of HFA-fluids on the basis of practice use results

Attempt of evaluation of these products representing different systems without detailed observations on specific conditions leads to a dangerously simplistic conclusions. However, the most important properties of the three types of HFA-fluids are shown in the following table: solution, emulsion, partially-synthetic emulsion and standard emulsion appear here side by side, evaluated on the basis of the results obtained in practice:

Property	Hydraulic fluid			Evaluation		
	Solution	Emulsion HFA-E		Solution	Emulsion HFA-E	
	HFA-S	Low content of mineral oil	High content of mineral oil	HFA-S	Low content of mineral oil	High content of mineral oil
Freezing temperature	> -5°C	> -15°C	< -25 °C	-	(+)	+
pH of (2% at water type Z)	9.3	9.5–11	8.9–7.5	-	-	+
Zinc compatibility	poor	low	good	-	-	+
Stability at water type Z	not for use	poor	good	not for use	-	+
Corrosion protection	good	good	good	+	+	+
Memory effect	very good	poor	very good	-	-	+
Corrosion at storage	very high	high	low	-	-	+
Acting at the enclosure movement	nd	nd	6-9 month without corrosion	nd	nd	+
Non-act caused by HFA-fluid	yes	nd	no	-	nd	+
Lubricating properties	poor	satisfied	good	-	(+)	+
Compatibility with sealing	satisfied	satisfied	good	-	-	+
Microbiological stability	high	high	satisfied	+	+	-
Need to control	high	high	high	-	-	-
Refractometer Correction Calculator	1.8–3.0	1.2–1.5	0.9–1.0	-	(-)	+
Hazard for environment	very low	very low	very low	+	+	+
Water hazard class (WGK)	WGK 1	WGK 1	WGK 1	+	+	+
Recognition of losses due to leakage	impossible, need to colour	results from composition of product	results from composition of product	-	(+)	+
Density (15°C)	1.0–1.1	approx. 1.0	0.9	-	-	+
Total cost (use)	high	medium	low	-	(±)	+
Cost of product	high	low	high	-	+	-

nd: no data (keine Erfahrungen)
+ : Positive value
(+): Mostly positive value
(-): Mostly negative value
- : Negative value (expieriences)

We of course recognize that in this evaluation were included only some of the properties, therefore it cannot in any way pretend to be a comprehensive assessment. Despite this, however, one can gain insight based on experience gained at the use of its products worldwide.

REFERENCES

[1] DIN 51 502: Kurzbezeichnung der Schmierstoffe und Kennzeichnung der Schmierstoffbehälter, Schmiergeräte und Schmierstellen

[2] DIN 24 230: Nonflammable hydraulic fluids ISO 6743 cz. 4: Lubricants, industrial oils and related products. Family H - Hydraulic systems

[2a] DIN ISO 12922: Schmierstoffe, Industrieöle und verwandte Erzeugnisse (Klasse L) – Familie H (Hydraulische Systeme). Requirements for category HFAE, HFAS, HFB, HFC, HFDR a HFDU; 1999

[3] DIN ISO 15380: Schmierstoffe, Industrieöle und verwandte Produkte (Klasse L) – Familie H (Hydraulische Systeme) Requirements for category HETG, HEPG, HEES a HEPR; 2004

[4] DIN 51 385: Schmierstoffe, Kühlschmierstoffe, Definitionen. Juni 1991

[5] Anforderungen und Prüfungen schwerentflammbarer Hydraulikflüssigkeiten zur hydrostatischen und hydrokinetischen Kraftübertragung und Steuerung; Luxemburger Bericht 7. Ausgabe, April 1994

[6] K. Bauer, Wasserhaltige Hydraulikflüssigkeiten, Ölhydraulik und Pneumatik, 47 (2003), Nr. 11-12 (K. Bauer,

[7] Einstufung von HFA-E Konzentraten in die WGK; Gutachten des Hygiene-Institutes des Ruhrgebietes, Gelsenkirchen vom 25.10.2006, A-144347-06

[8] Bewertung der Umwelt-/ Wassergefährdung von HFA-E-Gebrauchsemulsionen (1-3%) Gutachten des Hygiene-Institutes des Ruhrgebietes, Gelsenkirchen vom 27.07.2007, A-153392-06

[9] K. Bauer, HFA-Hydraulikflüssigkeiten für den Strebausbau, Glückauf 138 (2002), Nr. 5 S. 208-212

[10] Untersuchungen der DSK, F&E-Studie der Ruhranalytik, Studie 144 vom 01.02.2001: Untersuchung der Metallseifenbildung in Hydraulikkreisläufen mit dem Ziel der Verminderung von Filterverstopfungen

[11] REACH: Regulation (EC) No 1907/2006 of the European Parliament and of the Council from 18.12.2006 concerning the Registration, Evaluation, Authorization and Restriction from Chemicals (REACH)

[12] Gesetz zum Schutz vor gefährlichen Chemikalien (Chemikaliengesetz ChemG)

[13] Verordnung zum Schutz vor gefährlichen Stoffen (Gefahrstoffverordnung GefStoffV) Anhang IV, Nr. 19

[14] 76/769/EHS Guideline: Guideline on the Prohibition of Chemical

[15] Technische Regeln für Gefahrstoffe 611: Verwendungsbeschränkungen für wassermischbare bzw. wassergemischte Kühlschmierstoffe, bei deren Einsatz N-Nitrosamine auftreten können, 2002

[16] H. Dwuletzki, B. Pfaender, V. Preis, G. Tollkamp; Untersuchungen zur Bestimmung von Notlaufeigenschaften von HFA-Fluiden und Nachweis des „Memory-Effektes" von HFA-Emulsions-Systemen; Glückauf 144 (2008) Nr.1/2

[17] Deutsche Montan Technologie (DMT), Untersuchungsbericht Nr. 1535NN14 vom 14.08.2002 (Organizace Deutsche Montan Technologie (DMT)

International Mining Forum 2011, Dyczko et al. (eds) © 2011 Taylor & Francis Group, London, UK. ISBN 978-0-415-68938-0

Author Index